AF595078

New Insights into Food Fermentation

New Insights into Food Fermentation

Editors

Valentina Bernini
Juliano De Dea Lindner

MDPI • Basel • Beijing • Wuhan • Barcelona • Belgrade • Manchester • Tokyo • Cluj • Tianjin

Editors
Valentina Bernini
Food and Drug
University of Parma
Parma
Italy

Juliano De Dea Lindner
Food Science and Technology
Federal University of Santa Catarina (UFSC)
Florianópolis
Brazil

Editorial Office
MDPI
St. Alban-Anlage 66
4052 Basel, Switzerland

This is a reprint of articles from the Special Issue published online in the open access journal *Foods* (ISSN 2304-8158) (available at: www.mdpi.com/journal/foods/special_issues/food_fermentation).

For citation purposes, cite each article independently as indicated on the article page online and as indicated below:

LastName, A.A.; LastName, B.B.; LastName, C.C. Article Title. *Journal Name* **Year**, *Volume Number*, Page Range.

ISBN 978-3-0365-3920-1 (Hbk)
ISBN 978-3-0365-3919-5 (PDF)

Contents

About the Editors

Valentina Bernini

Valentina Bernini is Professor in Food Microbiology and Predictive Microbiology at the Department of Food and Drug at the University of Parma, Italy. She graduated and achieved a Ph.D. in Food Science and Technology at the University of Parma. She actually teaches in the Interuniversity International Master Degree in Food Safety and Food Risk Management (University of Parma), in Food Science and Technology Master Degree (University of Parma), and in Master Internacional en Tecnologia de los Alimentos (MITA) (University di Parma –Universidad de Buenos Aires). She is the Scientific Coordinator of University of Parma Culture Collection (UPCC) and a Member of the Scientific Committee of the Join Research Unit MIRRI-IT. She is a Member of the Food Science Ph.D. board at the University of Parma. She is co-founder of SIIIA (Società Italiana per l'Innovazione nell'Industria Alimentare s.r.l.), an University of Parma Spin-off. She collaborates with Consortia for the Protection of typical products, and national and international companies for the development of research and innovation in food production.

Juliano De Dea Lindner

Juliano De Dea Lindner is Professor in Food Microbiology and Quality Management in the Food Industry at the Department of Food Science and Technology at the Federal University of Santa Catarina, Brazil. He graduated in Pharmacy Science and Biochemistry at the Catholic University of Paraná, Brazil. He found the food companies Incorpore Foods Ltda in Brazil and Nutriplus S.r.l. in Italy. He is currently Professor in the Postgraduate Programs in Food Science and Biotechnology. He is a member of the International Dairy Federation (FIL-IDF) on the Permanent International Committee on Microbiological Hygiene. Consultant for the food industry in developing processes and products of a technical nature with more than 60 SKUs currently on the world market. He is Research Productivity Scholarship from the Brazilian National Council for Scientific and Technological Development linked to the Ministry of Science, Technology, Innovations, and Communications.

Preface to "New Insights into Food Fermentation"

This reprint presents a print version of the Special Issue of the journal {Foods} dedicated to new insights into food fermentation. Food fermentation has been used for thousands of years for food preservation. At present, fermented foods remain appreciated by consumers thanks to the high-quality standards achieved and the improvements in terms of nutritional and organoleptic characteristics. The production processes, type of raw material, microbial cultures, etc., can affect these products' quality and safety characteristics. A vast array of microorganisms can be found in fermented foods, and microbial succession during fermentation, as well as during ripening, contributes to the desired properties of these foods. In addition to the sensory and safety aspects, microorganisms present in fermented foods can positively affect people's health due to their potential probiotic nature and the production of beneficial metabolites such as vitamins and antioxidant compounds.

The goal of this Special Issue was to broaden the current knowledge on advanced approaches concerning food fermentation, gathering studies on conventional and unconventional food matrix fermentation, functional compounds obtained through fermentation, fermentations increasing quality and safety standards, as well as papers presenting innovative approaches shedding light on the microbial community that characterizes fermented foods.

In the 13 papers collected in this volume, interested readers will find a collection of scientific contributions providing a sample of the state-of-the-art and forefront research in food fermentation. Among the articles published in the Special Issue, the geographic distribution of the studies is wide enough to attract the interest of an international audience of readers.

The editors would like to thank the authors for their collaboration and commitment to publishing their high-quality scientific articles.

Valentina Bernini and Juliano De Dea Lindner
Editors

MDPI

Editorial

New Insights into Food Fermentation

Juliano De Dea Lindner [1,*] and Valentina Bernini [2,*]

[1] Food Technology and Bioprocess Research Group, Department of Food Science and Technology, Federal University of Santa Catarina (UFSC), Florianópolis 88034-000, Brazil
[2] Department of Food and Drug, University of Parma (UNIPR), 43124 Parma, Italy
* Correspondence: juliano.lindner@ufsc.br (J.D.D.L.); valentina.bernini@unipr.it (V.B.)

Citation: Lindner, J.D.D.; Bernini, V. New Insights into Food Fermentation. *Foods* **2022**, *11*, 283. https://doi.org/10.3390/foods11030283

Received: 12 January 2022
Accepted: 17 January 2022
Published: 21 January 2022

Food fermentation has been used for thousands of years for food preservation. At present, fermented foods remain appreciated by consumers thanks to the high-quality standards achieved, and the improvements in terms of nutritional and organoleptic characteristics. The production processes, type of raw material, microbial cultures, etc., can affect the quality and safety characteristics of these products. A vast array of microorganisms can be found in fermented foods, and microbial succession during fermentation, as well as during ripening, contributes to the desired properties of these foods. In addition to the sensory and safety aspects, microorganisms present in fermented foods can positively affect the health of people due to their potential probiotic nature and the production of beneficial metabolites such as vitamins and antioxidant compounds.

The goal of this Special Issue was to broaden the current knowledge on advanced approaches concerning food fermentation, gathering studies on conventional and unconventional food matrix fermentation, functional compounds obtained through fermentation, fermentations increasing quality and safety standards, as well as papers presenting innovative approaches shedding light on the microbial community that characterizes fermented foods. This Special Issue brings a series of 13 articles related to new insights into food fermentation.

Two review articles provide information concerning traditional fermented beverages. The first presents an updated view on Mexican beverages, systematizing information on the diversity, spatial distribution, and cultural history of beverages. Ojeda-Linares et al. [1] identified 16 Mexican fermented beverages (named *mescal*, *pulque*, *tejuino*, *pozol*, *chorote*, *colonche*, *saká*, *sendechó*, *balché*, *atole agrio*, *pox*, *sambudia*, *tesgüino*, *tepache*, *tuba*, and *taberna*) and 143 plant species involved in their production as substrates for fermentation. The authors highlight that microbial communities from backslopping have only scarcely been covered in studies of beverages such as *mescal*, *pulque*, and *atole agrio*. The second review addresses the properties, processing methods, microorganisms, and microbial dynamics issues regarding Ethiopian alcoholic beverages. In Ethiopia, the preparation and consumption of cereal- and fruit-based spontaneous, natural, and uncontrolled fermented alcoholic beverages are common. Cereal-based (*tella*, *borde*, *shamita*, *korefe*, *cheka*, and *keribo*) and fruit-based (*tej*, *ogol*, and *booka*) are described by Getachew Fentie et al. [2] as being the popular beverages in the country. Yeasts and lactic acid bacteria (LAB) are the predominant microorganisms encountered during fermentation. These beverages also contain significant amounts of total polyphenols and antioxidants. The alcohol contents and pH values of these beverages range from 1.53% to 21.7% and 2.9 to 4.9, respectively.

Many foods are still produced following traditional practices, although novel approaches to food fermentation have also attracted the interest of researchers and industries. Innovative technological and biological processes, as well as novel approaches of investigation, deeply interact to steer traditional products into modern diets and to open perspectives for the fermentation of unconventional substrates and food byproducts. The three papers submitted that dealt with functional foods addressed issues related to fruit and cereal fermentation. The research conducted in Italy by Maisto et al. [3] evaluated the addition of four different potential probiotic strains (*Lactiplantibacillus plantarum* subsp. *plantarum* ATCC14917, *Lactobacillus delbrueckii* subsp. *bulgaricus* ATCC 11842, *Lactobacillus acidophilus* ATCC 4356, and *Lacticaseibacillus rhamnosus* ATCC 7469) to date fruit-based fermented products (extruded snacks). After fermentation, changes in the polyphenol profile in terms of increased free phenolic compounds and related activity were observed. These results may be attributed to the metabolism of lactobacilli in catalyzing both the release of bioactive compounds from the matrix and the remodeling of polyphenolic composition in favor of more bioaccessible molecules. These positive effects were more evident when the snack was fermented with *L. rhamnosus*. The authors consider that the fermented snack may be proposed as a prototype of functional food, mainly indicated for athletics nutrition and supplementation.

The next two papers dealt specifically with the analysis of fermented cereals with heart health functional claims. Moon and Chang [4] fermented rice bran using a culture of *L. plantarum* EM, which exhibited significant cholesterol removal (45–68%) and strong antimicrobial activity against foodborne pathogenic bacteria and food spoilage fungi in vitro. Phytate levels were significantly reduced during fermentation by 53% due to the phytase activity of *L. plantarum*, indicating that fermented rice bran does not present nutrient-deficiency issues. The authors state that fermented rice bran is a promising low-cost functional food candidate and appears to satisfy consumer demands for environmental requirements concerning the re-utilization of biological byproducts. Yang et al. [5] explored the effects of different fermentation parameters on the quality of *natto* fermented with *Bacillus subtilis* GUTU09 and *Bifidobacterium animalis* subsp. *lactis* BZ25. Nattokinase activity, free amino nitrogen content, and sensory score were increased compared with control tests. The plate thrombolytic area and nattokinase activity increased significantly as the fermentation time increased, indicating that the *natto* exhibited strong thrombolytic action in vitro. Nattokinase is a fibrinolytic enzyme released by the *B. subtilis natto* bacteria during the fermentation process.

Despite recent innovations on "omic" techniques, including genomic approaches using high-throughput sequencing in combination with advanced metabolomics, there are many fermented products with limited information about their microbial diversity and dynamics (succession). This is particularly true for traditional products, which represent a rich niche for discoveries involving microorganisms more robust for industrial production. For example, the unknown microbiota can reveal new molecular mechanisms of quorum sensing and new bioactive molecules with beneficial effects.

The next four papers deal specifically with the analyses of the microbiome of fermented foods. The composition of the microbiota has an important impact on the quality and safety of products due to the growth and interaction between microorganisms during processing and ripening. Therefore, much effort has been made by the authors to applicate molecular tools and omic systems to investigate the microbial community composition of fermented foods. The paper presented by Jiang et al. [6] evaluated the effects of raw materials and fermentation periods on the microbial ecology of Chinese *paocai* using metagenomic analyses. *Paocai* is typically made by fermenting red radish or cabbage with brine for 6–10 days. *Secundilactobacillus paracollinoides* and *Furfurilactobacillus siliginis* were the characteristic bacteria in red radish *paocai*, whereas 15 species of characteristic microbes were found in cabbage *paocai*. Furthermore, the study also enabled the identification of volatile organic compounds (VOCs) and the establishment of their relationship to the microorganisms, which provided insights into the microbial flavor profiles of these fermented vegetables which are popular and traditional in Asian countries.

The microbiota of Protected Designation of Origin (PDO) cheeses play essential roles in defining their quality and typicity, and could be applied to protect these products from counterfeiting. Zago et al. [7] applied DNA meta-analysis to study the possible role of microbiota in distinguishing Italian *Grana Padano* (GP) cheese from generically hard cheeses (HCs). The microbial structure and the genotypic fingerprinting of the bacterial taxa of 119 GP samples were evaluated by 16S rRNA gene sequencing (DNA metabarcoding) and RAPD-PCR from total cheese DNA (RAPD-PCR metafingerprinting), respectively, and compared with 49 samples of generically HC from retail. The obtained metagenotypes were evaluated as possible tools to differentiate the two sampling groups, assuming, unlike metataxonomic analysis, that this technique was able to identify strain- or group-specific differences within complex microbial communities. Metataxonomic and metafingerprinting data were then used as inputs to train and validate a two-class (GP vs. HC) classifier based on a neural network as a computational model. Although metataxonomic data did not allow for reliable classification, the discriminatory power of metafingerprinting enabled the building of a robust model (very high binary accuracy). When trained with metagenotyping data, the model correctly classified the samples. The molecular (meta)fingerprint of the entire microbial community could be promising to assist GP authentication and to distinguish it from imitation products.

Degenhardt et al. [8] studied the presence of hepatitis E virus (HEV) and rotavirus-A (RV-A), as well as fungal and bacterial communities, using metagenomics and culture-dependent methods in artisanal *colonial* salami-type dry-fermented sausages from Brazil. LAB and yeasts dominated the microbiome. *Lactilactobacillus sakei* and *Debaryomyces hansenii* were ubiquitous and the most abundant species. The characteristics of the raw material and hygiene of the manufacturing process resulted in high loads of beneficial microorganisms and the absence of HEV and RV-A viruses, as determined by RT-qPCR assays. The wide array of traditional fermented sausages worldwide represents a reservoir of microbial biodiversity and can be an important source of new biotechnological strains able to preserve typical features lost due to the introduction of commercial cultures. Particularly, in recent decades, the use of starter cultures has been introduced in the meat industry to guide fermentation, enhancing product safety but losing biodiversity and peculiar characteristics to the products in terms of both technological and organoleptic traits. Barbieri et al. [9] analyzed 15 artisanal salamis from the Mediterranean area (Italy, Spain, Croatia, and Slovenia) to evaluate the microbiota composition through culture-dependent and culture-independent techniques. LAB and coagulase-negative cocci were the dominant populations, with the highest LAB presence in Croatian and Italian samples. Metagenomic analysis showed high variability in microbial composition: *L. sakei* was the dominant species, but *Companilactobacillus* spp. was present in high amounts (45–55% of the total amplicon sequence variant) in some Spanish sausages. *Staphylococcus epidermidis*, *Staphylococcus equorum*, *Staphylococcus saprophyticus*, *Staphylococcus succinus*, and *Staphylococcus xylosus* were detected. The growth and survival of different microbial groups in the sausages reflect in the first-instance safety characteristics (i.e., biogenic amine concentration). In addition, the volatilome, and consequently, the peculiar sensory traits of traditional products, are dependent on the complexity of the microbiota. In many sausages considered in this paper, increased microbial biodiversity caused VOCs to be more complex, both qualitatively and quantitatively.

Due to the increasing occurrences of worldwide food-borne disease outbreaks caused by biogenic amines, food safety has received more concern in the production of fermented meat products. The composition of the microbiota has a direct impact on the safety of products. Using high-throughput sequencing, culture-dependent, and HPLC methods, Ma et al. [10] investigated the contribution and regulation of biogenic amines (BAs) by dominant microorganisms during traditional fish sauce fermentation in China. *Tetragenococcus* (40.65%), *Lentibacillus* (9.23%), *Vagococcus* (2.20%), *Psychrobacter* (1.80%), *Pseudomonas* (0.98%), *Halomonas* (0.94%), and *Staphylococcus* (0.16%) were the dominant genera observed. The contents of BAs gradually increased as the fermentation progressed. After 12 months of fermentation, the histamine content (55.59 mg/kg) exceeded the toxic dose recommended by the United States Food and Drug Administration (FDA). Correlation analysis showed that the microbiota made a considerable contribution to the accumulation of BAs. *Staphylococcus nepalensis* 5-5 and *S. xylosus* JCM 2418 strains with a high BA reduction ability were screened out of 44 low BA-producing dominant isolated strains and might be potential functional cultures for BA control in meat. Yu et al. [11] explored the influences of thyme

(*Thymus vulgaris* L.) on the growth, gene expression, and histamine accumulation by *Proteus bacillus* isolated from Xinjiang smoked spontaneous fermentation horsemeat sausage, a popular appetizer in China. RT-qPCR was employed to evaluate the gene expression level of histidine decarboxylase (HDC) cascade-associated genes. Histamine accumulation was suppressed by inhibitory effects of the thyme microcapsule on histamine-producing bacteria and reductions in the transcription of hdcA and hdcP genes. Furthermore, the addition of thyme microcapsules in Xinjiang smoked horsemeat sausage inhibits the potential spoilage and pathogenic microbial growth (e.g., Enterobacteria).

Jiang et al. [12] investigated the effect of lipase addition on *suanzhayu*, a Chinese traditional solid fermented fish product, which is produced by mixing rice powders with seasonings and fresh fish meat in a sealed long fermentation condition. The addition of lipase had little effect on the structure of the microbiome, but promoted the growth of *Proteus* and the formation of VOCs, especially aldehydes and esters. The correlation analysis showed that *Lactobacillus*, *Enterococcus*, and *Proteus* played an important role in the safety of the product, inhibiting the potential pathogenic *Escherichia-Shigella*. The addition of lipase could be used as a novel means to enhance the quality of *suanzhayu*. Finally, the last paper evaluated lactic acid fermentation of *Arthrospira platensis* biomass, focusing on changes in the aromatic profile of this cyanobacterium widely used in food formulations and mainly consumed as a food supplement because of its high nutritional value. Martelli et al. [13] applied two different stabilization treatments on the biomass (UV light treatment and sterilization) prior to solid-state fermentation with *Lacticaseibacillus casei* 2240 and *L. rhamnosus* GG. The fermentation process was useful for off-flavor reduction. In particular, the fermentation process significantly influenced the concentration of those compounds responsible for aldehydic/ethereal, buttery/waxy (acetoin and diacetyl), alkane, and fermented aromatic notes (isoamyl alcohol). Fermentation with LAB can be an interesting tool to obtain a lyophilized spirulina powder with more pleasant sensory properties for potential use as a food supplement or in food formulations.

Author Contributions: All authors have made a substantial, direct, and intellectual contribution to the work, and approved it for publication. All authors have read and agreed to the published version of the manuscript.

Funding: This research received no external funding.

Acknowledgments: *In memoriam* of Giuliano Ezio Sansebastiano and Roberto Massini. The authors would like to recognize through this acknowledgment the friendship and teachings received from the extraordinary professors and friends.

Conflicts of Interest: The authors declare no conflict of interest.

References

1. Ojeda-Linares, C.; Álvarez-Ríos, G.D.; Figueredo-Urbina, C.J.; Islas, L.A.; Lappe-Oliveras, P.; Nabhan, G.P.; Torres-García, I.; Vallejo, M.; Casas, A. Traditional Fermented Beverages of Mexico: A Biocultural Unseen Foodscape. *Foods* **2021**, *10*, 2390. [CrossRef] [PubMed]
2. Getachew Fentie, E.; Admassu Emire, S.; Dessalegn Demsash, H.; Worku Dadi, D.; Shin, J. Cereal- and Fruit-Based Ethiopian Traditional Fermented Alcoholic Beverages. *Foods* **2020**, *9*, 1781. [CrossRef]
3. Maisto, M.; Annunziata, G.; Schiano, E.; Piccolo, V.; Iannuzzo, F.; Santangelo, R.; Ciampaglia, R.; Tenore, G.C.; Novellino, E.; Grieco, P. Potential Functional Snacks: Date Fruit Bars Supplemented by Different Species of *Lactobacillus* spp. *Foods* **2021**, *10*, 1760. [CrossRef]
4. Moon, S.-H.; Chang, H.-C. Rice Bran Fermentation Using *Lactiplantibacillus plantarum* EM as a Starter and the Potential of the Fermented Rice Bran as a Functional Food. *Foods* **2021**, *10*, 978. [CrossRef] [PubMed]
5. Yang, Y.; Lan, G.; Tian, X.; He, L.; Li, C.; Zeng, X.; Wang, X. Effect of Fermentation Parameters on Natto and Its Thrombolytic Property. *Foods* **2021**, *10*, 2547. [CrossRef] [PubMed]
6. Jiang, L.; Xian, S.; Liu, X.; Shen, G.; Zhang, Z.; Hou, X.; Chen, A. Metagenomic Study on Chinese Homemade Paocai: The Effects of Raw Materials and Fermentation Periods on the Microbial Ecology and Volatile Components. *Foods* **2022**, *11*, 62. [CrossRef] [PubMed]

7. Zago, M.; Rossetti, L.; Bardelli, T.; Carminati, D.; Nazzicari, N.; Giraffa, G. Bacterial Community of Grana Padano PDO Cheese and Generical Hard Cheeses: DNA Metabarcoding and DNA Metafingerprinting Analysis to Assess Similarities and Differences. *Foods* **2021**, *10*, 1826. [CrossRef] [PubMed]
8. Degenhardt, R.; Sobral Marques Souza, D.; Acordi Menezes, L.A.; de Melo Pereira, G.V.; Rodríguez-Lázaro, D.; Fongaro, G.; De Dea Lindner, J. Detection of Enteric Viruses and Core Microbiome Analysis in Artisanal Colonial Salami-Type Dry-Fermented Sausages from Santa Catarina, Brazil. *Foods* **2021**, *10*, 1957. [CrossRef] [PubMed]
9. Barbieri, F.; Tabanelli, G.; Montanari, C.; Dall'Osso, N.; Šimat, V.; Smole Možina, S.; Baños, A.; Özogul, F.; Bassi, D.; Fontana, C.; et al. Mediterranean Spontaneously Fermented Sausages: Spotlight on Microbiological and Quality Features to Exploit Their Bacterial Biodiversity. *Foods* **2021**, *10*, 2691. [CrossRef]
10. Ma, X.; Bi, J.; Li, X.; Zhang, G.; Hao, H.; Hou, H. Contribution of Microorganisms to Biogenic Amine Accumulation during Fish Sauce Fermentation and Screening of Novel Starters. *Foods* **2021**, *10*, 2572. [CrossRef] [PubMed]
11. Yu, H.; Huang, Y.; Lu, L.; Liu, Y.; Tang, Z.; Lu, S. Impact of Thyme Microcapsules on Histamine Production by *Proteus bacillus* in Xinjiang Smoked Horsemeat Sausage. *Foods* **2021**, *10*, 2491. [CrossRef] [PubMed]
12. Jiang, C.; Liu, M.; Yan, X.; Bao, R.; Liu, A.; Wang, W.; Zhang, Z.; Liang, H.; Ji, C.; Zhang, S.; et al. Lipase Addition Promoted the Growth of *Proteus* and the Formation of Volatile Compounds in *Suanzhayu*, a Traditional Fermented Fish Product. *Foods* **2021**, *10*, 2529. [CrossRef] [PubMed]
13. Martelli, F.; Cirlini, M.; Lazzi, C.; Neviani, E.; Bernini, V. Solid-State Fermentation of *Arthrospira platensis* to Implement New Food Products: Evaluation of Stabilization Treatments and Bacterial Growth on the Volatile Fraction. *Foods* **2021**, *10*, 67. [CrossRef] [PubMed]

Review

Traditional Fermented Beverages of Mexico: A Biocultural Unseen Foodscape

César Ojeda-Linares [1], Gonzalo D. Álvarez-Ríos [1], Carmen Julia Figueredo-Urbina [2], Luis Alfredo Islas [1], Patricia Lappe-Oliveras [3], Gary Paul Nabhan [4], Ignacio Torres-García [5], Mariana Vallejo [6,*] and Alejandro Casas [1,*]

[1] Instituto de Investigaciones en Ecosistemas y Sustentabilidad, Universidad Nacional Autónoma de México, Antigua Carretera a Pátzcuaro 8701, Col. San José de la Huerta, Morelia 58190, Mexico; cojeda@cieco.unam.mx (C.O.-L.); galvarez@cieco.unam.mx (G.D.Á.-R.); lislas@cieco.unam.mx (L.A.I.)
[2] Cátedras CONACYT-Laboratorio de Genética, Área Académica de Biología, Instituto de Ciencias Básicas e Ingeniería, Universidad Autónoma del Estado de Hidalgo, Mineral de la Reforma 42184, Mexico; figueredocj@gmail.com
[3] Instituto de Biología, Tercer Circuito Exterior, S/N Ciudad Universitaria, Coyoacán, Ciudad de México 04510, Mexico; lappe@ib.unam.mx
[4] The Southwest Center of University of Arizona, 1401 E. First St., Tucson, AZ 85721, USA; gpnabhan@arizona.edu
[5] Escuela Nacional de Estudios Superiores Unidad Morelia, Universidad Nacional Autónoma de México, Antigua Carretera a Pátzcuaro No. 8701, Col. Ex Hacienda de San José de la Huerta C.P., Morelia 58190, Mexico; itorresg@enesmorelia.unam.mx
[6] Jardín Botánico, Instituto de Biología, Universidad Nacional Autónoma de México, Tercer Circuito Exterior, S/N Ciudad Universitaria, Coyoacán, Ciudad de México 04510, Mexico
* Correspondence: mariana.vallejo@ib.unam.mx (M.V.); acasas@cieco.unam.mx (A.C.); Tel.: +52-55-5622-9045 (M.V.); +52-55-5322-2703 (A.C.)

Citation: Ojeda-Linares, C.; Álvarez-Ríos, G.D.; Figueredo-Urbina, C.J.; Islas, L.A.; Lappe-Oliveras, P.; Nabhan, G.P.; Torres-García, I.; Vallejo, M.; Casas, A. Traditional Fermented Beverages of Mexico: A Biocultural Unseen Foodscape. *Foods* **2021**, *10*, 2390. https://doi.org/10.3390/foods10102390

Academic Editors: Valentina Bernini and Juliano De Dea Lindner

Received: 9 September 2021
Accepted: 7 October 2021
Published: 9 October 2021

Abstract: Mexico is one of the main regions of the world where the domestication of numerous edible plant species originated. Its cuisine is considered an Intangible Cultural Heritage of Humanity and ferments are important components but have been poorly studied. Traditional fermented foods are still diverse, but some are endangered, requiring actions to promote their preservation. Our study aimed to (1) systematize information on the diversity and cultural history of traditional Mexican fermented beverages (TMFB), (2) document their spatial distribution, and (3) identify the main research trends and topics needed for their conservation and recovery. We reviewed information and constructed a database with biocultural information about TMFB prepared and consumed in Mexico, and we analyzed the information through network approaches and mapped it. We identified 16 TMFB and 143 plant species involved in their production, species of Cactaceae, Asparagaceae, and Poaceae being the most common substrates. Microbiological research has been directed to the potential biotechnological applications of *Lactobacillus*, *Bacillus*, and *Saccharomyces*. We identified a major gap of research on uncommon beverages and poor attention on the cultural and technological aspects. TMFB are dynamic and heterogenous foodscapes that are valuable biocultural reservoirs. Policies should include their promotion for conservation. The main needs of research and policies are discussed.

Keywords: ethnobiology; ethnozymology; fermentation; Mesoamerican biocultural heritage; traditional food systems

1. Introduction

The consumption of ferments by humans probably started unintentionally since fermentation occurs in nature as a common process [1]. However, whatever its origin, the consumption of fermented foods has accompanied people since ancient times [2]. The obtaining of fermented products with desirable attributes and the conditions determining them were eventually recognized and deliberately selected and favored [3]. Together with roasting, fermentation is perhaps one of the oldest practices and techniques used by humans to process plant and animal materials to prepare food [4]. Fermented products commonly include foods, among them beverages, but dyes, fibers, and other products involving fermentation have also been recorded since remote past times [5–7]. At some point in the human cultural history, knowledge over the fermentation process increased, hygienic practices were developed, and new techniques were adopted, outstandingly, the use of inoculums; subsequently, the manufacturing of fermented foods and beverages became the responsibility of skilled craftspersons who were mainly responsible for developing and improving the technologies for making fermented foods. Those crafters have been recognized in the literature as traditional fermentation managers [7].

Currently, a small fraction of species of organisms are predominantly used in the human diet worldwide, and this fact makes the hegemonic food systems vulnerable [8]. Fermentation plays an important role in improve the shelf life of several products, as well as to increase the diversity of food supplies and their nutritional value [9]. Therefore, the goals of diversifying food systems of the world have a key support in ferments. Most traditional ferments represent keystone products in several countries because they are expressions of interrelationships between biological and cultural diversity, and dynamic biocultural knowledge changing according to adaptive processes [9,10].

Fermented products are outputs of complex interactions of three main components, substrates, microorganisms causing fermentation, and humans who drive the process and value the final products [11,12]. Humans have observed and directed the fermentation process by choosing substrates, and selecting specific attributes of their beverages looking for pleasant-tasting, nicely textured, and long-lasting products [7,12–14]. Through time, humans have improved their techniques by experimenting with local substrates, adding others, testing variations in the general conditions of the process, and evaluating the product resulting from that process [11–16]. These biocultural practices have generated thousands of food types, including beverages, throughout human history and it is a process that continues evolving to the present day.

The fermented foods and beverages were used long before any awareness of microbiology and biochemistry [17], but now these sciences provide research tools that make it possible to identify the crucial role of microorganism assemblages in fermentation, as well as the techniques to direct fermentation according to human purposes. Microorganisms are present in substrates (plant or animal) or could be added as consortiums of microorganisms known as starter cultures. These starters may simply be small batches of previous fermentations, stored and added to new substrates to guarantee a faster and homogeneous fermentation that yields specific sensorial attributes [18–20]. Moreover, these may be microorganisms that remain in the containers where previous fermentation processes occurred, and people leave them there for new fermentation [21]. Environmental conditions are variable during fermentation, which confers variable quality and sensorial characteristics to the final product. However, traditional producers perform different practices to reduce the effect of environmental heterogeneity during the fermentation processes [7,15,22,23], while commercially fermented products are highly technified and the outcome is standardized.

Fermented beverages are among the most iconic fermented foods worldwide; these are essential components of local diets in many cultures and are mainly prepared from plant substrates [24–27]. It has been reported that through fermentation, foods may be improved, preserved, and their organoleptic properties enhanced [11]. Fermented products can have an important sociocultural role [15,24–27], consumed during holidays, ceremonies, and rituals, and linked to fruiting seasons and the harvest of agricultural products [27]. Both consumers and producers decide if the quality of a fermented product is acceptable considering their attributes and determine if it is kept or not in their diet [11–14,25–27]. Despite the importance of fermented products around the world, efforts for tracking their loss and actions to recover their production are uncommon. However, there are iconic and inspiring cases such as that in northeastern Poland, where researchers registered signs of recovering the tradition of producing juniper beers [28].

The traditional ecological knowledge involved in the preparation of fermented products appears to be at risk because of the overreliance on the massive campaigns of commercially produced beverages, which reach rural regions where the traditional products are prepared [29,30]. This fact, as well as the decline of knowledge transfer and gaps in documenting the traditional know-how of local practices to manage the microbiota, ingredients, and fermentation process have favored the marginalization and disappearance of homemade fermented products. This phenomenon contrasts with the increasing interest in the nutritional value of ferments and their economic profits in markets, especially for the trendy fermented beverages, also called functional products, for instance, kombucha [29–31].

The human diet worldwide is vulnerable since it is mostly limited to 10 to 50 plant species that provide about 95% of the world's caloric intake [32–35]. In contrast, ethnobotanists and anthropologists have documented thousands of edible species throughout the world. Only in Mexico, several studies have reported more than 2000 edible plant species, [36–40]. Among them, numerous wild species provide vegetables, fruits, nuts, tubers, and other edible products. However, little is known about the diversity of plant substrates that are employed for fermented products, their distribution, and if they are handled for specific fermentation purposes.

In 2010, the UNESCO recognized food as an intangible cultural heritage and included the Mexican gastronomy in the representative list of "Intangible Cultural Heritage of Humanity" [41,42]. However, the traditional fermented beverages are a neglected food group of the cultural heritage of the Mexican cuisine, and little is known about its conservation and biocultural status, permanence, or possible recovery. Several authors [43–46] have defined the traditional products considering the following criteria: (1) The key production steps of a traditional food product is performed in a certain national, regional, or local area, (2) the traditional food product must be authentic in its recipe, origin of raw material, and/or production process, (3) the traditional food product could have been exchanged locally for several years, available often through barter or only consumed by the family core, and (4) it is part of the gastronomic heritage. The tradition provides materials (raw substrates and tools) that allow groups to be rooted in their past and move and rebuild themselves in the present [43–46]. Therefore, it has been recognized that traditional products are dynamic.

In Mexico, a broad spectrum of traditional fermented foods, including beverages, is produced from different raw materials by Indigenous and mestizo people in different regions, the processing methods and microbiota varying among regions, localities, and producers [47–50]. A previous work by Godoy and collaborators [51] reported 66 types of fermented beverages, 4 of them registered only in historical books but not consumed anymore. Most of the beverages reported by Godoy are a combination of fermented agave sap and several fruits, known as curados. Pineda [52] described over 75 types of fermented beverages, including a few distilled for preparing spirits, listing numerous herbs, fruits, roots, and stalks that were used to infuse those beverages. Bruman [53] conducted an important work in the 1930s when he visited several regions of Mexico and recorded several traditional beverages; therefore, in this study we consider his work fundamental to characterize a previous distribution of the TMFB and to compare it with the present. The previous works help to identify historically important beverages prepared with wild or cultivated plants and to determine whether the plants or the traditional knowledge and practices have become rare or lost.

Several studies have been directed to characterize the microbial communities associated with some TMFB [47–50], their possible benefits to human health [47–50], and their biotechnological and functional applications as the probiotic activity or the production of antimicrobial activity [47–50]. However, to our knowledge, none of them have been under detailed clinical trials to corroborate their benefits for human health [47–50] However, few studies have been directed to characterize the rich legacy and diversity of fermented beverages as biocultural reservoirs of diversity of organisms and the variety of practices to elaborate a culturally accepted product.

The characterization of microbial communities of Mexican fermented products has been commonly performed by culture-dependent approaches. Early classifications of products were conducted based on the physiological and biochemical criteria and microscopic inspection [51]. Nowadays, molecular markers such as 16S and ITS have been used to accurately characterize pure strains of bacteria and fungi. Nevertheless, the scope for such approaches is limited, time-consuming, and the results might dismiss the identification of uncultivable microorganisms. Culture-independent approaches have been used in other studies; for instance, denaturing gradient gel electrophoresis (DGGE) has been used to characterize the microbial communities of TMFB such as **mescal** [47,48], **pulque** [54], and **pozol** [55]. Although culture-independent techniques give insights of the composition of microbial communities, these are also time-consuming and do not provide information on rare species.

More recently, some studies have used high throughput sequencing (HTS) techniques, such as Illumina MiSeq paired-end sequencing of barcoded polymerase chain reaction (PCR) amplicons, but these studies are still scarce in TMFB. However, studies with shotgun metagenomics have recently been conducted to characterize and infer the dynamics of microbial communities of **pulque** [56,57], **atole agrio** [58], and **tuba** [59]. A few studies have been performed under a polyphasic approach but most of these studies have been used to characterize the bacterial communities, while fungal communities have not been deeply studied. The study of microbial communities in fermented products is crucial since it allows for monitoring the processes and practices and gives insights of the distinctive autochthonous peculiarities of a product. Microbial diversity is a determinant for understanding the chemistry and nutritional properties for health, yield, and quality of the plant substrates employed for fermentation [60].

This review started with the assumption that the TMFB are the outcome of the synthesis of the vast biological and cultural diversity. In particular, we expected to identify a high diversity of TMFB, which is related to the similarly high cultural diversity represented by the number of Indigenous groups that are distributed throughout the different regions and ecosystems of Mexico. We expected that those more widely distributed and more studied groups from ethnobiological perspectives (the Nahua, Maya, Mixtec, and Zapotec in central and southern Mexico, as well as the Rarámuri, Seri, Pápago, or O'odham and the Tepehuan in northern Mexico [61]) are also those with more records of TMFB. The processes of cultural erosion have been documented in different regions, which have led to the loss of traditional cultural elements; we supposed these include TMFB. However, it is relevant to identify which ones and how much are endangered because of their decreasing availability or consumption. We, in addition, expected that those beverages based on substrates with a broader distribution and higher rooting would be the most frequent and more studied. Finally, we expected to identify a core of components of microbial communities in the diverse beverages since most of the TMFB pass through a lactic and alcoholic fermentation stage. The aims of this review are therefore to: (1) Characterize the diversity of the TMFB to provide an overview of the research on fermented beverages, the cultural groups that produce them, the plant substrates used, and the microorganisms identified; (2) visualize the spatial distribution of these beverages in the country; (3) identify their presence or absence in the foodscape and the status of conservation policies promoted by the Mexican authorities; and (4) identify the main trends of studies conducted on TMFB and the topics that are needed for a research agenda towards sustainable use of traditional ferments of Mexico. We emphasize the importance of the fermented beverages as reservoirs of biological diversity, and their relevance as Mexican biocultural heritage conferring identity to cultural groups as unique and diverse foodscapes. We, in addition, aspire to provide helpful information for designing strategies for conserving and recovering such a valuable heritage.

2. Materials and Methods

2.1. Literature Review

We conducted a search of peer-reviewed literature in Scopus, Google Scholar, Google, and the Web of Science databases to visualize the current state of research on TMFB. In addition, since numerous studies are not covered in these databases, we reviewed local journals, technical reports, books, and Ph.D. theses from regional universities of Mexico to complement the information. We also consulted references from gastronomic literature and governmental sources information. Through this search, we identified 328 peer-reviewed articles plus 197 other references related with TMFB. A search of peer-reviewed literature was performed with the following keywords: Mexican fermented beverages, traditional fermented beverages, **tepache**, **pulque**, **mescal**, **colonche**, **jobo** or **hobo**, **colonche**, **nawait**, **pozol**, **tejuino**, **tesgüino**, **piznate**, **taberna**, **cocoyol**, **tuba**, Mexican palm wine, **balché**, **xtabentún**. As these names are mainly in Spanish, we used the Boolean operators OR and AND. For example, in the Web of Science, the following search string was used: "Agave" AND "fermentation", "tradicional" AND "bebidas", "traditional" OR "ancestral" to improve the research parameters. We extended the search in local databases using keywords such as Mexican fermented beverages, bebidas fermentadas mexicanas, fermented beverages in Mexico, bebidas fermentadas en México, fermented products in Mexico, productos fermentados en México, **tepache**, **pulque**, **colonche**, **hobo**, Mexican wines, vinos mexicanos, **colonche**, **nawait**, **pozol**, **sidra**, **tejuino**, **tesgüino**, **taberna**, **cocoyol**, **tuba**, Mexican palm wine, and **balché**.

Most of the beverages recorded are wine-like beverages produced by the fermentation of several fruit species; however, we found numerous cases in which the producers use to add sugar cane alcohol or spirit beverages to confer specific flavors, and these beverages were not considered in our analysis. We found different names for beverages produced with **pulque** mixed with several fruits known as *pulques curados*; in this case, we only considered the name **pulque** for all of these beverages. In a broad sense, in this review, we only considered the main substrates for fermentation and secondary substrates employed during the fermentation process.

To characterize the current state of research on TMFB, the information of the title and the abstract of each article were used as inputs in the software VOSviewer 1.6.15 [62,63]. The construction and visualization of the resulting conceptual networks were performed with the full counting method with at least three occurrences of a term, and a consideration of 60% of relevance. The layout attraction and layout repulsion parameters were scored 1 and 0, respectively, the distance between nodes represents the connectivity among concepts, and the nodes size represents the number of mentions of the concept. The clustering resolution and minimum cluster size parameters were set to 1.25 and 5, respectively.

The information from this search was systematically stored in a database using Access with the following fields: (1) Fermented product name; (2) plant substrates (both scientific and local names; (3) microorganisms (both bacteria and yeasts, and the techniques employed to identify them, e.g., culture-dependent or culture-independent); (4) ethnic groups producing and using the beverage (the dosage of consumption, related customs, and rituals; whether it is still consumed or extinct); (5) sensorial features associated to beverages (sensorial attributes reported as sweet, acid, etc.); and (6) medicinal (if it has been reported to improve human health). Searches in the databases considered studies from 1960 to August 2020. In addition, to document the conservation policies over biological resources and traditional knowledge, we searched in governmental agencies databases.

2.2. Map Construction

A map was constructed to visualize the documented distribution of fermented beverages based on the data collected in the literature; the area of the municipalities identified was considered as the documented distribution area. Another map was constructed based on the estimated distribution regions proposed by Bruman [53]. In addition, to visualize the distribution of TMFB documented and that of the cultural groups, we considered the distribution of the Indigenous languages of Mexico reported by the Comisión Nacional para el Conocimiento y Uso de la Biodiversidad (CONABIO) [64]. The map design and analysis were made with the Qgis free software [65]. Finally, a map was constructed with the overlapped data of the documented distribution and Bruman's proposal to visualize a potential area of distribution of the TMFB and the gaps that have not been documented.

2.3. Network Analysis

To visualize the uniqueness, diversity, and interaction of the substrates employed to produce fermented beverages, and the co-occurrence of specialist or generalist species in microbial communities, we performed an exploration through the bipartite network approach. The network's theory offers powerful tools to describe complex communities and the distribution of species specificity within them. Moreover, this approach was performed with the microorganisms identified in the literature to each traditional fermented beverage. The data from the network structure were used as descriptors of diversity and uniqueness. The analysis was performed through the Rstudio software [66] with the igraph package [67].

3. Results

3.1. Traditional Mexican Fermented Beverages

Based on a previous review performed by Godoy and collaborators [51], 66 types of fermented beverages were registered. Nevertheless, most of them are beverages prepared with pulque whose main substrate is the fermented sap from *Agave* plants and the addition of fruits of native and introduced plant species. By the current review, we identified 16 names for traditional fermented beverages produced from several substrates such as seeds, stems, fruits, tree barks, fruit pulps, and sap. Stems are the dominant substrates used for preparing traditional fermented beverages. We identified 140 plant species used as main substrates for fermentation or as promoters of fermentation.

Through the peer-reviewed literature, we identified 10 traditional fermented beverages and we extended the number of studies based on the information from theses and the local literature, from which we identified 6 additional beverages; therefore, we documented a total of 16 fermented beverages. The most studied beverages are **mescal**, **pulque**, **tejuino**, **pozol**, **chorote**, **colonche**, **saká**, **sendechó**, **balché**, **atole agrio**, **pox**, **sambudia**, **tesgüino**, **tepache**, **tuba**, and **taberna** (Table 1).

Since we did not find recent studies on other beverages, we cannot confirm what is their actual status of production and distribution. Clearly, further field studies should be directed to document what is currently happening with these beverages. It can be visualized that the current research tendency is markedly directed towards the beverages grouped into the category of *Agave* spirits or *Agave* distillates referred to in this study as **mescal**, compared with the rest of the traditional fermented beverages in peer-reviewed literature.

The current state of research in peer-reviewed articles on TMFB can be visualized in the network of Figure 1, which displays nine conceptual clusters that highlight the main topics related to the beverages. Cluster (1) (in red) is related to the application, evaluation, and sensory properties of the fermented products. This cluster resembles the classical biotechnological approach from the early 1990s, most of the research being directed to improve the sensory properties of **mescal**. It also includes new trends related to the application of defined starter cultures. Cluster (2) (in green) is the **mescal** cluster, in which *Agave potatorum* appears as the most studied species, and *Saccharomyces cerevisiae* as the most commonly characterized microorganism. This cluster highlights the economic importance of **mescal** production, and the interest in improving the fermentation profile using several *S. cerevisiae* strains. Cluster (3), the yeast cluster (in blue), emphasizes the use of non-*Saccharomyces* yeast species to improve the aroma profile and optimize the ethanol production in **mescal**. Cluster (4) is the pathogens cluster (in olive) and refers to the effects of pathogens such as *Scyphophorus acupunctatus*, which mainly affect the *Agave* groups, decreasing the production of beverages related to these plants. Cluster (5) is the *Agave* *salmiana* cluster (in purple), which is mainly related to the influence of environmental factors in the development of plants and the fermentation process.

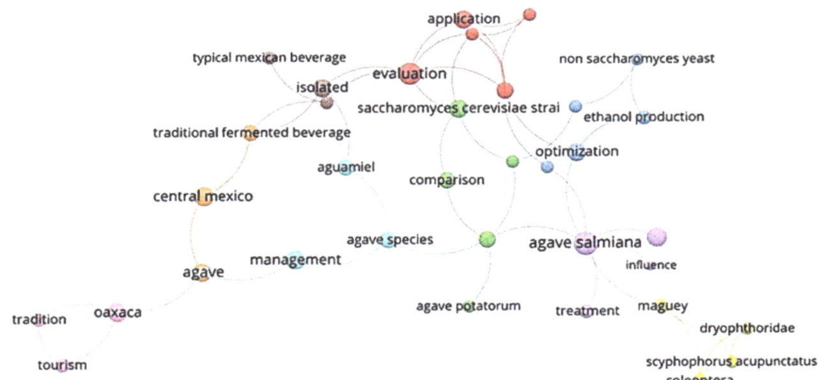

Figure 1. Clusters representing different research topics reported in the literature. The red cluster • is related to biotechnological applications in traditional fermented beverages; the green cluster • represents studies related to the mescal production; the blue cluster • are those on the application of yeasts through the production of traditional fermented beverages; the olive cluster • includes studies on pathogens affecting the development of *Agave*; the purple cluster • comprises the research on environmental factors influencing the fermentation process; the light blue cluster • are studies on management practices of *Agave* and sap extraction; the orange cluster • indicates the reports on the presence of traditional fermented beverages in central Mexico, mainly **pulque** and **mescal**; the brown cluster • shows the research that has been directed to maize beverages and the possible probiotic features associated with this microbiota; the pink cluster • shows the importance of cultural groups of related to production of fermented beverages, but it is only related with mescal production.

The following clusters are related to the most traditional beverages and human groups related to them: Cluster (6), the management cluster (in light blue), groups the management practices on *Agave* species, mainly those involved in the production of **pulque** and the implications of management of agaves for sap extraction. Cluster (7), the one of traditional fermented beverages (in orange), features the less common traditional fermented beverages prepared in central Mexico, those with *Agave* species such as **pulque** or **mescal** that occur mainly in central Mexico. Cluster (8), the maize group (in brown), mainly accounts for research directed to isolate and characterize the bacteria and yeasts associated to possible probiotic attributes in **tejuino** and **pozol**. Cluster (9), the cultural cluster (in light purple), groups studies related to traditions and tourism, mainly in the state of Oaxaca. It can be highlighted that this cluster is also related to **mescal** production.

Table 1. Main traditional Mexican fermented beverages. Main substrates for fermentation, microorganisms described, and cultural groups associated to its production.

Beverages	Main Substrate	Main Microorganisms Recorded in the Literature	Cultural Groups Associated	Literature
Pozol	*Zea mays* (grains)	**Bacteria**: *Aerobacter, Acetobacter, Achromobacter, Agrobacterium, Alcaligenes, Bacillus, Bifidobacterium, Clostridium, Enterobacter, Enterococcus, Escherichia, Exiguabacterium, Klebsiella, Kosakonia Lactobacillus, Lactococcus, Leuconostoc, Paracolobactrum, Pediococcus, Pseudomonas, Propionibacterium, Streptococcus, Weissella.* **Yeasts**: *Candida, Cyberlindera, Debaryomyces, Kluyveromyces, Galactomyces, Meyerozyma, Pichia, Rhodotorula, Trichosporon.* **Fungi**: *Cladosporium, Monilia, Mucor, Phoma, Penicillium.*	Chol, Chontal, Lacandon, Mam, Maya, Tojolabal, Tzeltal, Tzotzil, Zapotec, Zoque, Mestizo	[52–55,61,68–70] [71–79]
Atole agrio	*Zea mays* (grains)	**Bacteria**: *Acetobacter, Aerococcus, Bacillus, Enterococcus, Clostridium, Lactobacillus, Lactococcus, Leuconostoc, Pediococcus, Streptococcus,* Weissella. **Yeasts:** *Candida, Cryptococcus, Clavispora, Debaryomyces, Hanseniaspor, Pichia Saccharomycesa.* **Fungi:** *Aspergillus, Fusarium, Penicillium.*	Maya, Mazatec, Nahua, Purepecha, Totonac, Tzeltal, Tzotzil, Wixarika, Mestizo	[58,72,73,80–83]
Saká	*Zea mays* (grains)	ND	Maya, Tzotzil, Tzeltal	[84]
Tejuino	*Zea mays* (grains)	**Bacteria**: *Acetobacter, Bacillus, Brochothrix, Chyseobacterium, Kurthia, Lactobacillus, Leuconostoc, Pantoea, Pseudomoonas, Strotococcus, Weissella.* **Yeasts**: *Candida, Galactomyces, Lachancea, Meyerozyma, Saccharomyces, Wickehamomyces.***Fungi**. *Aspergillus, Penicillium*	Mestizo	[85,86]
Tesgüino	*Zea mays* (grains)	**Bacteria**: *Bacillus, Lactobacillus, Bacillus, Leuconostoc, Pediococcus, Streptococcus.* **Yeasts**: *Brettanomyces, Candida, Clavispora, Cryptococcus, Kluyveromyces, Lachancea, Metschnikowia, Meyerozyma, Pichia, Saccharomyces, Wicherhamomyces.* **Fungi**: *Aspergillus, Penicillium.*	Guajiro, Pame, Pima, Tarahumara, Tepehuan, Tubar, Wixarika, Yaqui, Zapotec	[46–49,87–92]
Pulque	*Agave* spp. (sap)	**Bacteria**: *Acetobacter, Acetobacterium, Acinetobacter, Acrobacter, Adlercreutzia, Ardescatena, Bacillus, Commensalibacter, Citrobacter, Cellulomonas, Cellulosimicrobium, Chelativorum, Chryseobacterium, Chryseomonas, Clostridium, Comensalbacter, Corynebacterium, Devosia, Dysgonomonas, Enterobacter, Erwinia, Escherichia, Euzebia, Flavobacterium, Fructobacillus, Gluconobacter, Hafnia, Halomicronema, Kluyvera, Klebsiella, Kokuria, Komagataeibacter, Lactobacillus, Lactococcus, Luteomicrobium, Leuconostoc, Marivitia, Macrococcus, Mesorhizobium, Micrococcus, Microbacterium, Micrococcus, Novosphingobium, Providencia, Pediococcus, Pseudomonas, Rhodobacter, Rhodovulum, Ruminococcus, Sacrcina, Salinibacterim. Sarcandra, Serratia, Sphaerotilus, Sphingomonas, Sphingopyxis, Streptococcus, Streptomyces, Sulfuropirillum, Synechococcus, Tanticharoenia, Trochococcus, Weissella, Zymomonas.* **Yeasts**: *Bullera, Candida, Clavispora, Cryptococcus, Cystofilobasidium, Debaryomyces, Dekkera, Galactomyces, Hanseniaspora, Kazachstania, Kluyveromyces, Lipomyces, Meyerozyma, Pichia, Rhodotorula, Saccharomyces, Schwanniomyces. Torulaspora, Westerdykella, Wickerhamomyces, Zygosaccharomyces.* **Fungi**: *Aureobasidium. Aspergillus, Cladosporium. Penicillium, Rhizopus.*	Hñähñu, Ixcatec, Mazahua, Mixtec, Nahua, Ngiwa, Purhepecha, Triqui, Zapotec, Mestizo	[23,45–48,93–97]

Table 1. *Cont.*

Beverages	Main Substrate	Main Microorganisms Recorded in the Literature	Cultural Groups Associated	Literature
Tuba	*Cocos nucifera* (sap)	**Bacteria**: *Bacillus, Cronobacter, Enterococcus, Erwinia, Fructobacillus, Gluconacetobacter, Klebsiella, Lactobacillus, Lactococcus, Leuconostoc, Micrococcus, Serratia, Sphingomonas, Vibrio, Zymomonas.* **Yeasts**: *Candida, Cryptococcus, Hanseniaspora, Saccharomyces.*	Mestizo	[96,98–115]
Taberna	*Acrocomia acuelata* (sap)	**Bacteria**: *Aerobacter, Acetobacter, Bacillus, Brevundimonas, Citrobacter, Enterobacter, Enterococcus, Fructobacillus, Gluconobacter, Klebsiella, Kluyvera, Lactobacillus, Lactococcus, Pantoea, Sphingomonas, Zymomonas.* **Yeasts**: *Candida, Hanseniapora, Issatchenkia, Kazachstania. Meyerozyma, Pichia. Rhodotorula, Saccharomyces, Schizosaccharomyces*	Zapotec, Mestizo	[98–100]
Tepache	*Ananas comosus* (fruit)	***Bacteria***: *Acetobacter, Acinetobacter, Bacillus, Escherichia, Enterobacter, Enterococcus, Gluconobacter, Klebsiella, Lactobacillus, Lactococcus, Leuconostoc, Micrococcus, Pediococcusa, Weissella.* ***Yeasts***: *Candida, Cryptococcus, Hanseniaspora, Meyerozyma, Pichia, Rhodotorula, Saccharomyces.* **Fungi**: *Penicillium.*	Mestizo	[49,50,116–118]
Colonche	*Opuntia* spp. (fruits), *Pacchycerus, Stenocereus*	**Bacteria**: *Enterococcus, Lactobacillus, Leuconoctoc, Pediococcus, Weissella.* **Yeasts**: *Candida, Hanseniaspora, Pichia, Saccharomyces.*	Chichimecan groups, Pame, Zapotec, Mestizo	[22,50,52,119]
Mescal	*Agave* spp.	**Bacteria**: *Acetobacter, Acinetobacter, Acetobacterium, Bacillus, Citrobacter,* Enterobacter, Erwinia, Chryseobacterium, *Gluconobacter, Kluyvera, Kokuria, Komagataeibacter, Lactobacillus, Lactococcus, Leuconostoc, Microbacterium, Providencia, Oenococcus, Pediococcus, Pseudomonas Serratia, Weissella, Zymomonas.* **Yeasts**: *Candida, Citeromyces, Clavispora, Cryptococcus, Debaryomyces, Dekkera, Diutinia, Hanseniaspora, Issatchenkia, Kazachstania, Kluyveromyces, Meyerozyma, Millerozyma, Naganishia, Ogataea, Pichia, Pseudozyma, Rhodosporidiobolus, Rhodotorula, Saccharomyces, Saturnispora, Schizosaccharomyces Sporidiobolus, Torulaspora, Trichosporon, Wickerhamomyces, Yamadazyma, Zygosaccharomyces.*	Cahiti, Guasave, Ixcatec, Pima, Tepehuan, Warohiro, Wixarika, Mestizo	[116,120–126]
Chorote	*Zea mayz (grains) Theobroma cacao* rosted beans),	*Fructobacillus, Lactobacillus, Leuconostoc, Gluconacetobacter, Sphingomonas, Vibrio,*	Maya, Mestizo	[127]
Balché	*Lonchocarpus* spp. *(bark and flowers)* and honeybee	**Yeasts**: *Saccharomyces.*	Lacandon, Maya	[84,128–133]
Pox	*Saccharum officinarum* and *Zea mays (stems)*	ND	Chol, Tzeltal, Tzotzil Mestizo	[53]

Table 1. *Cont.*

Beverages	Main Substrate	Main Microorganisms Recorded in the Literature	Cultural Groups Associated	Literature
Sambudia	*Ananas comosus* (fruit)	ND	Mazahua, Mestizo	[134–136]
Sendechó	*Zea mays (grains)* and *Capsicum* sp.	**Bacteria**: *Acetobacter, Bacillus, Enterococcus, Klebsiella, Kocuria, Lactobacillus, Leuconostoc, Micrococcus, Pediococcus, Pseudomionas, Staphylococcus, Zymomonas.* **Yeasts**: *Candida, Clavispora, Cryptococcus, Galactomyces, Kluyveromyces, Rhodotorula, Saccharomyces, Torulaspora, Wickerhamomyces, Zygosaccharomyces.* **Fungi**: *Aureobasidium, Cladosporium, Epicoccum, Fusarium, Paecilomyces, Penicillum, Phima, Sclerotium, Verticillium.*	Mazahua, Hñähñu	[48–50,137–141]

ND: No data recorded.

The network shows that most of the current research is directed to study the **mescal** production process and its optimization. This is not surprising because of the national and international rising market boom of **mescal**. However, it is lower than research on tequila, which is the fourth largest export product of Mexico [142]. After **mescal**, **pulque** is the most studied beverage, particularly biotechnological and management aspects involved in its production.

3.2. Current State of the Conceptual Overview

The main microorganism referred to in the studies is *S. cerevisiae*, which is a common species in fermented products around the world and it is frequently found in alcoholic beverages. Surprisingly, it is possible to see in the network that the non-*Saccharomyces* species form a node; this is a trend around several industrialized beverages, such as beer [70], wine [143], and recently in cocoa fermentation [144], which looks for flavors and aromas in the final products. Research has been directed to characterize bacteria in the traditional fermented beverages, mainly because most of these beverages pass through a lactic fermentation stage, but perhaps the most common purpose is the isolation and evaluation of these bacteria to improve human health [145].

Two major trends can be identified in the current research on traditional fermented beverages in Mexico. The first one includes biotechnological approaches, which could be visualized in the first three clusters referred to above, which is related to the worldwide interest of the dairy industry to promote products with probiotic and prebiotic compounds [146]. This topic has been constitutively addressed in numerous research centers for several beverages around the world [147]. In the markets, there is an increasing demand for functional dairy products and healthy food [148,149] that pushes this major trend, which is reinforced by the search for options to improve nutrition in developing countries by using biotechnology [87,150]. In Mexico, this trend is clear in fermented beverages such as **pozol**, **pulque**, **tepache** and **tejuino**, whose microbiota and potential benefits to health have been majorly characterized.

The second research trend is related to the *Agave* beverages such as **pulque** and **mescal,** but also beverages such as **pozol**, which entails biotechnological approaches and the traditional management of sources of substrates, particularly *Agave* species. The rising market of **mescal** appears to be leading this trend, while in the case of **pulque**, the possible food functionalities appear to be the main drivers of the current research programs. It is possible to see the great cultural node related to **mescal** production, but few studies focused on the cultural groups in other beverages.

There is a huge gap in the research agendas on traditional fermented beverages. It is notorious the major interest that *Agave* products have but almost null on the rest of the traditional fermented beverages. As mentioned, there is high interest in biotechnological research programs specially to explore the central region of Mexico, which could be seen in Figure 4A. The remaining beverages are not only marginalized in markets, but also in the scientific agendas. Greater efforts to maintain them should be performed.

3.3. Plant Diversity Used in Mexican Traditional Fermented Beverages

We identified 143 plant species used as main and secondary substrates for fermentation, as promoters of the fermentations process, and as additives to improve the shelf life or flavor (see Table S1). The network analysis reported a low connectance, a low number of links between the species and beverages analyzed, which elucidates the high specialization of ingredients, preparation, and assemblages of the beverages as can be seen in Figure 2, in which the clusters are relatively distant from each other (see also Table 2). This pattern is corroborated with the low niche overlap and the mean of shared partners in the network. The latter value also explains the low number of generalist species involved in the network; species such as *Z. mays (Z)*, *A. salmiana (A)*, *S. officinarum (S)*, and *Cinnamomum verum (C)* can be seen involved in the production of different fermented beverages. *S. officinarum* is a species frequently employed to strengthen fermentation as it is used as an external supply of sugars, commonly added as processed brown sugar. *C. verum* is used to add different flavors to the final product but not during the fermentation process. Cinnamon has been recorded to have antimicrobial and pathogens inhibitors properties, but it has not been reported on its use to avoid spoilage. Sugarcane and cinnamon are non-native species from Mexico but are strongly integrated into the traditional fermented beverages production as enhancers or additives.

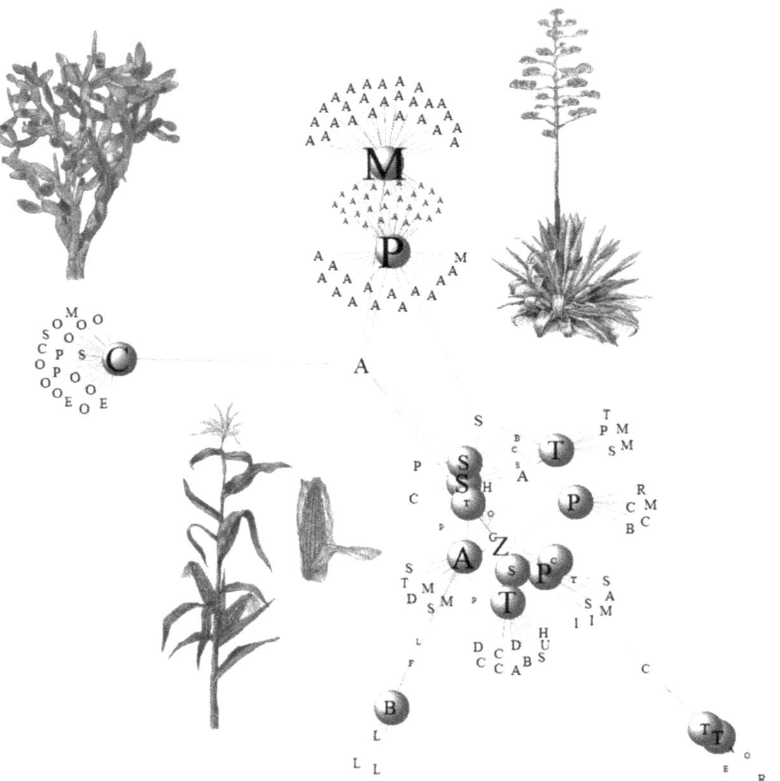

Figure 2. Network of substrates. Capital letters in the grey circles represent the TMFB, the size of circles and letters are related to the number of substrates employed; the letters without circles represent the species employed in fermentation, their size represents the number of uses in different fermented beverages. The M and P clusters are related to the substrates used for preparing **mescal** and **pulque**, respectively; the A represents the *Agave* species employed. The A in the middle of **pulque** and **mescal** cluster represents shared species, while in the external position represents exclusive agave species. The C circle is the **colonche** production, it is prepared with several cacti fruits, and shares *A. salmiana* for its production as other beverages in which the A is in the center of the network. The centered cluster is related to maize produced beverages (**pozol**, **tesgüino**, **tejuino**, **atole**, **chorote**, **saká**) in which *Z. mays* (Z in the middle) is the core substrate for these fermented beverages. The B cluster is the **balché** group, in which L letters refer to *Lonchocarpus* spp.; the T cluster is related to beverages prepared mainly by the fermentation of palm sap such as **taberna** and **tuba**. Species and beverages are listed in the Table S1. Illustration credits to Rosa Jeannine Xochicale Solís.

Table 2. Network values for plant substrates and microorganisms.

Network	Connectance	Links per Species	Niche Overlap	Mean of Shared Partners
Plant substrates	0.085	1.29	0.20	0.31
Microorganisms	0.23	1.75	0.99	7.21

The network of substrates of the traditional fermented beverages of Mexico is diverse (S = 5.21), with a minimal connectance, which indicates the uniqueness of most of the beverages with a high specificity in the substrates. The documentation and protection of those unique substrates should be attended to maintain and guarantee their production. The network allows for visualizing that maize is the most generalist substrate, which is not strange since maize is a keystone of the Mexican food systems and cultures and is prepared in great diverse ways.

3.3.1. Cluster of Maize Beverages

Eight TMFB involve the use of maize as a main or secondary substrate, most of them are prepared with maize grains, including: **atole agrio**, **saká**, **tejuino**, **tesgüino**, **sendechó**, **chorote**, and **pozol**. **Pox** was usually prepared with maize stems as the main substrate, but sugarcane stem is more commonly used now. For processing most of these products, people employ specific races of maize; we identified 21 races used to produce them, and the beverages receive different names according to the maize race and locality where they are produced. Beverages produced with maize have a deeply and strong relationship with ethnic groups. Through this review, we identified that maize beverages are the most consumed by 21 cultural groups of Mexico.

Tesgüino is produced and consumed mainly by the Rarámuri or Tarahumara people and other cultural groups such as the Yaqui and the Wixarika, who live in northern Mexico [46–49,87–90]. This beverage is commonly mentioned as the Mexican beer or maize beer, and locally named batári by the Tarahumara. It is consumed for several cultural purposes related to celebrations such as weddings, funerals, rituals associated to practices of the agricultural cycle, baptisms, and for maintaining the cohesion of the communities in the celebrations called tesgüinadas [90]. It is a beverage mainly elaborated by women by fermenting sprouted maize kernels. About five days after germination, kernels are grounded and boiled for approximately 12 h, then cooled, and the mass obtained is placed in a container known as tesgüinera where fermentation takes place. It is a spontaneous fermentation where producers commonly add leaves and other parts of local weeds and trees such as *Stevia serrata, Chimaphila maculata, Datura meteloides, Hamaemelum nobile,* and *Cojoba arborea,* all of them identified with local names [90–92]. *Usnea dillenius*, a lichen, is used as a catalyst [90].

It is commonly assumed that **tejuino** and **tesgüino** are the same beverage since the fermentation core substrate in both cases is the grounded kernels of maize (a dough); however, the process, the plant species, and the varieties added, locations, and the cultural groups are different. **Tejuino** is a TMFB prepared with the dough of one variety of maize and the addition of brown sugar. It is mostly consumed in the states of Colima and Jalisco by mestizo people, and non-special containers are used for fermentation. The **atole agrio** is a similar beverage. It is the fermentation of maize dough, and no sugar is added. It has been recorded in southern Mexico and its consumption involves the use of different local races of maize among the localities.

Pozol and **saká,** are non-alcoholic fermented beverages from the Mayan region, prepared with dry corn kernels, boiled with calcareous stones and then grinded into a dough. This is a process of nixtamalization that confers a particular consistency and flavor to the dough and adds nutrients such as calcium oxide and others that become bioavailable [54,81,151]. This maize dough can be prepared with several maize races available in the region. For both beverages, the process is clearly similar, but **pozol** is more common among the mestizo communities and **saká** is associated to Indigenous communities in religious contexts. For its production, maize dough is stored into small balls in banana leaves until fermentation, a process that varies depending on the environmental conditions and the producers' preferences. The dough balls are dissolved in water and consumed as a daily beverage [54,81–83,91,92,151]. In different locations, **pozol** dough can be mixed with batata, toasted cocoa, coconut, cinnamon, or vanilla [72–76]. These beverages are considered a source of nutrients and an essential food in southern Mexico, where its consumption is associated with traditional ceremonies, or consumed as a daily drink. It is also consumed as remedy for gastrointestinal illnesses in some localities [54,76–79].

Sendechó is a beverage that can be prepared with several maize races (five reported), depending on the producers' preferences, so its final color might vary from light red to purple, light yellow, white, and black. Through the germinating grains step, *Buddleja americana* is mixed with the grains [48,49]. **Pulque**, called ixquini in the region, is added as a starter inoculum [137,138]. The production process is similar to those maize beverages mentioned above; sprouted maize grains are ground, in this step chili peppers could be added [138], and then a maize dough is boiled with water, it is filtered with a cloth, and then spilled in a clay pot where *Agave* sap or fermented sap can be added as a starter of fermentation. It is produced in areas close to Mexico City and is still consumed by the Mazahua, Otomi, and mestizo groups for several cultural purposes.

3.3.2. Cluster of Agave Beverages

The cluster of *Agave* beverages is a highly diverse group, with 68 agave species involved in the fermentation of **pulque** and **mescal** as the main substrate to produce both beverages. The fermented sap of *A. salmiana* is the most frequently employed to produce **pulque** and it is also added as an inoculum to other beverages. Although *Agave* is a diverse group, we recorded low connectivity among its components, which suggests unique fermented products. Since species and varieties of *Agave* are not explicitly recorded in some studies, we might be underestimating their distinctiveness; there are several local names for the local varieties, suggesting high variation, but unfortunately, their taxonomic identity has been insufficiently studied and indicated.

Mescal appears to be the predominant beverage of *Agave* throughout the Mexican territory. It is in fact the beverage that employs a major number of species to get a final product, most of them are locally distributed and different species are often mixed to produce a unique sensorial profile. We consider **mescal** as a TMFB because most of the species employed for its production are endemic to Mexico, and most of them wild but some have gone through a domestication process or are under different levels of domestication by people of this country. Moreover, because these products are locally and/or regionally produced, their production is mostly unique for each producer, are part of the local gastronomic heritage, and include local and foreign technologies as mentioned before [45–48]. This group is highly variable in the production process among regions and producers, using different substrates, fermentative spaces, types of containers, and fermentation techniques. In fact, local names for mescal spirits are diverse and depend mostly on the region and substrate employed. The basic fermentative substrates are the cooked *Agave* stems and foliar bases whose products are then distilled; however, the species used are numerous and variable among the regions.

As mentioned, we did not consider other plant or animal products that are often mixed with pulque consumption which are generally called pulques curados [93–95]. We did not consider them because the main substrate is the fermented agave sap and fruits are added after **pulque** is produced. Moreover, we did not consider the species that are used as tools, such as *Lagenaria siceraria*, to store or transport the sap of *Agave* species. However, the number of *Agave* species employed as main substrates for **pulque** production (42 species) is high. **Pulque** results from the fermentation of *Agave* sap, and it has been recorded as spontaneous fermentation [91,92], but also as the product of using inoculums, which have been characterized in different regions [23,93–96]. **Pulque** preparation starts with the collection of the agave sap called aguamiel directly from the agave stem, which is mostly performed under non-sterile conditions [95,96]. The fermentation process varies depending on the agave species, the quality of the sap, the region, and other factors. There is not one single type of **pulque** because of the plethora of practices, species, and conditions for its production. These practices might change the fermentation process in time, the assemblages of yeasts and bacteria, and the final sensorial attributes.

Pulque has been used historically for festivities, ceremonies, agricultural rituals, births, and funerals [93]. There is a specific regulatory system for **pulque** in Mexican law (NMX-V-022.1972), which defines quality standards and sensorial attributes. In addition to the health benefits of **pulque** or its biotechnological applications, the diversity of plants employed and the practices performed reflect a historical know-how developed by humans in interaction with their local environments. However, this production system has been marginalized.

3.3.3. Cluster of Cacti Beverages

Colonche is a traditional fermented beverage that can be prepared with fruits of at least 17 cacti species. **Colonche** is the common name of this beverage in the region of the Altiplano central region of Mexico, where it is mainly prepared with several *Opuntia* species, although the most common and preferred by the producers and consumers is that prepared with *O. streptacantha* [22,50]. **Nawait** is another common name for a fermented beverage mainly produced by the fermentation of *Carnegiea gigantea* fruits, whose distribution covers the Sonoran Desert in northern Mexico and the southwestern U.S.A. [52]. In this region, a **colonche**-like beverage is also prepared with *Pachycereus pringlei.* In the Tehuacán-Cuicatlán Valley in south-central Mexico, **colonche** is known as nochoctli or pulque rojo. There, the fruits of several cacti are employed to produce this fermented beverage, the most common are those of *Pachycereus weberi*, *Escontria chiotilla*, and *Stenocereus* spp. This beverage is produced in different seasons depending on the availability of the substrates. For instance, in south-central Mexico, colonche is produced in April–May and in August–September, when fruits of the different species of columnar cacti and *Opuntia* used to produce it are available. The production and fermentations practices of **colonche** are variable in the different regions [22]. In most regions, fermentation occurs spontaneously; nonetheless, inoculums are employed by some producers by the addition of pulque from *A. salmiana* or by using inoculums from previous fermentations of cacti fruits. Most of the fermentations occur in clay pots continually used and that have been maintained by several generations of the managers [22].

3.3.4. Cluster of Palm Beverages

The beverages **tuba** and **taberna** form the palm cluster. These are prepared by the fermentation of the sap extracted from different palm trees, similarly as it is carried out with other palm wines consumed around the world, such as legmi in Africa, or kallu in southern India, as well as several Asian palm wines [98,99]. The main differences of these beverages are the palm species and the palm parts from which sap is obtained [99,100]. These beverages were adopted in Mexico by the Philippine influence during the Spanish colonial period and are currently produced in the Pacific coastal zones of Mexico. The species used for producing **taberna** is *Acrocomia aculeata* known as coyol, and it is produced in localities of the southern state of Chiapas [100,139–141,152–155]. The production occurs by the deliberate cut of the shoot apical meristem of the palm tree, then by making a cavity where sap is accumulated and fermented. The cavity is covered to avoid contamination by insects and then the sap is collected and consumed fresh. Sap for preparing **tuba** is obtained from the inflorescences of *Cocos nucifera*, which is collected and then stored in plastic containers, although the iconic containers are those manufactured with *L. siceraria* fruits, which are called bules [98–100,139–141,152–155].

3.3.5. Balché Cluster Beverages

Balché is an exceptionally important TMFB since it is the sacred drink for the Mayan, consumed in several ceremonies. It is the result of the fermentation of bark from several *Lonchocarpus* species, among them *L. punctatus* or *L. violaceus*, and *L. longistylus*, which is mixed with honey (either from *Apis mellifera* or melliponini bees). It has a particular pink color and sweet taste, and has been consumed since pre-Hispanic times [84,128–132]. To produce this beverage, some producers boil the bark of the tree to remove the compounds that confer a bitter flavor, thus the bark releases its characteristic color and fragrance. Then, it is dried to later be boiled with virgin water, which is collected from cenotes or rivers. After that, the bark, honey, and water are mixed and fermented spontaneously in a hole made inside a tree; later, the hole is sealed with banana or palm leaves for two or three days. Hitherto, *S. cerevisiae* has been the only microorganism recorded in the fermentation of this beverage [133], but more studies are clearly needed.

3.3.6. The Cluster of Tepache and Sambudia

Tepache can be prepared with at least 10 plant species as a substrate. Nowadays, in central Mexico the most common substrate is pineapple (*Ananas comosus*), but its etymology in the Náhuatl language derives from tepitl, a beverage made with maize in the past and in few localities [49,50], which suggests that this beverage is a derivation from an ancient maize beverage. In western Mexico, **tepache** is also produced by the fermentation of other fruits that belong to the Bromeliaceae family, such as *Bromelia karatas*, known as tumbiriche or timbiriche. It can also be prepared with fruits of plants introduced into Mexico such as apple, orange, and guava [117,118]. In the case of **tepache** prepared with pineapple, the process starts by peeling the infrutescences and adding the rind into a wooden container known as tepacheras, then brown sugar is added. It is a spontaneous fermentation where the microbiota is mainly associated with the sorosis epidermis [49,50]. Production of **tepache** with other fruits can be fermented in plastic containers; this beverage is commonly homemade, and its quality varies from kitchen to kitchen.

Sambudia is a beverage prepared with several substrates, including maize, rice, and barley. **Sambudia** is the name used for different beverages produced in the state of Mexico; a first way to produce it is using pineapple peel as a substrate [50,134–136] through a process similar to that of **tepache**. A second way to prepare it is with ground grains of rice, or barley; in this case, cinnamon, cloves, pepper, roasted and ground maize leaves, and pulque are added in a container; brown sugar is added to sweeten the mixed elements and the mixture is fermented for at least one day [136]. Moreover, it can be prepared with the fermentation of *B. karatas* fruits by adding pineapple peel, maize leaf, and ground maize. Fermentation occurs by inoculating the remnants of past fermentations in the containers, which are clay pots called sambudieras [134].

3.4. Traditional Fermented Beverages as Dynamic Systems: The Addition of New Substrates

The historical records indicate that practices and techniques for manufacturing fermented foods evolved independently in every hemisphere and were developed based on the resources available in the local environments [12]. In Mexico, the fermented beverages, practices, and techniques are applied on the three main groups of substrates referred to above. However, the inclusion of foreign species or technologies to diversify the diet are not an exception in the Mexican cuisine. Therefore, rather than cultural erosion, the contact with new species, ingredients, and techniques of the Old World appears to have enriched the Mexican fermented beverages. These historical processes support a diversification hypothesis [101].

As humans migrated from region to region, food cultures and production practices moved as well [12,101–105]. Although the most common plants used as main substrates to produce fermented beverages are maize, agaves, and columnar cacti, the inclusion of new species replaced some substrates. A clear example is the traditional fermented beverage **pox** prepared with the stems of *Z. mays* nowadays, mainly prepared with *S. officinarum* stems [53], but also the several beverages with added brown sugar from this crop originating in Southeast Asia [106]. *C. verum*, also originated in Asia [107–109], is commonly used as an additive to flavor several Mexican beverages.

The fermentation of numerous palms sap is common in countries of Asia and Africa [96,98–111] and, in Mexico, **taberna** and **tuba** are examples of the adoption of new technologies and instruments, including distillers for the Philippine coconut spirits distillation technique [112–114]. Nowadays, there is a debate about possible pre-Columbian distillation, but the fact is that the Philippine distillation technique is commonly used in several localities for producing **mescal** and other beverages [115].

3.5. Traditional Knowledge and Microbial Communities in Fermented Beverages: How Do Traditional Fermenters Promote Microbial Reservoirs and Microbial Diversity

The most common practice to promote fermentation worldwide is the inoculation of a substrate with starter cultures. For instance, in Korea, soy sauce is prepared using meju (solely fermented soybeans), whereas in Japan and China it is prepared using koji (a fermented mixture of soybeans, wheat flour, and wheat) [156]. Similar examples exist around the world and, recently, the use and characterization of starter cultures has been a main trend of research that could be used to improve fermentations and prevent the possible risk of spoilage [157–159]. This is, for instance, the case of traditional and industrial production of wines [160–164]. However, there are still few studies identifying and characterizing traditional starter cultures for traditional products in Mexico.

Through this review, we identified names such as tibicos, castaña, xinaiste, jinaiste, zinaiste, el pie, el pie de pulque, asiento, ixquini, semilla, xaxtle, and nancle among the most frequent names that fermenters give to those mixed starter cultures for traditional fermented beverages. The study and characterization of these microbial communities have only been scarcely covered in studies of beverages such as **mescal** [97,165], **pulque** [61,166,167], and recently in **atole agrio** [21]. This technique has been used for years and the methods to prepare them should be considered in further studies.

A commonly accepted assumption in traditional fermented products is that spontaneous fermentations have inconsistent or heterogeneous quality outcomes. However, fermentation managers procure to simplify the diversity of the environmental conditions and practices to decrease such heterogeneity. The practice of selecting fermentative environments or controlled facilities is common; for instance, people procure fermenting inside the house or special areas to control the external environmental heterogeneity in light incidence, temperature, external contaminants, or pathogens. Moreover, there is a cultural selection of the person who performs the fermentation and the substrate's quality [22,23,97,163,164]. A commonly overlooked aspect is how the microbial communities of starters are selected, prepared, conceived, stored, and used by local fermenters looking for the most favorable composition to improve the quality of their final products. Documenting such a process would give insights about how these communities of microorganisms are managed.

The interrelationships between the native microbial strains, substrates, techniques, tools, and fermentative environments of TMFB have been little studied. We identified studies that characterized the clay pots or tinas de fermentación for pulque [164] and **tesgüino** [85]. In fact, for **tesgüino** clay pots, fermenters use the specific name of tesgüineras and if the clay pot is broken, the pieces are trembled inside a new one, thus maintaining the microbiota from the old one. Nevertheless, few studies have attended to the importance of these containers as reservoirs of microbiota and their relevance in the fermentation process.

How Diverse Are the Inconspicuous Microbial Environments?

As mentioned, most studies on microbial environments in fermented beverages have been conducted with culture-dependent methods, isolating the most common microorganisms, among them *S. cerevisiae*, as the main microorganism responsible for alcoholic fermentation. These are invariably present in **mescal** and **pulque**. Nevertheless, for maize fermented beverages, the isolation and characterization of microorganisms have been directed to bacteria to characterize the possible probiotic functions of lactic acid bacteria (LAB) and, more recently, their function as a source of nitrogen fixation bacteria [163].

We found that only 10 beverages have been studied to characterize their microorganism communities, identifying 255 species of microorganisms. **Pozol** (Poz), **pulque** (Pul), and **mescal** (Mes) are the beverages with more species recorded (Table S2). This could be partially explained because a larger number of studies have been carried out on these beverages as also shown in the conceptual network (Figure 2). The network analysis reveals low connectivity that may be explained because there is high specificity in microorganisms in some of the beverages (Figure 3). In addition, it shows numerous links per species since genera such as *Lactobacillus* (Lac), *Bacillus* (Bac), and *Saccharomyces* (Sac) are shared among most groups of beverages. The network values are shown in Table 2.

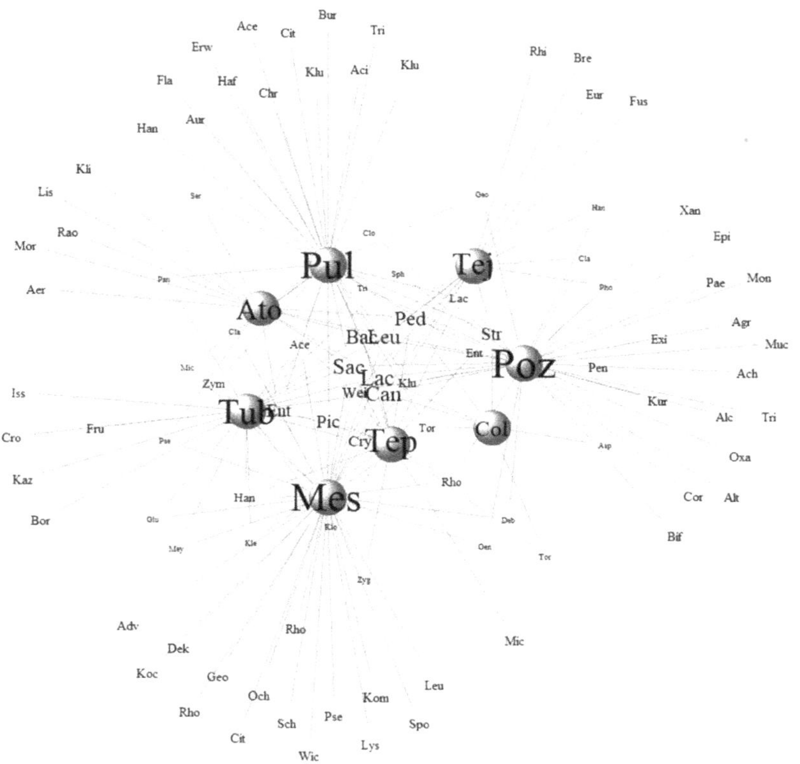

Figure 3. Network of microorganisms. Grey circles represent the names of the TMFB, Tep is for **tepache,** Col is for **colonche**, Mes is **mescal**, Ato for **atole agrío**, Pul for **pulque**, Tub for **tuba**, and **Tej** for tejuino. The size of the letters represents the number of microorganisms recorded for each beverage. The letters in the middle represent the microorganisms that are shared commonly in the studied beverages. *Saccharomyces* (Sac), *Lactobacillus* (Lac), and *Candida* (Can), among others, are common shared microorganisms. **Mescal**, **pulque**, and **pozol** are the most studied beverages in which a major group of microorganisms has been recorded. **Colonche** displays the lowest connectivity due to the few studies on this beverage. The genera of microorganisms and the names of beverages are displayed in the Table S2.

Through the nested analysis, we can see that **mescal**, **pulque**, **atole agrio**, and **pozol** are the beverages with the highest number of microorganism species identified, while **colonche** is the beverage with fewer genera identified. In general, six modules of microorganisms in TMFB were detected: (1) **mescal**; (2) **pulque**; (3) **atole agrio**; (4) **pozol**; (5) **colonche**, and (6) palm and **tepache**. This analysis shows that 99 species occur in the most studied fermented beverages, which might be the generalist core of microorganisms intervening in these fermented products. Interestingly, with this analysis, we identified that **pozol** has 24, **mescal** 23, **pulque** 22, **atole agrio** 10, and **colonche** 2 genera with low shared similarities with other beverages, and those could be specialist genera in each traditional beverage. Nevertheless, as the production practices and substrates vary for each event of production and ferment type, further analyses are needed for more precise information.

Forthcoming technologies such as CRISPR/cas9 [167–174] or the improvement of specific strains [115,156] could play important roles to improve these products as it has happened with yeast from *S. cerevisiae* in the fermentation of several products [172,173]. Nevertheless, careful studies should be performed about how and when these microorganisms are or can be intentionally used in these beverages because there is not a clear panorama about the ecological implications of using these microorganisms in replacing local native microbiota.

3.6. Uses of Fermented Beverages by Human Cultural Groups and Future Directions

Mestizo people produce and consume most of the TMFB, mainly those prepared with maize and agave as the main substrates. This could be explained because mestizo people are the most numerous in Mexican society and have recovered Indigenous traditions; also, because maize plays the key role in Mexican gastronomy, it is widespread, and fundamental in the local communities' nutrition [175,176]. These facts also explain that maize is the most generalist substrate of TMFB and the key role of these beverages in social cohesion, festivities, and ceremonies, such as tesgüinadas [177,178]. Likewise, many of the TMFB that have been recorded are used as medicine to prevent diarrhea, reduce infections, constipation, and, in general, to improve health. These beverages are currently part of daily life food, refreshing beverages during the working hours, and part of local ceremonies.

Indigenous cultural groups are interested in looking forward and renewing their connections with their lands and cultural heritage and recovering their traditional foodways to regain cultural strength and personal and community health [30,179]. This could be visualized in numerous forms of research that have identified how communitarian leaders and elders have encouraged youth to learn about harvesting and how to prepare their traditional foods around the world. Combining knowledge from different sources and epistemic systems is necessary to understand the diversity within and across ecological, social, and cultural systems, which are important factors underpinning conservation and natural resource management strategies [179–181]. This should be a particular concern in the case of traditional fermented beverages around the world, not only in Mexico.

It is important to highlight the marked level of segregation of the Mexican Indigenous people, which has been verified by numerous anthropological and socio-demographic studies based on synthetic demographic and poverty indexes [182–185]. Figure 4C shows a clear correspondence of TMFB with cultural groups. Most of them correspond to the mestizo group in the central region of Mexico. It should be considered that Indigenous people are not a homogeneous sector. The censuses of the Mexican population report gradients from monolingual to bilingual speakers [185], and numerous studies show that most Mexican people have extraordinary mixtures of Indigenous and non-Indigenous cultural aspects. However, it is generally recognized that Indigenous people are one of the sectors living in extreme poverty with no access to education and health services. The lack of monetary income and household goods and social marginality are factors eroding Indigenous cultures, pushing people to migrate from their original areas to cities, where Indigenous languages and culture are often discriminated [185]. This process endangers the maintenance of general culture and the transmission of the traditional knowledge on the production of fermented beverages.

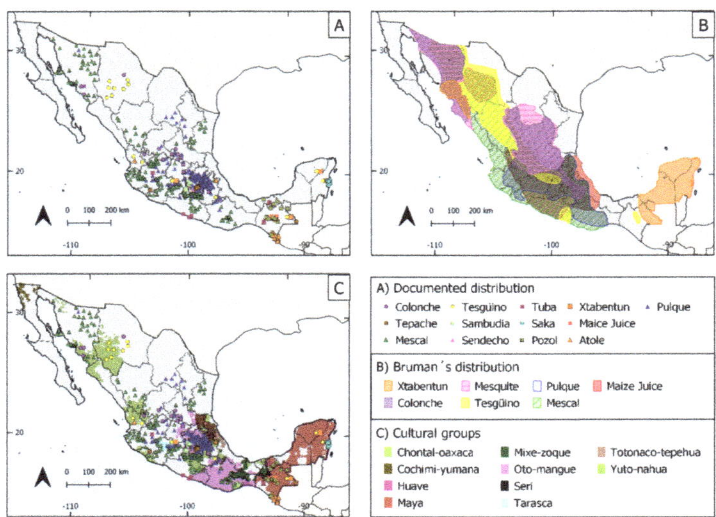

Figure 4. Maps of distribution of TMFB: (**A**) documented distribution of the TMFB; labels are indicated in the low right rectangle; (**B**) distribution proposed by Bruman; (**C**) distribution of TMFB according to the current review and cultural groups grouped by language.

Efforts should be directed to characterize and document aspects of TMFB, such as their uniqueness, human cultural specificity, their heritage status, the dietary patterns, their role in cultural identity, practices in the agricultural production of raw materials for fermentation, dishes and gastronomic innovation, preparation techniques, recipes, food traditions, symbolic dimensions, and material aspects such as utensils and dishware, among other biocultural topics. In addition, promotion activities would be relevant. For instance, promoting events in the communities, sponsored through schools, councils, and cultural centers, can be effective to illustrate, highlight, and demonstrate the values of traditional food systems [186,187]. These activities could help to maintain this marginalized biocultural heritage.

3.7. Reviewed Distribution of the TMFB

We compared the distribution of TMFB documented by Bruman in 1993 (Figure 4B) with information documented in this review (Figure 4A). We overlapped both maps to visualize the localities where these beverages could be produced nowadays and, accordingly, should be attended as priorities in further studies (Figure 4). It can be noticed that Bruman reported 7 beverages and we found 15, adding specific sites of **tepache**, **sendechó**, **sambudia**, **atole agrio**, **saká**, **tuba,** and **pozol** distribution.

The most striking result is the current absence of mention of a traditional fermented beverage made with mesquite pods reported by Bruman and a beverage named **bingarrote** or **bingui** from central Mexico, reported by Pineda [52], which was prepared with underground cooked and then crushed maguey stems; this material was placed to ferment into a pulque container, and afterwards the fermented liquid was distilled in an alembic. The ferment is called **binguí** or **benjuí** and the spirit **bingarrote**. It is a sort of culinary hybrid between pulque and mescal, but it is unclear which agave species was used. This information could be crucial to recover this apparently lost food, which could be a lost knowledge to humanity [188–191].

A first sight of maps in Figure 4 suggests that there has been a dramatic reduction in the production and consumption of the traditional fermented beverages, since Bruman defined areas in the early 1990s until the current research. However, it is important to emphasize that Bruman's methodology was mainly based on his expert knowledge and data that he collected through fieldtrips. We identify that the locations recorded in our review represent shorter areas due to the following reasons: (1) there is a gap in the research on uncommon traditional beverages contrasting with that on **pulque** and **mescal**, (2) there is a systematic oversampling in the same localities for most of the studies; for instance, five **pozol** studies are performed in the same locality and this fact does not allow characterizing the potential area of this beverage, and (3) there has not been regular monitoring of the production of some of these beverages, and, consequently, it has not been attended before.

We therefore propose that deeper studies are still necessary to identify, characterize, and monitor the actual and potential production areas of TMFB. This task will require fieldwork, information about the distribution of the substrates related to each beverage, and a careful characterization of localities where it is produced, using approaches from ethnobiology and ethnography that could bring insights about the possible processes of cultural erosion and the effects of market pressures and cultural discrimination over these marginal products. Such studies could also be the way to promote maintaining or the revival of these products [28,192]. The loss of traditional food systems will result in decreasing culture-specific food activities, thus influencing the decrease of dietary diversity [193]. Moreover, TMFB could be part of touristic development, reliable small-scale food markets, and local health strategies, but most of them remain unexplored and some others are in danger of extinction.

3.8. How Are We Attending the Maintenance of This Intangible Biocultural Heritage?

In recent years, the awareness to protect traditional relationships between humans and environments has promoted the development of systems of legal protection and biocultural management around the world [194]. Numerous challenges have arisen to protect genetic resources, intellectual property rights, and natural resources, among other important topics related to traditional local knowledge [195,196]. Studies have been directed to document the local knowledge systems in several countries to define what is local and global, traditional, and Indigenous, and the authenticity and origin, as well as different ways to interact with the environment and how to attend to environmental crisis issues [197,198]. Traditional food systems can play a key role in the strategy to combat malnutrition while ensuring sustainable development [199]. The Food and Agricultural Organization (FAO) now recognizes that food items such as fermented products must be considered important in areas where malnutrition is evident. Use and conservation are commonly associated issues; conserving traditional food is a powerful way to conserve biocultural heritage.

Food systems are part of complex bodies of knowledge, most of them constructed empirically [200] and transmitted in multiple verbal and non-verbal ways, such as know-how or learning-by-doing [200], and fermented products are not exceptions. These products have been embedded as part of the daily lives of many people, including those currently marginalized rural or Indigenous groups [201]. The diversity of fermented products is an outstanding reservoir of genetic resources that has high potential to obtain secondary products such as extracts, enzymes, dyes, and others compound that can be involved in global markets and could help to solve problems such as hunger [202,203] and poverty and may play a key role to reinforce cultural identity [12]. Nevertheless, several cases of the commodification and commercial use of local knowledge have been recorded, which commonly decontextualize and make inappropriate retribution, and contribute to disarticulate local communities [204–207]. Therefore, analyzing the conditions for fair trade are indispensable when analyzing sustainable ways of using these traditional resources and products. This is a particular concern in relation to the industrial **mescal** production, for instance.

The Mexican government signed the Nagoya Protocol in 2011 and was ratified in 2012. This is the major international regulation system about access to genetic resources and fair and equitable sharing of the benefits derived from their use, which emerged from the Convention on Biological Diversity. However, it was not until 2014 that the Nagoya Protocol was enacted as a supreme law in article 133 of the Mexican Constitution.

The main aim of this protocol is favoring incentives for the conservation of biological diversity, the sustainable use of its components, and the prevention of misappropriation of genetic resources and traditional knowledge on them, which is relevant for a bioculturally mega-diverse country such as Mexico.

Since Indigenous peoples hold traditional food system knowledge, interinstitutional and intersectoral initiatives would make more significant contribution to increase the potential of these resources, but local people should be protagonists. The documentation of traditional food systems is urgent since knowledge of food harvesting and preparation is rapidly disappearing. It is therefore crucial to enhance the public policies to support the maintenance of this traditional knowledge.

The conservation of species and varieties involved is equally important. Cultural historians, ethnobiologists, and the general public should become aware of the distinctive varieties of plants used as substrates for fermentation, their conservation, and recovery. For instance, the varieties of *Agave rhodacantha* classified by the Mayo people as San Antoneña and El Chino were used in the fermentation on an agave beverage called **yocogihua** in southern Sonora and northern Sinaloa. Most commercial production of mescal in this region blinked out during the Ley Seca that began in Sonora in 1916, but remarkably the persistence of these varieties was documented off and on from 1888 to 1965 when the last plantation of them was closed. Nevertheless, the Mayo who worked in the plantation rescued or dispersed them to rancherias neighboring the plantation in Masiaca, Sonora, thus several fencerow plantings of *A. rhodocantha* persist nearby, even though it is otherwise rare except for one other location in Sonora. The recovery of these cultivars and possible reintegration into the contemporary production of culturally distinctive fermented beverages in Sonora is now being undertaken [208,209]. As mentioned, **bingarrote** and **bingui** remain used for an analogous beverage today that is rarely produced in the state of Guanajuato where *A. salmiana* varieties are still in use for both pulque and mezcal production [209]. Identifying through community elders which varieties or species are currently associated with sporadic **bingarrote** and **bingui** production near San Miguel de Allende could help to rescue both the plant genetic resources and the traditional cultural practices historically linked to them.

For local knowledge associated with microorganisms, the situation might be overly complex and difficult to attend, as certain microorganisms are not cultivable and not easy to identify. The lack of a microorganism's accessions in collections, the low characterization of the selection mechanisms, and the implications of this selection over genetic and phenotypic traits in microorganisms can make the establishment of protective laws or conservation policies on this topic challenging. To understand the crucial roles of microorganisms on the TMFB process, we need to know their ecological niches, population dynamics, and relationships between microbiome and environments and microbiome and the selection process.

4. Conclusions

Mexico has a unique gastronomic culture characterized by its multiethnic, high biocultural diversity and long cultural history. Biological diversity is used in traditional Mexican cuisine, and, undoubtedly, Mexican fermented beverages are intimately associated with plant products that have been managed and domesticated in Mexico, as it can be seen in the relevance of fermented beverages based on maize, agave sap and stems, or cacti fruits. However, this is also the case of dozens of wild species that are involved in traditional beverages. Foreign technologies, tools, and plants have been integrated into the Mexican foodscapes, and this fact has promoted a diversification of beverages and a diversification of sensory qualities acceptable to a wide range of palates. TMFB are clear examples that traditional knowledge is a dynamic process that is constantly changing, adjusting, and evolving.

Traditional Mexican cuisine is an outstanding reservoir of genetic resources such as plants, animals, fungi, and, generally less considered, microorganisms such as yeast and bacteria which could play a relevant role for future nourishment applications. The TFMB are reservoirs of cultural diversity, practices, and worldviews of communities. However, studies have focused mainly on the biotechnological applications and on beverages of the *Agave* group, leaving apart other beverages that are rarely studied, thus limiting the action to combat the endangered permanence of these products and the local food systems. The study of fermented beverages under transdisciplinary approaches is fundamental to provide information to construct regulatory frameworks or proposals to protect the diversity that TMFB involves in maintaining the complex interactions between humans, plants, and microorganisms. We expect that these examples demonstrate how biocultural conservation and restoration can be tangibly integrated into the protection and promotion of the Mesoamerican gastronomic patrimony recognized by the UNESCO and the routes of research needed to achieve it.

Supplementary Materials: The following are available online at https://www.mdpi.com/article/10.3390/foods10102390/s1, Table S1: Plants involved in the production of TMFB registered in the literature. Grey boxes represent the presence of the specie and white boxes the absence. ***Ato*** refers to ***atole agrio***, ***Bal*** to ***balché***, ***Col*** to colonche, ***Mes*** to ***mescal***, ***Pox*** to ***pox***, ***Poz*** to ***Pozol***, ***Sak*** to ***saká***, ***Sam*** to ***sambudia***, ***Sen*** to ***sendechó***, ***Tab*** to ***taberna***, ***Tub*** to ***tuba***, ***Tep*** to ***tepache***, ***Tej*** to ***tejuino***, ***Tes*** to ***tesgüino***, ***Cho*** to ***chorote***, ***Pul*** to ***pulque***. Table S2. Microorganism's genera previously registered in the literature. Grey boxes represent the presence of the genera and white boxes the absence. **Ato** refers to **atole agrio**, **Col** to **colonche**, **Mes** to **mescal**, **Poz** to **Pozol**, **Pul** to **pulque**, **Tej** to **tejuino**, **Tep** to **tepache**, **Tub** to **tuba**. The first three letters of the genera are displayed in the Figure 3.

Author Contributions: Conceptualization, C.O.-L., M.V., and A.C.; methodology, C.O.-L., A.C. and L.A.I.; formal analysis, C.O.-L. and L.A.I.; investigation, C.O.-L., G.D.Á.-R., I.T.-G., C.J.F.-U., A.C. and M.V.; resources, A.C.; data curation, C.O.-L., G.D.Á.-R., C.J.F.-U., L.A.I., I.T.-G., M.V., G.P.N., A.C. and P.L.-O.; writing—original draft preparation, C.O.-L., M.V., and A.C.; writing—review and editing, C.O.-L., A.C., M.V., I.T.-G., G.D.Á.-R., L.A.I., G.P.N., and C.J.F.-U.; visualization, C.O.-L., A.C. and L.A.I.; supervision and funding acquisition, A.C., M.V. and I.T.-G. All authors have read and agreed to the published version of the manuscript.

Funding: This research was funded by the Consejo Nacional de Ciencia y Tecnología (CONACyT), Mexico for academic and financial support of postgraduate studies of the first author. Moreover, to CONACYT, research project A1-S-14306 and the Cátedra project 1245, and PAPIIT, DGAPA, UNAM, research projects IN206520, AI203321 and IA207721, and CONABIO RG-023 for financial support of field and technical work.

Institutional Review Board Statement: Not applicable.

Informed Consent Statement: Not applicable.

Data Availability Statement: Data supporting the reported results are systematized in a database available by request to the first author.

Acknowledgments: The authors thank the Posgrado en Ciencias Biológicas, UNAM and the CONACYT, Mexico for financial support of postgraduate studies of the first author. Moreover, to CONACYT, research project A1-S-14306 and the Cátedra project 1245, as well as PAPIIT, DGAPA, UNAM, research projects IN206520, AI203321 and IA207721, and CONABIO RG023 for financial support of the field and technical work. We also thank Rosa Jeannine Xochicale Solis for helping in the illustrations work and we mainly thank the Mexican fermenters for keeping their practices alive.

Conflicts of Interest: The authors declare no conflict of interest.

References

1. Blandino, A.; Al-Aseeri, M.E.; Pandiella, S.S.; Cantero, D.; Webb, C. Cereal-based fermented foods and beverages. *Food Res. Int.* **2003**, *36*, 527–543. [CrossRef]
2. Dudley, R. The Fruits of Fermentation. In *The Drunken Monkey*; University of California Press: Berkley, CA, USA, 2014; pp. 11–33.
3. Gibbons, J.G.; Rinker, D.C. The genomics of microbial domestication in the fermented food environment. *Curr. Opin. Genet. Dev.* **2015**, *35*, 1–8. [CrossRef]
4. Lévi-Strauss, C. The Culinary Triangle. In *Levi-Strauss: Structuralism and Dialectics*; Paidos: Buenos Aires, Argentina, 1968; pp. 39–57.
5. Achi, O.K. The potential for upgrading traditional fermented foods through biotechnology. *Afr. J. Biotechnol.* **2005**, *4*, 375–380.

6. Marsh, A.J.; Hill, C.; Ross, R.P.; Cotter, P.D. Fermented beverages with health-promoting potential: Past and future perspectives. *Trends Food Sci. Technol.* **2014**, *38*, 113–124. [CrossRef]
7. Nabhan, G.P. Ethnobiology for a Diverse World: Microbial Ethnobiology and the Loss of Distinctive Food Cultures. *J. Ethnobiol.* **2010**, *30*, 181–183. [CrossRef]
8. Godfray, H.C.J.; Crute, I.R.; Haddad, L.; Lawrence, D.; Muir, J.F.; Nisbett, N.; Pretty, J.; Robinson, S.; Toulmin, C.; Whiteley, R. The future of the global food system. *Philos. Trans. R. Soc. B Biol. Sci.* **2010**, *365*, 2769–2777. [CrossRef]
9. Battcock, M. *Fermented Fruits and Vegetables: A Global Perspective*; No. 134; FAO: Rome, Italy, 1998.
10. Pieroni, A.; Vandebroek, I. *Traveling Cultures and Plants: The Ethnobiology and Ethnopharmacy of Migrations*; Berghahn Books: New York, NY, USA, 2007; Volume 7.
11. Steinkraus, K.H. Nutritional significance of fermented foods. *Food Res. Int.* **1994**, *27*, 259–267. [CrossRef]
12. Tamang, J.P.; Cotter, P.D.; Endo, A.; Han, N.S.; Kort, R.; Liu, S.Q.; Mayo, B.; Westerik, N.; Hutkins, R. Fermented foods in a global age: East meets West. *Compr. Rev. Food Sci. Food Saf.* **2019**, *19*, 184–217. [CrossRef] [PubMed]
13. Steinkraus, K.H. Classification of fermented foods: Worldwide review of household fermentation techniques. *Food Control* **1997**, *8*, 311–317. [CrossRef]
14. Tamang, J.; Samuel, D. Dietary Cultures and Antiquity of Fermented Foods and Beverages. In *Fermented Foods and Beverages of the World*; Tamang, J.P., Kailasapathy, K., Eds.; CRC Press: New York, NY, USA, 2010; pp. 1–40.
15. Legras, J.L.; Merdinoglu, D.; Cornuet, J.M.; Karst, F. Bread, beer and wine: Saccharomyces cerevisiae diversity reflects human history. *Mol. Ecol.* **2007**, *16*, 2091–2102. [CrossRef]
16. Valamoti, S.M. Brewing beer in wine country? First archaeobotanical indications for beer making in Early and Middle Bronze Age Greece. *Veg. Hist. Archaeobot.* **2018**, *27*, 611–625. [CrossRef]
17. Campbell-Platt, G. Fermented foods—A world perspective. *Food Res. Int.* **1994**, *27*, 253–257. [CrossRef]
18. Hassan, A.N.; Frank, J.F. Starter cultures and their use. In *Applied Dairy Microbiology*; CRC Press: Plymouth, UK, 2001; pp. 171–226.
19. Zorba, M.; Hancioglu, O.; Genc, M.; Karapinar, M.; Ova, G. The use of starter cultures in the fermentation of boza, a traditional Turkish beverage. *Process. Biochem.* **2003**, *38*, 1405–1411. [CrossRef]
20. Kebede, A.; Viljoen, B.; Gadaga, T.; Narvhus, J.; Lourens-Hattingh, A. The effect of container type on the growth of yeast and lactic acid bacteria during production of Sethemi, South African spontaneously fermented milk. *Food Res. Int.* **2007**, *40*, 33–38. [CrossRef]
21. Väkeväinen, K.; Hernández, J.; Simontaival, A.-I.; Severiano-Pérez, P.; Díaz-Ruiz, G.; von Wright, A.; Wacher-Rodarte, C.; Plumed-Ferrer, C. Effect of different starter cultures on the sensory properties and microbiological quality of Atole agrio, a fermented maize product. *Food Control* **2019**, *109*, 106907. [CrossRef]
22. Ojeda-Linares, C.I.; Vallejo, M.; Lappe-Oliveras, P.; Casas, A. Traditional management of microorganisms in fermented beverages from cactus fruits in Mexico: An ethnobiological approach. *J. Ethnobiol. Ethnomed.* **2020**, *16*, 1–12. [CrossRef]
23. Álvarez-Ríos, G.D.; Figueredo-Urbina, C.J.; Casas, A. Physical, Chemical, and Microbiological Characteristics of Pulque: Management of a Fermented Beverage in Michoacán, Mexico. *Foods* **2020**, *9*, 361. [CrossRef]
24. Gadaga, T.H.; Mutukumira, A.N.; Narvhus, J.A.; Feresu, S.B. A review of traditional fermented foods and beverages of Zimbabwe. *Int. J. Food Microbiol.* **1999**, *53*, 1–11. [CrossRef]
25. Tamang, J. Diversity of Fermented Beverages and Alcoholic Drinks. In *Fermented Foods and Beverages of the World*; Routledge: New York, NY, USA, 2010; pp. 85–125. [CrossRef]
26. Tamang, J.P.; Thapa, S.; Tamang, N.; Rai, B. Indigenous fermented food beverages of Darjeeling hills and Sikkim: A process and product characterization. *J. Hill Res.* **1996**, *9*, 401–411.
27. Tamang, J.P.; Tamang, N.; Thapa, S.; Dewan, S.; Tamang, B.; Yonzan, H.; Kharel, N. Microorganisms and nutritional value of ethnic fermented foods and alcoholic beverages of Northeast India. *Indian J. Tradit. Knowl.* **2012**, *11*, 7–25.
28. Madej, T.; Pirożnikow, E.; Dumanowski, J.; Łuczaj, Ł. Juniper beer in Poland: The story of the revival of a traditional bever-age. *J. Ethnobiol.* **2014**, *34*, 84–103. [CrossRef]
29. Baschali, A.; Tsakalidou, E.; Kyriacou, A.; Karavasiloglou, N.; Matalas, A.-L. Traditional low-alcoholic and non-alcoholic fermented beverages consumed in European countries: A neglected food group. *Nutr. Res. Rev.* **2017**, *30*, 1–24. [CrossRef] [PubMed]
30. Medina, F.X.; Aguilar, A. Sustainable diets: Social and cultural perspectives. In *Sustainable Diets: Linking Nutrition and Food Systems*; CABI: Wallingford, UK, 2019; p. 131.
31. Morales, D. Biological activities of kombucha beverages: The need of clinical evidence. *Trends Food Sci. Technol.* **2020**, *105*, 323–333. [CrossRef]
32. Meyer, R.S.; DuVal, A.E.; Jensen, H.R. Patterns and processes in crop domestication: An historical review and quantitative analysis of 203 global food crops. *New Phytol.* **2012**, *196*, 29–48. [CrossRef]
33. Smýkal, P.; Nelson, M.N.; Berger, J.D.; Von Wettberg, E.J. The impact of genetic changes during crop domestication. *Agronomy* **2018**, *8*, 119. [CrossRef]
34. Gaut, B.S.; Seymour, D.K.; Liu, Q.; Zhou, Y. Demography and its effects on genomic variation in crop domestication. *Nat. Plants* **2018**, *4*, 512–520. [CrossRef]
35. Von Wettberg, E.; Davis, T.M.; Smýkal, P. Wild Plants as Source of New Crops. *Front. Plant Sci.* **2020**, *11*, 1426. [CrossRef]

36. Blancas, J.; Casas, A.; Moreno-Calles, A.I.; Caballero, J. Cultural motives of plant management and domestication. In *Ethnobotany of Mexico*; Springer: New York, NY, USA, 2016; pp. 233–255.
37. Caballero, J.; Casas, A.; Cortés, L.; Mapes, C. Patrones en el conocimiento, uso y manejo de plantas en pueblos indígenas de México. *Estud. Atacameños* **1998**, *16*, 181–196. [CrossRef]
38. Casas, A.; Torres-Guevara, J.; Parra, F. *Domesticación y en el Continente Americano. Manejo de Biodiveridad y Evolución Dirigida por las Culturas del Nuevo Mundo*; Universidad Nacional Autónoma de México; Universidad Nacional Agraria La Molina: Mexico City, Mexico, 2016; Volume 1, pp. 25–50.
39. Casas, A.; Parra, F.; Rangel-Landa, S.; Blancas, J.; Vallejo, M.; Moreno-Calles, A.I.; Guillén, S.; Torres-García, I.; Delgado-Lemus, A.; Pérez-Negrón, E.; et al. *Manejo y Domesticación de Plantas en Mesoamérica. Una Estrategia de Investigación y Estado del Conocimiento Sobre los Recursos Genéticos de México*; Universidad Nacional Autónoma de México; Universidad Nacional Agraria La Molina: Mexico City, Mexico, 2016; Volume 1.
40. Linares, M.E.; Bye, R. Las especies subutilizadas de la milpa. *Rev. Digit. Univ.* **2015**, *16*, 22.
41. Scovazzi, T. The Definition of Intangible Cultural Heritage. In *Cultural Heritage, Cultural Rights, Cultural Diversity*; Brill Nijhoff: Leiden, The Netherlands, 2012; pp. 179–200.
42. Romaine, S.; Gorenflo, L.J. Linguistic diversity of natural UNESCO world heritage sites: Bridging the gap between nature and culture. *Biodivers. Conserv.* **2017**, *26*, 1973–1988. [CrossRef]
43. Gallagher, J.; McKevitt, A. Laws and Regulations of Traditional Foods: Past, Present and Future. In *Traditional Foods*; Springer: Cham, Switzerland, 2019; pp. 239–271.
44. Kühne, B.; Vanhonacker, F.; Gellynk, X.; Verbeke, W. Innovation in traditional food products in Europe: Do sector innovation activities match consumers' acceptance? *Food Qual. Prefer.* **2010**, *21*, 629–638. [CrossRef]
45. Gellynck, X.; Kühne, B. Innovation and collaboration in traditional food chain networks. *J. Chain Netw. Sci.* **2008**, *8*, 121–129. [CrossRef]
46. Guerrero, L.; Guàrdia, M.D.; Xicola, J.; Verbeke, W.; Vanhonacker, F.; Zakowska-Biemans, S.; Sajdakowska, M.; Sulmont-Rossé, C.; Issanchou, S.; Contel, M.; et al. Consumer-driven definition of traditional food products and innovation in traditional foods. A qualitative cross-cultural study. *Appetite* **2009**, *52*, 345–354. [CrossRef]
47. Romero-Luna, H.E.; Hernández-Sánchez, H.; Dávila-Ortiz, G. Traditional fermented beverages from Mexico as a potential probiotic source. *Ann. Microbiol.* **2017**, *67*, 577–586. [CrossRef]
48. Ramírez-Guzmán, K.N.; Torres-León, C.; Martinez-Medina, G.A.; de la Rosa, O.; Hernández-Almanza, A.; Alvarez-Perez, O.B.; Araujo, R.; González, L.R.; Londoño, L.; Ventura, J.; et al. Traditional Fermented Beverages in Mexico. In *Fermented Beverages*; Woodhead Publishing: Sawston, UK, 2019; pp. 605–635.
49. Pérez-Armendáriz, B.; Cardoso-Ugarte, G. Traditional fermented beverages in Mexico: Biotechnological, nutritional, and functional approaches. *Food Res. Int.* **2020**, *136*, 109307. [CrossRef]
50. Robledo-Márquez, K.; Ramírez, V.; González-Córdova, A.; Ramírez-Rodríguez, Y.; García-Ortega, L.; Trujillo, J. Research opportunities: Traditional fermented beverages in Mexico. Cultural, microbiological, chemical, and functional aspects. *Food Res. Int.* **2021**, *147*, 110482. [CrossRef]
51. Godoy, A.; Herrera, T.; Ulloa, M. *Más allá del Pulque y el Tepache: Las Bebidas Alcohólicas no Destiladas Indígenas de México*; Unam: Mexico City, Mexico, 2003; Volume 1.
52. Wilson, I.; Pineda, A. Pineda's Report on the Beverages of New Spain. *Arizona West* **1963**, *5*, 79. Available online: http://www.jstor.org/stable/40167046 (accessed on 18 August 2021).
53. Bruman, H.J. *Alcohol in Ancient Mexico*; University of Utah Press: Salt Lake City, UT, USA, 2000.
54. Enríquez-Salazar, M.I.; Veana, F.; Aguilar, C.N.; De la Garza-Rodríguez, I.M.; López, M.G.; Rutiaga-Quiñones, O.M.; Morlett-Chávez, J.A.; Rodríguez-Herrera, R. Microbial diversity and biochemical profile of aguamiel collected from Agave salmiana and A. atrovirens during different seasons of year. *Food Sci. Biotechnol.* **2017**, *26*, 1003–1011. [CrossRef] [PubMed]
55. Ben Omar, N.; Ampe, F. Microbial Community Dynamics during Production of the Mexican Fermented Maize Dough Pozol. *Appl. Environ. Microbiol.* **2000**, *66*, 3664–3673. [CrossRef]
56. Rocha-Arriaga, C.; Espinal-Centeno, A.; Martinez-Sánchez, S.; Caballero-Pérez, J.; Alcaraz, L.D.; Cruz-Ramírez, A. Deep microbial community profiling along the fermentation process of pulque, a biocultural resource of Mexico. *Microbiol. Res.* **2020**, *241*, 126593. [CrossRef]
57. Chacón-Vargas, K.; Torres, J.; Giles-Gómez, M.; Escalante, A.; Gibbons, J.G. Genomic profiling of bacterial and fungal communities and their predictive functionality during pulque fermentation by whole-genome shotgun sequencing. *Sci. Rep.* **2020**, *10*, 1–13. [CrossRef]
58. Pérez-Cataluña, A.; Elizaquível, P.; Carrasco, P.; Espinosa, J.; Reyes, D.; Wacher, C.; Aznar, R. Diversity and dynamics of lactic acid bacteria in Atole agrio, a traditional maize-based fermented beverage from South-Eastern Mexico, analysed by high throughput sequencing and culturing. *Antonie Van Leeuwenhoek* **2018**, *111*, 385–399. [CrossRef]
59. Astudillo-Melgar, F.; Ochoa-Leyva, A.; Utrilla, J.; Huerta-Beristain, G. Bacterial Diversity and Population Dynamics During the Fermentation of Palm Wine from Guerrero Mexico. *Front. Microbiol.* **2019**, *10*, 531. [CrossRef]
60. Villarreal-Morales, S.L.; Montañez-Saenz, J.C.; Aguilar-González, C.N.; Rodriguez-Herrera, R. Metagenomics of traditional beverages. In *Advances in Biotechnology for Food Industry*; Academic Press: Cambridge, MA, USA, 2018; pp. 301–326.

61. Toledo, V.M.; Alarcón-Chaires, P.; Moguel, P.; Olivo, M.; Cabrera, A.; Leyequien, E.; Rodríguez-Aldabe, A. El atlas etnoeco-lógico de México y Centroamérica: Fundamentos, métodos y resultados. *Etnoecológica* **2001**, *6*, 7–41.
62. Van Eck, N.J.; Waltman, L. Software survey: VOSviewer, a computer program for bibliometric mapping. *Scientometrics* **2009**, *84*, 523–538. [CrossRef] [PubMed]
63. Zahedi, Z.; Van Eck, N.J. Visualizing readership activity of Mendeley users using VOSviewer. In Proceedings of the Altmetrics14: Expanding Impacts and Metrics, Workshop at Conference, Bloomington, IN, USA, 24–26 June 2014; Volume 1041819.
64. Blomberg, Á.; de Norma, A.; Moreno, D. *Distribución de las Lenguas Indígenas de México*; Jardín Etnobotánico de Oaxaca—Comisión Nacional para el Conocimiento y Uso de la Biodiversidad: Mexico City, Mexico, 2008; Volume 1.
65. QGIS Development Team. QGIS Geographic Information System. Open Source Geospatial Foundation Project. Available online: http://qgis.osgeo.org (accessed on 8 March 2021).
66. R Team. *RStudio: Integrated Development for R*; RStudio, Inc.: Boston, MA, USA, 2020; Available online: http://www.rstudio.com (accessed on 17 March 2021).
67. Csardi, G.; Nepusz, T. The igraph software package for complex network research. *Int. J. Complex Syst.* **2006**, *1695*, 1–9.
68. Nuraida, L.; Wacher, M.C.; Owens, J.D. Microbiology of pozol, a Mexican maize dough. *World J. Microbiol. Biotechnol.* **1995**, *11*, 567–571. [CrossRef] [PubMed]
69. Ray, P.; Sanchez, C.; O'sullivan, D.J.; McKay, L.L. Classification of a bacterial isolate, from pozol, exhibiting antimicrobial activity against several gram-positive and gram-negative bacteria, yeasts, and molds. *J. Food Prot.* **2000**, *63*(8), 1123–1132. [CrossRef]
70. Larroque, M.; Carrau, F.; Fariña, L.; Boido, E.; Dellacassa, E.; Medina, K. Effect of Saccharomyces and non-Saccharomyces native yeasts on beer aroma compounds. *Int. J. Food Microbiol.* **2021**, *337*, 108953. [CrossRef]
71. Wacher, C.; Cañas, A.; Bárzana, E.; Lappe, P.; Ulloa, M.; Owens, J.D. Microbiology of Indian and Mestizo pozol fermentations. *Food Microbiol.* **2000**, *17*, 251–256. [CrossRef]
72. De la Luz, S. Memoria Alimentaria y transición cultural en la comunidad Mazahua de San Pedro de los baños, Ixtlahuaca, Estado de México. Bachelor's Thesis, Universidad Autónoma del Estado de México, Ciudad de México, Mexico, 2019.
73. Jiménez Vera, R.; Gonzalez Cortes, N.; Magaña Contreras, A.; Corona Cruz, A. Microbiological and sensory evaluation of fermented white pozol, with cacao (Theobroma cacao) and coconut (Cocos nucifera). *Rev. Venez. Cienc. Tecnol. Aliment.* **2010**, *1*, 70–80.
74. Velázquez-López, A.; Covatzin-Jirón, D.; Toledo-Meza, M.D.; Vela-Gutiérrez, G. Fermented drink elaborated with lactic acid bacteria isolated from chiapaneco traditional pozol. *Cienc. UAT* **2018**, *13*, 165–178.
75. Torres-León, C.; Ramírez-Guzmán, N.; Ascacio-Valdes, J.; Serna-Cock, L.; dos Santos Correia, M.T.; Contreras-Esquivel, J.C.; Aguilar, C.N. Solid-state fermentation with Aspergillus niger to enhance the phenolic contents and antioxidative activity of Mexican mango seed: A promising source of natural antioxidants. *LWT* **2019**, *112*, 108236. [CrossRef]
76. Trabanino, F.; Meléndez, L. El ajkum sa'o'pozol de camote'-una bebida entre los mayas palencanos del Clásico Tardío. *Ketzalcalli* **2016**, *2*, 3–21.
77. Wacher, C.; Canas, A.; Cook, P.E.; Bárzana, E.; Owens, J.D. Sources of microorganisms in pozol, a traditional Mexican fer-mented maize dough. *World J. Microbiol. Biotechnol.* **1993**, *9*, 269–274. [CrossRef]
78. Díaz-Ruiz, G.; Guyot, J.; Ruiz-Teran, F.; Morlon-Guyot, J.; Wacher, C. Microbial and physiological characterization of weak-ly amylolytic but fast-growing lactic acid bacteria: A functional role in supporting microbial diversity in pozol, a Mexican fermented maize beverage. *Appl. Environ. Microbiol.* **2003**, *69*, 4367–4374. [CrossRef] [PubMed]
79. Escobar-Ramírez, M.C.; Jaimez-Ordaz, J.; Escorza-Iglesias, V.A.; Rodríguez-Serrano, G.M.; Contreras-López, E.; Ramírez-Godínez, J.; González-Olivares, L.G. Lactobacillus pentosus ABHEAU-05: An in vitro digestion resistant lactic acid bacte-rium isolated from a traditional fermented Mexican beverage. *Rev. Argent. Microbiol.* **2020**, *52*, 305–314. [CrossRef]
80. Guadarrama Orozco, K.D. Tipificación de Bacterias Lácticas Aisladas del Axocotl, Atole Agrio de la Sierra Norte de Puebla, por Medio de ADRA. Bachelor's Thesis, Facutlad de Química Universidad Nacional Autónoma de México, Mexico City, Mexico, 2007.
81. Ampe, F.; ben Omar, N.; Moizan, C.; Wacher, C.; Guyot, J.-P. Polyphasic Study of the Spatial Distribution of Microorganisms in Mexican Pozol, a Fermented Maize Dough, Demonstrates the Need for Cultivation-Independent Methods To Investigate Traditional Fermentations. *Appl. Environ. Microbiol.* **1999**, *65*, 5464–5473. [CrossRef] [PubMed]
82. Escobar Colmenares, S. *Reivindicación de la dieta de la milpa y otros alimentos que la complementan en la región los Altos de Chiapas*; Universidad de Ciencias y Artes de Chiapas: Tuxtla Gutiérrez, Chiapas, Mexico, 2019.
83. Silva, L.M.L.; Trabanino, F. Preparación de pozol de camote con maíz en Palenque, Chiapas. *Antrópica Rev. Cienc. Soc. Humanid.* **2018**, *4*, 159–169.
84. McGee, R.J. The balché ritual of the Lacandon Maya. *Estud. Cult. Maya* **2013**, *18*, 439–457.
85. Lappe, P.; Ulloa, M.; Gómez, J. Microbial and chromatographic study of tejuino from Jalisco, Mexico. *Rev. Mex. Microbiol.* **1989**, *5*, 181–203.
86. Valdez, A.V.; Garcia, L.E.S.; Kirchmayr, M.; Rodríguez, P.R.; Esquinca, A.G.; Coria, R.; Mathis, A.G. Yeast communities associated with artisanal mezcal fermentations from Agave salmiana. *Antonie Van Leeuwenhoek* **2011**, *100*, 497–506. [CrossRef]
87. Lappe, P.; Ulloa, M. *Estudios Étnicos, Microbianos y Químicos del Tesgüino Tarahumara*; UNAM: Mexico City, Mexico, 1989; Volume 1.
88. Cruz, S.; Ulloa, M. *Alimentos Fermentados de Maíz Consumidos en México y Otros Países Latinoamericanos*; UNAM: Mexico City, Mexico, 1989; Volume 1.
89. Castro, J.T. Bebidas fermentadas indígenas: Cacao, pozol, tepaches, tesgüino y tejuino. In *Conquista y Comida: Consecuencias del Encuentro de dos Mundos*; Universidad Nacional Autónoma de México: Mexico City, Mexico, 1997; pp. 437–448.

90. Moreno-Fuentes, Á.; Aguirre-Acosta, E.; Pérez-Ramírez, L. Conocimiento tradicional y científico de los hongos en el estado de Chihuahua, México. *Etnobiología* **2004**, *4*, 89–117.
91. Väkeväinen, K.; Valderrama, A.; Espinosa, J.; Centurión, D.; Rizo, J.; Reyes-Duarte, D.; Plumed-Ferrer, C. Characterization of lactic acid bacteria recovered from atole agrio, a traditional Mexican fermented beverage. *LWT* **2018**, *88*, 109–118. [CrossRef]
92. Gutiérrez Rojas, D. Por el Camino Real de la Costa: Apuntes sobre la tradición mariachera en la costa de Michoacán. *Relac. Estud. Hist. Soc.* **2014**, *35*(139), 281–304. [CrossRef]
93. Escalante, A.; López Soto, D.R.; Velázquez Gutiérrez, J.E.; Giles-Gómez, M.; Bolívar, F.; López-Munguía, A. Pulque, a traditional Mexican alcoholic fermented beverage: Historical, microbiological, and technical aspects. *Front. Microbiol.* **2016**, *7*, 1026. [CrossRef] [PubMed]
94. Rojas-Rivas, E.; Rendón-Domínguez, A.; Felipe-Salinas, J.A.; Cuffia, F. What is gastronomy? An exploratory study of social representation of gastronomy and Mexican cuisine among experts and consumers using a qualitative approach. *Food Qual. Preference* **2020**, *83*, 103930. [CrossRef]
95. Yañez, S.B.; Ortíz, H.T.; Ortega, A.E.; Bordi, I.V. Venta informal de pulque como estrategia de reproducción social. Evidencias del centro de México. *Rev. Geogr. Agrícola* **2019**, *62*, 49–67.
96. Chandrasekhar, K.; Sreevani, S.; Seshapani, P.; Pramodhakumari, J. A review on palm wine. *Int. J. Res. Biol. Sci.* **2012**, *2*, 33–38.
97. Ruiz-Terán, F.; Martínez-Zepeda, P.N.; Geyer-de la Merced, S.Y.; Nolasco-Cancino, H.; Santiago-Urbina, J.A. Mezcal: In-digenous Saccharomyces cerevisiae strains and their potential as starter cultures. *Food Sci. Biotechnol.* **2019**, *28*, 459–467. [CrossRef] [PubMed]
98. Santiago-Urbina, J.A.; Ruíz-Terán, F. Microbiology and biochemistry of traditional palm wine produced around the world. *Int. Food Res. J.* **2014**, *21*, 1261–1269.
99. Santiago-Urbina, J.A.; Arias-García, J.A.; Ruiz-Terán, F. Yeast species associated with spontaneous fermentation of taberna, a traditional palm wine from the southeast of Mexico. *Ann. Microbiol.* **2015**, *65*, 287–296. [CrossRef]
100. Sánchez, D.G.; Ríos, G.L. Manejo de la palma de coco (Cocos nucifera L.) en México. Revista Chapingo. Serie ciencias forestales y del ambiente. *Rev. Chapingo Ser. Cienc. For. Ambiente* **2002**, *8*(1), 39.
101. De Albuquerque, U.P. Re-examining hypotheses concerning the use and knowledge of medicinal plants: A study in the Caatinga vegetation of NE Brazil. *J. Ethnobiol. Ethnomed.* **2006**, *2*, 1–10. [CrossRef] [PubMed]
102. Vandebroek, I.; Balick, M.J. Globalization and loss of plant knowledge: Challenging the paradigm. *PLoS ONE* **2012**, *7*(5), e37643. [CrossRef] [PubMed]
103. Ceuterick, M.; Vandebroek, I.; Torry, B.; Pieroni, A. Cross-cultural adaptation in urban ethnobotany: The Colombian folk pharmacopoeia in London. *J. Ethnopharmacol.* **2008**, *120*, 342–359. [CrossRef]
104. Ceuterick, M.; Vandebroek, I.; Pieroni, A. Resilience of Andean urban ethnobotanies: A comparison of medicinal plant use among Bolivian and Peruvian migrants in the United Kingdom and in their countries of origin. *J. Ethnopharmacol.* **2011**, *136*, 27–54. [CrossRef] [PubMed]
105. Van Andel, T.; Westers, P. Why Surinamese migrants in the Netherlands continue to use medicinal herbs from their home country. *J. Ethnopharmacol.* **2010**, *127*, 694–701. [CrossRef] [PubMed]
106. Paterson, A.H.; Moore, P.H.; Tew, T.L. The Gene Pool of Saccharum Species and Their Improvement. In *Genomics of the Saccharinae*; Springer: New York, NY, USA, 2013; pp. 43–71.
107. Bennett, B.; Prance, G.T. Introduced plants in the indigenous Pharmacopoeia of Northern South America. *Econ. Bot.* **2000**, *54*, 90–102. [CrossRef]
108. Kumarathilake, D.M.H. Study on the Extinction Risks, Conservation and Domestication of Endemic Wild Cinnamomum Species in Sri Lanka. Master's Thesis, University of Ruhuna, Ruhuna, Sri Lanka, 2012.
109. Ranasinghe, P.; Pigera, S.; Premakumara, G.S.; Galappaththy, P.; Constantine, G.R.; Katulanda, P. Medicinal properties of 'true'cinnamon (*Cinnamomum zeylanicum*): A systematic review. *BMC Complem. Altern. Med.* **2013**, *13*, 275. [CrossRef] [PubMed]
110. Titilayo, F.; Temitope, A. Microbiological and Physicochemical Changes in Palm Wine Subjected to Spontaneous Fermentation During Storage. *Int. J. Biotechnol.* **2019**, *8*, 48–58.
111. Djeni, T.N.; Kouame, K.H.; Ake, F.D.M.; Amoikon, L.S.T.; Dje, M.K.; Jeyaram, K. Microbial Diversity and Metabolite Profiles of Palm Wine Produced from Three Different Palm Tree Species in Côte d'Ivoire. *Sci. Rep.* **2020**, *10*, 1–12. [CrossRef]
112. Zizumbo-Villarreal, D. History of coconut (*Cocos nucifera* L.) in Mexico: 1539? *Genet. Resour. Crop. Evol.* **1996**, *43*, 505–515. [CrossRef]
113. Zizumbo-Villarreal, D.; Colunga-Garcíamarín, P. Early coconut distillation and the origins of mezcal and tequila spirits in west-central Mexico. *Genet. Resour. Crop. Evol.* **2007**, *55*, 493–510. [CrossRef]
114. Zizumbo-Villarreal, D.; González-Zozaya, F.; Olay-Barrientos, A.; Almendros-López, L.; Flores-Pérez, P.; Colunga-Garcíamarín, P. Distillation in Western Mesoamerica before European Contact. *Econ. Bot.* **2009**, *63*, 413–426. [CrossRef]
115. Machuca, P. *El Vino de Cocos en la Nueva España. Historia de una Transculturación en el Siglo XVII*; Colegio de Michoacán: Zamora, Mexico, 2018.
116. Corona-González, R.; Ramos-Ibarra, J.; Gutiérrez-González, P.; Pelayo-Ortiz, C.; Guatemala-Morales, G.; Arriola-Guevara, E. The use of response surface methodology to evaluate the fermentation conditions in the production of tepache. *Rev. Mex. Ing. Qum.* **2013**, *12*, 19–28.

117. Moreno-Terrazas, R.; Reyes-Morales, H.; Huerta-Ochoa, S.; Guerrero-Legarreta, I.; Vernon-Carter, E.J. Note. Consumer awareness of the main sensory attributes of tepache, a traditional fermented fruit beverage. *Food Sci. Technol. Int.* **2001**, *7*, 411–415. [CrossRef]
118. Velázquez-Quiñones, S.E.; Moreno-Jiménez, M.R.; Gallegos-Infante, J.A.; González-Laredo, R.F.; Álvarez, S.A.; Rosales-Villarreal, M.C.; Rocha-Guzmán, N.E. Apple Tepache Fermented with Tibicos: Changes in Chemical Profiles, Antioxidant Activity and Inhibition of Digestive Enzymes. *J. Food Process. Preserv.* **2001**, *45*, e15597.
119. Rodríguez-Lerma, G.K.; Gutiérrez-Moreno, K.; Cárdenas-Manríquez, M.; Botello-Álvarez, E.; Jiménez-Islas, H.; Rico-Martínez, R.; Navarrete-Bolaños, J.L. Microbial ecology studies of spontaneous fermentation: Starter culture selection for prickly pear wine production. *J. Food Sci.* **2011**, *76*, 6. [CrossRef]
120. Lappe-Oliveras, P.; Moreno-Terrazas, R.; Arrizón-Gaviño, J.; Herrera-Suárez, T.; García-Mendoza, A.; Gschaedler-Mathis, A. Yeasts associated with the production of Mexican alcoholic non-distilled and distilled Agave beverages. *FEMS Yeast Res.* **2008**, *8*, 1037–1052. [CrossRef]
121. Moreno-Terrazas, R. Determinación de las Características Microbiológicas, Bioquímicas Fisicoquímicas y Sensoriales para la Estandarización del Procesode Elaboración del Tepache. Iztapalapa, México. Ph.D. Thesis, Universidad Autónoma Metropolitana, Mexico City, Mexico, 2005.
122. Kirchmayr, M.; Segura, L.E.; Lappe-Oliveras, P.; Moreno-Terrazas, R.; de la Rosa, M.; Gschaedler Mathis, A. Impact of environmental conditions and process modifications on microbial diversity, fermentation efficiency and chemical profile during the fermentation of Mezcal in Oaxaca. *LWT—Food Sci. Technol.* **2017**, *79*, 160–169. [CrossRef]
123. Páez-Lerma, J.B.; Arias-García, A.; Rutiaga-Quiñones, O.M.; Barrio, E.; Soto-Cruz, N.O. Yeasts Isolated from the Alcoholic Fermentation of Agave duranguensis during Mezcal Production. *Food Biotechnol.* **2013**, *27*, 342–356. [CrossRef]
124. Pérez, E.; González-Hernández, J.C.; Chávez-Parga, M.C.; Cortés-Penagos, C. Caracterización fermentativa de levaduras productoras de etanol a partir de jugo de Agave cupreata en la elaboración de mezcal. *Rev. Mex. Ing. Quím.* **2013**, *12*, 451–461.
125. Walker, G.M.; Lappe-Oliveras, P.; Moreno-Terrazas, R.; Kirchmayr, M.; Arellano-Plaza, M.; Gschaedler-Mathis, A.C. Chapter 16 Yeasts Associated with the Production of Distilled Alcoholic Beverages. In *Yeasts in the Production of Wine*; Romano, P., Ciani, M., Fleet, G.H., Eds.; Springer Nature: New York, NY, USA, 2019; pp. 478–515.
126. Narváez-Zapata, J.A.; Rojas-Herrera, R.A.; Rodríguez-Luna, I.C.; Larralde-Corona, C.P. Culture-Independent Analysis of Lactic Acid Bacteria Diversity Associated with Mezcal Fermentation. *Curr. Microbiol.* **2010**, *61*, 444–450. [CrossRef]
127. Castillo-Morales, M.; Wacher-Rodarte, C.; Hernández-Sánchez, H. Preliminary studies on chorote a tradicional Mexican fermented product. *World J. Microbiol. Biotechnol.* **2005**, *21*, 293–296. [CrossRef]
128. Bernard Menna, A.; Lozano Cortés, M. "Las Bebidas Sagradas Mayas: El Balché Y El Saká." Sincronía. Available online: http://sincronia.cucsh.udg.mx/mennacortes03.html (accessed on 16 June 2021).
129. Aguirre-Dugua, X.; Pérez-Negrón, E.; Casas, A. Phenotypic differentiation between wild and domesticated varieties of Crescentia cujeteL. and culturally relevant uses of their fruits as bowls in the Yucatan Peninsula, Mexico. *J. Ethnobiol. Ethnomed.* **2013**, *9*, 76. [CrossRef]
130. Chuchiak, J.F., IV. "It is Their Drinking That Hinders Them": Balché and the Use of Ritual Intoxicants among the Colonial Yucatec Maya. *Estud. Cult. Maya* **2013**, *24*, 1550.
131. Aroche, D.S. Con el diablo adentro. El consumo medicinal y ritual del balche'entre los mayas de Yucatán visto desde una perspectiva etnohistórica. *Hist. Conoc. Hist. Clave Digit.* **2016**, *10*, 42–55.
132. Cox-Tamay, L.D.; Yamasaki, E.; Heredia-Campos, E.B. Plantas Utilizadas En La Ceremonia Maya: Ch'a Cháak. *Desde El Herbario CICY* **2016**, *8*, 167. Available online: https://www.academia.edu/29506808/Plantas_utilizadas_en_la_Ceremonia_Maya_Cha_ch%C3%A1 (accessed on 2 February 2021).
133. Menezes, C.; Vollet-Neto, A.; Contrera, F.A.F.L.; Venturieri, G.C.; Imperatriz-Fonseca, V.L. The role of useful microorganisms to stingless bees and stingless beekeeping. In *Pot-Honey*; Springer: New York, NY, USA, 2013; pp. 153–171.
134. Quintero-Salazar, B.; Bernáldez Camiruaga, A.I.; Dublán-García, O.; Barrera García, V.D.; Favila Cisneros, H.J. Consumo y conocimiento actual de una bebida fermentada tradicional en Ixtapan del Oro, México: La sambumbia. *Alteridades* **2012**, *22*, 115–129.
135. Wacher Rodarte, C. La biotecnología alimentaria antigua: Los alimentos fermentados. *Rev. Digit. Univ.* **2014**, *15*, 121–136.
136. Romero, E.B.S. *El Sjendechjø: Bebida Mazahua de Maíz Fermentado*; Consejo Estatal para el Desarrollo Integral de los Pueblos Indígenas del Estado de México: Mexico City, Mexico, 1995.
137. Lorence-Quiñones, A.; Wacher-Rodarte, C.; Quintero-Ramírez, R. Cereal fermentations in Latin American countries. *FAO Agric. Serv. Bull.* **1999**, *138*, 99–114.
138. Unai. "Sendechó, la Cerveza Precolombina". 2019. Available online: https://www.delgranoalacopa.com/sendecho-la-cerveza-precolombina/ (accessed on 30 April 2021).
139. De la Fuente-Salcido, N.M.; Castañeda-Ramírez, J.C.; García-Almendárez, B.E.; Bideshi, D.K.; Salcedo-Hernández, R.; Barboza-Corona, J.E. Isolation and characterization of bacteriocinogenic lactic bacteria from M-Tuba and Tepache, two traditional fermented beverages in México. *Food Sci. Nutr.* **2015**, *3*, 434–442. [CrossRef]
140. Flores-Gallegos, A.; Vázquez-Vuelvas, O.; López-López, L.; Sainz-Galindo, A.; Ascacio-Valdes, J.; Aguilar, C.N.; Rodriguez-Herrera, R. Tuba, a Fermented and Refreshing Beverage from Coconut Palm Sap. In *Non-Alcoholic Beverages*; Woodhead Publishing: Sawston, UK, 2019; pp. 163–184.
141. Coutiño, B.; Rodríguez, R.; Belmares, R.; Aguilar, C.; Ruelas, X. Selección de la bebida "taberna" obtenida de la palma Acro-comia aculeata y análisis químico proximal. *Multiciencias* **2015**, *15*, 397–409.

142. Available online: https://www.gob.mx/agricultura\T1\textbar{}jalisco/articulos/alcanza-tequila-ventas-por-mas-de-568-millones-de-dolares-en-primer-semestre-del-ano-138868 (accessed on 12 February 2021).
143. Borren, E.; Tian, B. The Important Contribution of Non-Saccharomyces Yeasts to the Aroma Complexity of Wine: A Review. *Foods* **2020**, *10*, 13. [CrossRef]
144. Wahyuni, N.L.; Sunarharum, W.B.; Muhammad, D.R.A.; Saputro, A.D. Formation and development of flavour of cocoa (*Theobroma cacao* L.) cultivar Criollo and Forastero: A review. *IOP Conf. Ser. Earth Environ. Sci.* **2021**, *733*, 012078. [CrossRef]
145. De Filippis, F.; Pasolli, E.; Ercolini, D. The food-gut axis: Lactic acid bacteria and their link to food, the gut microbiome and human health. *FEMS Microbiol. Rev.* **2020**, *44*, 454–489. [CrossRef] [PubMed]
146. Kisan, B.S.; Kumar, R.; Ashok, S.P.; Sangita, G. Probiotic foods for human health: A review. *J. Pharmacogn. Phytochem.* **2019**, *8*, 967–971.
147. Neffe-Skocińska, K.; Rzepkowska, A.; Szydłowska, A.; Kołożyn-Krajewska, D. Trends and possibilities of the use of probiotics in food production. In *Alternative and Replacement Foods*; Academic Press: Cambridge, MA, USA, 2018; pp. 65–94.
148. Homayouni, A.; Alizadeh, M.; Alikhah, H. Functional dairy probiotic food development: Trends, concepts, and products. In *Immunology and Microbiology "Probiotics"*; IntechOpen: London, UK, 2012.
149. Torres-León, C.; Ramírez-Guzman, N.; Londoño-Hernandez, L.; Martinez-Medina, G.A.; Díaz-Herrera, R.; Navarro-Macias, V.; Aguilar, C.N. Food waste and byproducts: An opportunity to minimize malnutrition and hunger in developing countries. *Front. Sustain. Food Syst.* **2018**, *2*, 52. [CrossRef]
150. Rad, A.H.; Abbasi, A.; Kafil, H.S.; Ganbarov, K. Potential Pharmaceutical and Food Applications of Postbiotics: A Review. *Curr. Pharm. Biotechnol.* **2020**, *21*, 1576–1587. [CrossRef] [PubMed]
151. Sanchez Cortes, M.S.; Verdugo Valdez, A.G.; Lopez Zuñiga, E.J.; Vela Gutierrez, G. *El Atole Agrio: Su Presencia y Modo de Preparación en la Cultura Alimentaria de Chiapas*; UNICACH: Mexico City, Mexico, 2019.
152. Bueno López, B.P. Evaluación del Potencial Probiótico de Bacterias Ácido Lácticas (BAL), Aisladas de una Bebida Fermentada Tradicional de Chiapas (Taberna). Doctoral Dissertation, Universidad de Ciencias y Artes de Chiapas, Tuxtla Gutiérrez, Chiapas, Mexico, 2013.
153. Trujillo Jiménez, C.A. Evaluación de la capacidad productiva de etanol utilizando levaduras aisladas de agave y taberna. Doctoral dissertation, Universidad de Ciencias y Artes de Chiapas, Tuxtla Gutiérrez, Chiapas, Mexico, 2013.
154. Alcántara-Hernández, R.J.; Rodríguez-Álvarez, J.A.; Valenzuela-Encinas, C.; Gutiérrez-Miceli, F.A.; Castañón-González, H.; Marsch, R.; Dendooven, L. The bacterial community in 'taberna'a traditional beverage of Southern Mexico. *Lett. Applied Microbiol.* **2010**, *51*, 558–563. [CrossRef]
155. Ambrocio Ríos, J.A. Ecología de Levaduras Asociadas a la Taberna, Bebida Extraída de la Palma de Coyol (*Acrocomia aculeata* (Jacq.) Lodd. ex Mart). Master's Thesis, Universidad de Ciencias y Artes de Chiapas, Tuxtla Gutiérrez, Chiapas, Mexico, 2018.
156. Devanthi, P.V.P.; Gkatzionis, K. Soy sauce fermentation: Microorganisms, aroma formation, and process modification. *Food Res. Int.* **2019**, *120*, 364–374. [CrossRef]
157. Pakuwal, E.; Manandhar, P. Production of Rice Based Alcoholic Beverages and their Quality Evaluation. *J. Food Sci. Technol. Nepal* **2020**, *12*, 37–48. [CrossRef]
158. Cano, A.N.H.; Suárez, M.E. Ethnobiology of algarroba beer, the ancestral fermented beverage of the Wichí people of the Gran Chaco I: A detailed recipe and a thorough analysis of the process. *J. Ethn. Foods* **2020**, *7*, 1–12. [CrossRef]
159. Daryaei, H.; Coventry, J.; Versteeg, C.; Sherkat, F. Combined pH and high hydrostatic pressure effects on Lactococcus start-er cultures and Candida spoilage yeasts in a fermented milk test system during cold storage. *Food Microbiol.* **2010**, *27*, 1051–1056. [CrossRef] [PubMed]
160. Garofalo, C.; Arena, M.P.; Laddomada, B.; Cappello, M.S.; Bleve, G.; Grieco, F.; Beneduce, L.; Berbegal, C.; Spano, G.; Capozzi, V. Starter Cultures for Sparkling Wine. *Fermentation* **2016**, *2*, 21. [CrossRef]
161. Roudil, L.; Russo, P.; Berbegal, C.; Albertin, W.; Spano, G.; Capozzi, V. Non-Saccharomyces commercial starter cultures: Scientific trends, recent patents and innovation in the wine sector. *Recent Pat. Food Nutr. Agric.* **2020**, *11*, 27–39. [CrossRef] [PubMed]
162. Valadez-Blanco, R.; Bravo, G.; Santos, N.; Velasco, S.; Montville, T. The artisanal production of pulque, a traditional beverage of the Mexican highlands. *Probiotics Antimicrob. Proteins* **2012**, *4*, 140–144. [CrossRef]
163. Rizo, J.; Rogel, M.A.; Guillén, D.; Wacher, C.; Martinez-Romero, E.; Encarnación, S.; Rodríguez-Sanoja, R. Nitrogen fixation n pozol, a traditional fermented beverage. *Appl. Environ. Microbiol.* **2020**, *86*, e00588-20. [CrossRef]
164. Nuñez-Guerrero, M.E.; Salazar-Vázquez, E.; Páez-Lerma, J.B.; Rodríguez-Herrera, R.; Soto-Cruz, N.O. Physiological characterization of two native yeasts in pure and mixed culture using fermentations of agave juice. *Int. J. Agric. Nat. Resour.* **2019**, *46*, 1–11. [CrossRef]
165. Escalante-Minakata, P.; Blaschek, H.; de la Rosa, A.B.; Santos, L.; De León-Rodríguez, A. Identification of yeast and bacteria involved in the mezcal fermentation of Agave salmiana. *Lett. Appl. Microbiol.* **2008**, *46*, 626–630. [CrossRef] [PubMed]
166. Gutiérrez-Uribe, J.A.; Figueroa, L.M.; Martín-del-Campo, S.T.; Escalante, A. Pulque. In *Fermented Foods in Health and Disease Prevention*; Academic Press: Cambridge, MA, USA, 2017; pp. 543–556.
167. Jakočiūnas, T.; Jensen, M.K.; Keasling, J. CRISPR/Cas9 advances engineering of microbial cell factories. *Metab. Eng.* **2016**, *34*, 44–59. [CrossRef]

168. Jakočiūnas, T.; Bonde, I.; Herrgård, M.; Harrison, S.J.; Kristensen, M.; Pedersen, L.E.; Jensen, M.K.; Keasling, J.D. Multiplex metabolic pathway engineering using CRISPR/Cas9 in Saccharomyces cerevisiae. *Metab. Eng.* **2015**, *28*, 213–222. [CrossRef] [PubMed]
169. DiCarlo, J.; Norville, J.; Mali, P.; Rios, X.; Aach, J.; Church, G.M. Genome engineering in Saccharomyces cerevisiae using CRISPR-Cas systems. *Nucleic Acids Res.* **2013**, *41*, 4336–4343. [CrossRef] [PubMed]
170. Magalhães, F.; Krogerus, K.; Castillo, S.; Ortiz-Julien, A.; Dequin, S.; Gibson, B. Exploring the potential of Saccharomyces eubayanus as a parent for new interspecies hybrid strains in winemaking. *FEMS Yeast Res.* **2017**, *17*, 17. [CrossRef]
171. Nikulin, J.; Vidgren, V.; Krogerus, K.; Magalhães, F.; Valkeemäki, S.; Kangas-Heiska, T.; Gibson, B. Brewing potential of the wild yeast species Saccharomyces paradoxus. *Eur. Food Res. Technol.* **2020**, *246*, 2283–2297. [CrossRef]
172. Wagner, J.M.; Alper, H.S. Synthetic biology and molecular genetics in non-conventional yeasts: Current tools and future advances. *Fungal Genet. Biol.* **2016**, *89*, 126–136. [CrossRef] [PubMed]
173. Wu, D.; Xie, W.; Li, X.; Cai, G.; Lu, J.; Xie, G. Metabolic engineering of Saccharomyces cerevisiae using the CRISPR/Cas9 system to minimize ethyl carbamate accumulation during Chinese rice wine fermentation. *Appl. Microbiol. Biotechnol.* **2020**, *104*, 4435–4444. [CrossRef]
174. Vigentini, I.; Gebbia, M.; Belotti, A.; Foschino, R.; Roth, F. CRISPR/Cas9 System as a Valuable Genome Editing Tool for Wine Yeasts with Application to Decrease Urea Production. *Front. Microbiol.* **2017**, *8*, 2194. [CrossRef] [PubMed]
175. Fernandez Suarez, R.; Morales Chavez, L.A.; Galvez Mariscal, A. Importance of mexican maize landraces in the national diet: An essential review. *Rev. Fitotec. Mex.* **2013**, *36*, 275–283.
176. Valerino-Perea, S.; Lara-Castor, L.; Armstrong, M.E.G.; Papadaki, A. Definition of the Traditional Mexican Diet and Its Role in Health: A Systematic Review. *Nutrients* **2019**, *11*, 2803. [CrossRef]
177. Delgado, Á.A. La mujer en la cosmovisión y ritualidad rarámuri. *Bol. Antropol. Univ. Antioq.* **2007**, *21*, 41–63.
178. Delgado, Á.A.; Gómez, G.A. The rarámuri race as a metaphor of cultural resistance. *J. Hum. Sport Exerc.* **2009**, *4*, 6–19. [CrossRef]
179. Turner, B.L.; Klaus, H.D. *Diet, Nutrition, and Foodways on the North Coast of Peru*; Springer: Berlin/Heidelberg, Germany, 2020.
180. Lertzman, K. The Paradigm of Management, Management Systems, and Resource Stewardship. *J. Ethnobiol.* **2009**, *29*, 339–358. [CrossRef]
181. Salomon, A.K.; Lertzman, K.; Brown, K.; Wilson, K.B.; Secord, D.; McKechnie, I. Democratizing conservation science and practice. *Ecol. Soc.* **2018**, *23*, 23. [CrossRef]
182. Ham, P.F. Infant mortality in the indigenous population: Backwardness and contrasts. *Demos* **1993**, *6*, 12–13.
183. Serrano Carreto, E.; Embriz Osorio, A.; Fernández Ham, P. *Indicadores Socioeconómicos de los Pueblos Indígenas de México*; Instituto Nacional Indígenista: Mexico City, Mexico, 2002.
184. Lartigue, F.; Quesnel, A. *La Población Indígena Entre los Enfoques de Política Pública y las Categorías Antropodemográficas*; Alteridades, CIESAS: Mexico City, Mexico, 2003; pp. 11–34.
185. Barbary, O. Social Inequalities and Indigenous Populations in Mexico: A Plural Approach. In *Social Statistics and Ethnic Diversity*; Springer: Cham, Switzerland, 2015; pp. 209–228.
186. Cerjak, M.; Haas, R.; Brunner, F.; Tomić, M. What motivates consumers to buy traditional food products? Evidence from Croatia and Austria using word association and laddering interviews. *Br. Food J.* **2014**, *116*, 1726–1747. [CrossRef]
187. Vanhonacker, F.; Van Loo, E.J.; Gellynck, X.; Verbeke, W. Flemish consumer attitudes towards more sustainable food choices. *Appetite* **2013**, *62*, 7–16. [CrossRef]
188. Almansouri, M.; Verkerk, R.; Fogliano, V.; Luning, P.A. Exploration of heritage food concept. *Trends Food Sci. Technol.* **2021**, *111*, 790–797. [CrossRef]
189. Senos, R.; Lake, F.K.; Turner, N.; Martinez, D. Traditional ecological knowledge and restoration practice. In *Restoring the Pacific Northwest: The Art and Science of Ecological Restoration in Cascadia*; Apostol, D., Sinclair, M., Eds.; Island Press: Washington, DC, USA, 2006; Chapter 17; pp. 393–426.
190. Kuhnlein, H.V.; Receveur, O. Dietary change and traditional food systems of indigenous peoples. *Annu. Rev. Nutr.* **1996**, *16*, 417–442. [CrossRef]
191. Turner, N.J.; Turner, K.L. Traditional Food Systems, Erosion and Renewal in Northwestern North America. *Indian J. Tradit. Knowl.* **2007**, *6*, 57–68.
192. Quave, C.; Pieroni, A. Fermented Foods for Food Security and Food Sovereignty in the Balkans: A Case Study of the Gorani People of Northeastern Albania. *J. Ethnobiol.* **2014**, *34*, 28–43. [CrossRef]
193. Posey, D.A. Biodiversity, genetic resources and indigenous peoples in Amazonia: (Re) discovering the wealth of traditional resources of Native Amazonians. In *Amazonia at the Crossroads*; Oxford University Press: Oxford, UK, 2000; pp. 188–204.
194. Mauro, F.; Hardison, P.D. Traditional knowledge of indigenous and local communities: International debate and policy initiatives. *Ecol. Appl.* **2000**, *10*, 1263–1269. [CrossRef]
195. Drahos, P. Developing Countries and International Intellectual Property Standard-Setting. *J. World Intellect. Prop.* **2005**, *5*, 765–789. [CrossRef]
196. Hafstein, V.T. *The Making of Intangible Cultural Heritage: Tradition and Authenticity, Community and Humanity*; University of California: Berkeley, CA, USA, 2004.
197. Alexiades, M.N. Ethnobotany in the third millennium: Expectations and unresolved issues. *Delpinoa* **2003**, *45*, 15–28.

198. Rocha, J.A.; Boscolo, O.H.; Fernandes, L.R. Ethnobotany: A instrument for valorisation and identification of potential for the protection of traditional knowledge. *Interações Campo Grande* **2015**, *16*, 67–74. [CrossRef]
199. Mintz, S.W.; Du Bois, C.M. The Anthropology of Food and Eating. *Annu. Rev. Anthr.* **2002**, *31*, 99–119. [CrossRef]
200. Ingold, T. *The Perception of the Environment: Essays on Livelihood, Dwelling and Skill*; Routledge: London, UK, 2002.
201. Ellen, R.F.; Fischer, M.D. *Introduction: On the Concept of Cultural Transmission*; Berhahn Books: New York, NY, USA; Oxford, UK, 2013; pp. 1–54.
202. Pelto, G.H.; Vargas, L.A. Perspectives in dietary change. *Ecol. Food Nutr.* **1992**, *27*, 159–161. [CrossRef]
203. Marshall, E.; Mejia, D. *Traditional Fermented Food and Beverages for Improved Livelihoods*; FAO Diversification Booklet; FAO: Rome Italy, 2011; p. 21.
204. Pai, J.S. Applications of microorganisms in food biotechnology. *Indian J. Biotech.* **2003**, *2*, 382–386.
205. Alexiades, M. The cultural and economic globalisation of traditional environmental knowledge systems. In *Landscape, Process and Power: Reevaluating Traditional Environmental Knowledge*; Berghahn Books: New York, NY, USA; Oxford, UK, 2009; pp. 68–90.
206. Cunningham-Sabo, L.D.; Davis, S.M.; Koehler, K.M.; Fugate, M.L.; DiTucci, J.A.; Skipper, B.J. Food preferences, practices, and cancer-related food and nutrition knowledge of southwestern American Indian youth. *Cancer Interdiscip. Int. J. Am. Cancer Soc.* **1996**, *78*, 1617–1622.
207. Reyes-García, V.; Menendez-Baceta, G.; Aceituno-Mata, L.; Acosta-Naranjo, R.; Calvet-Mir, L.; Domínguez, P.; Pardo-de-Santayana, M. From famine foods to delicatessen: Interpreting trends in the use of wild edible plants through cultural eco-system services. *Ecol. Econ.* **2015**, *120*, 303–311. [CrossRef]
208. Holguín Balderrama, J. Yocogihua: La Última Fábrica de Aguardiente Alamense. Available online: https://infocajeme.com/general/2018/03/yocogihua-la-ultima-fabrica-de-aguardiente-alamense/ (accessed on 26 March 2018).
209. Nabhan, G.P.; Suro Pineda, D. *Agave Spirit World Past, Present and Future of Mezcals and Their Kin*; W.W. Norton: New York, NY, USA, 2021.

Review

Cereal- and Fruit-Based Ethiopian Traditional Fermented Alcoholic Beverages

Eskindir Getachew Fentie [1,2], Shimelis Admassu Emire [3], Hundessa Dessalegn Demsash [3], Debebe Worku Dadi [4] and Jae-Ho Shin [2,*]

1 College of Biological and Chemical Engineering, Addis Ababa Science and Technology University, Addis Ababa 16417, Ethiopia; eskench@gmail.com

2 Department of Applied Biosciences, Kyungpook National University, Daegu 41900, Korea

3 School of Chemical and Bio-Engineering, Addis Ababa Institute of Technology, Addis Ababa University, P.O. Box 385, King George VI Street, Addis Ababa 16417, Ethiopia; shimelisemire@yahoo.com (S.A.E.); hundessad@gmail.com (H.D.D.)

4 Department of Food Engineering and Postharvest Technology, Institute of Technology, Ambo University, Ambo 2040, Ethiopia; debeworku2010@gmail.com

* Correspondence: jhshin@knu.ac.kr; Tel.: +82-53-950-5716

Received: 29 October 2020; Accepted: 28 November 2020; Published: 1 December 2020

Abstract: Traditional fermented alcoholic beverages are drinks produced locally using indigenous knowledge, and consumed near the vicinity of production. In Ethiopia, preparation and consumption of cereal- and fruit-based traditional fermented alcoholic beverages is very common. *Tella*, *Borde*, *Shamita*, *Korefe*, *Cheka*, *Tej*, *Ogol*, *Booka*, and *Keribo* are among the popular alcoholic beverages in the country. These beverages have equal market share with commercially produced alcoholic beverages. Fermentation of Ethiopian alcoholic beverages is spontaneous, natural and uncontrolled. Consequently, achieving consistent quality in the final product is the major challenge. Yeasts and lactic acid bacteria are the predominate microorganisms encountered during the fermentation of these traditional alcoholic beverages. In this paper, we undertake a review in order to elucidate the physicochemical properties, indigenous processing methods, nutritional values, functional properties, fermenting microorganisms and fermentation microbial dynamics of Ethiopian traditional alcoholic beverages. Further research will be needed in order to move these traditional beverages into large-scale production.

Keywords: traditional alcoholic beverage; Ethiopia; processing; physicochemical; fermentative microorganisms

---

1. Introduction

Worldwide production and consumption of fermented beverages has a long history, and is believed to have started around 6000 BC [1,2]. Production techniques and consumption of these traditional beverages are very localized [3]. Ethiopia, like other parts of the world, produces and consumes a significant volume of traditional alcoholic beverages (Table 1). About eight million hectoliters of Ethiopian traditionally fermented alcoholic beverages are produced yearly. Commercially and traditionally produced alcoholic beverages have an almost equal market share [4] and annual per capital pure alcohol consumption in the country is about 2 L [5].

Traditional alcoholic drinks are widely produced and consumed in Asia and Africa [6]. Rwanda's ikigage [7], Nigeria's oti-oka [8], Uganda's kwete [9], Kenya's Busaa [10], Korea's makgeolli [11] and Mexico's pulque [12] are among the most common traditional alcoholic beverage that are consumed and produced in each respective country.

In Ethiopia, *Tella* [1], *Borde* [13], *Shamita* [14], *Korefe* [15], *Keribo* [16], *Cheka* [17], *Tej* [18], *Ogol* [19] and *Booka* [20] are very popular indigenous fermented alcoholic beverages. The total alcohol content of these beverage is in the range of 1.53–21.7% (*v*/*v*) [18,20]. All of these Ethiopian alcoholic beverages are produced at a small scale and sold by local alcohol venders from their homes. These traditional alcoholic beverages are classified under the category of acid-alcohol fermentation systems [21].

Scholars define wine and beer based on various perspectives. For instance, Herman [22] defined wines as alcoholic beverages made from sound ripe grapes, whereas Pederson [23] defined alcoholic beverages based on the kind of substrates: beers are produced from cereals whereas wines are produced from fruits. In addition, Steinkraus [24] defined wine as an alcoholic beverage that uses sugar as the principal source of fermentable carbohydrate. According to Steinkraus [24], beverages made from honey, sugar cane and palm are classified under the category of wine. Hence, *Tej*, *Ogol* and *Booka* can alternatively be called wines, since honey is used as a major substrate for the fermentation process.

"Gesho" (*Rhamnus prinoides* L.), also known as "dog wood", is the most common ingredient used to prepare Ethiopian alcoholic beverages, primarily as a flavoring and bittering agent. The substance β-sorigenin-8-O-β-D-glucoside (*"geshoidin"*) is the naphthalenic compound responsible for imparting bitterness [25,26]. In addition to this, it is also a source of fermentative microorganisms and plays a significant role during fermentation in regulating the microbial dynamic [27].

Due to the absence of standardized processes, back-slopping, and starter culture, Ethiopian beverages often have poor quality and failure to achieve their objective [28]. Moreover, preparation of these fermented alcoholic beverages is time-consuming and laborious [29]. As far as we know, this review is the first of its kind to address the research trends, significant research gaps and directions for future research outputs on Ethiopian traditional fermented alcoholic beverages. In particular, the raw materials, processing methods, physicochemical properties, nutritional values, functional properties, responsible fermenting microorganisms, fermentation microbial dynamics and storage stability of Ethiopian alcoholic beverages are the key points reviewed in the paper.

Table 1. Summary of cereal- and fruit-based Ethiopian traditional fermented alcoholic beverages.

Category of Beverages	Beverages	Raw Materials	Prominent Production and Consumption Regions	References
Beers	*Tella*	Barley (*Hordeum vulgare* L.), wheat (*Triticum aestivum* L.), maize (*Zea mays* L.), finger millet (*Eleusine coracana* L.), sorghum (*Sorghum bicolor* L.), *"teff"* (*Eragrostis tef* L.), "gesho" (*R. prinoides*)	Amhara, Oromia, Tigray, SNNP, Addis Ababa	[30,31]
	Borde	Maize (*Z. mays*), barley (*H. vulgare*), wheat (*T. aestivum*), finger millet (*E. coracana*), sorghum (*S. bicolor*)	SNNP	[13,32]
	Shamita	Roasted barley (*H. vulgare*) flour, salt, linseed (*Linum usitatissimum* L.) flour, chili pepper (*Capsicum annuum*)	SNNP, Addis Ababa	[13,33]
	Korefe	Malted and non-malted barley (*H. vulgare*), *"gesho"* (*R. prinoides*)	Amhara	[15]
	Keribo	Barley (*H. vulgare*), sugar, bakery yeast (*Saccharomyces cerevisiae*)	Oromia, Amhara, Addis Ababa	[16,34]
	Cheka	Sorghum (*S. bicolor*), maize (*Z. mays*), finger millet (*E. coracana*), vegetables, root of taro (*Colocasia esculenta* L.)	SNNP	[17,35,36]
	Areke	Barley (*H. vulgare*), "gesho" (*R. prinoides*), sorghum	Amhara, Oromia, Tigray, SNNP, Addis Ababa	[37–39]
Wine	*Tej*	Honey, "gesho" (*R. prinoides*)	Oromia, Amhara, Tigray, Addis Ababa	[18,39,40]
	Ogol	honey, barks of native tree (*Blighia unijungata* L.)	Gambella (Majangir)	[19]
	Booka	Honey, bladder of cow	Oromia (Gujii)	[20]

SNNP—Southern Nations, Nationalities, and Peoples Region.

2. Cereal-Based Traditional Alcoholic Beverages

2.1. *Tella*

Tella is the most consumed traditional fermented alcoholic beverage in Ethiopia. It is the most popular beverage in the Oromia, Amhara and Tigray regions (Table 1). Barley, wheat, maize, millet, sorghum, "*teff*" (*E. tef*) and "*gesho*" leaves (*R. prinoides*) along with naturally-present microorganisms are the ingredients used to produce *Tella* [1]. Even though the volume of production and consumption is high, the fermentation process is still spontaneous, uncontrolled and unpredictable [41].

The *Tella* making process and its raw materials vary among ethnic groups and economic and traditional situations [37]. Although there are minor changes in the process in different localities, the basic steps are similar throughout the country. The making of "*Tejet*", "*Tenses*" and "*Difdif*" are the fundamental steps in the *Tella* preparation process [1].

The *Tella* making process starts by soaking the barley in water for about 24 h at room temperature to produce a malt, locally called "*Bikil*". After 24 h, the moistened grain is covered by using fresh banana leaves and kept in a dry place for an additional three days [39]. Then, the germinated barley grain is sun-dried and ground to produce malt flour. At the same time "gesho" (*R. prinoides*) leaves and stems are sun-dried and ground. Then, "*Bikil*" flour and "gesho" powder are mixed with an adequate amount of water in a clean and smoked traditional bioreactor known as "*Insera*". This mixture is left to ferment for two days to form "*Tejet*" [31]. Subsequently, millet, sorghum and "teff" (*E. tef*) flours of equal proportion are mixed with water to form a dough. The dough is then baked to produced unleavened bread locally known as "*ye Tella kita*" [41], which is sliced into pieces and added to the earlier produced "*Tejet*". The mixture is then sealed tightly to ferment anaerobically for 5 to 7 days to turned into "*Tenses*" [30].

While the "*Tenses*" is fermenting, maize grain is soaked in water for about 3 d, and then it is dried, roasted and ground to make a dark maize flour called "*Asharo*". "*Asharo*" is the main ingredient that determines the color of *Tella* [31]. "*Asharo*" is then added to the previously produced "*Tenses*" and fermented anaerobically for a period of 10 to 20 days. After this period of fermentation, a thick mixture locally called "*Difdif*" is formed. Water is added to "*Difdif*" and left to ferment for an additional 5 to 6 h. Finally, solid residues are removed by filtration and served to consumers as *Tella*. In order to produce 25 to 28 L of pure *Tella*, 1 kg of "*gesho*" (*R. prinoides*) powder, 0.5 kg of "*Bikil*", 5 kg of "*ye Tella kita*", 10 kg of "*Asharo*" and 30 L of water are required [41].

Ingredients and utensils used to prepare *Tella* are the major source of microorganisms for the fermentation process [42]. As shown in Table 2, genera of *Saccharomyces*, *Lactobacillus* and *Acetobacter* are the most predominant fermenting microorganisms present in *Tella* [1,30,41]. The alcohol content and pH of *Tella* collected from different localities vary from 3.98–6.48% (*v*/*v*) and 1.52–4.99, respectively [43]. The alcohol content of *Tella* is greater than that of Rwanda's ikigage [7] and is very much lower than Korean makgeolli [11]. The electric conductivity, salinity and total dissolved solids (TDS) of *Tella* are 2359 μs/cm, 1.2% and 1180 mg/L, respectively [44].

Since the production of *Tella* is performed at the household level, it seriously lacks aseptic processing conditions. Consequently, the shelf life is no longer than 5 to 7 days at room temperature. Beyond that, the flavor becomes too sour to drink. *Acetobacter* species are mostly responsible for this sourness because they convert ethanol to acetic acid in the presence of oxygen [45].

Table 2. Physicochemical properties, microbial load, and storage stability of *Tella*.

Area of Investigation	Shelf Stability, Microbial and Physicochemical Properties	Concluding Remarks	References
Storage stability, and microbial dynamics for vacuum filtered (VF), pasteurized (P) and control *Tella*	• S. cerevisiae and Acetobacter xylinum (A. xylinum) are the dominating microorganisms; • pH of control sample decreased, while VF and P pH samples increased during storage time; • Turbidity of the control sample increased, while VF and P turbidity decreased or remained the same.	Pasteurization is an efficient method to extend the shelf life compared to vacuum filtration	[1]
Optimization of *Tella* production	• 3:1 (malt to "gesho" (*R. prinoides*)) showed lower pH after nine days of fermentation; • Fermentation rate increased with increasing malt to "gesho" (*R. prinoides*) ratio.	The optimum fermentation process parameters: • Temperature = 20–25 °C, average; • pH = 4.78; • Malt to "gesho" (*R. prinoides*) ratio = 1:3.	[46]
Isolation and characterization of *S. cerevisiae* from *Tella*	• Six *S. cerevisiae* strains were isolated and characterized phenotypically; • Isolates produce 10–15% mL/L of absolute ethanol; • Isolates showed 84% of viability at higher sugar concentration; • Isolates had an average 65% flocculation capacity.	Isolated strains have a good fermentative potential, especially for beer production	[30]
Physicochemical properties of fresh and matured *Tella*	• pH 4.67–3.87; • Alcohol content (%*v*/*v*) 3.04–3.75; • Specific extract 1.0056–1.0037; • Original extract 7.50–7.27; • CO_2 content (%) 0.24–0.034.	• Alcohol content increases with increasing maturation time; • pH, specific extract, original extract, and CO_2 content decrease with increasing maturation time.	[31]

2.2. Borde

Borde is a cereal-based Ethiopian traditional fermented low alcoholic beverage that uses maize (*Z. mays*), wheat (*T. aestivum*), finger millet (*E. coracana*) and sorghum (*S. bicolor*) interchangeably or sometimes proportionally as the main ingredients [29]. It is commonly produced and consumed in the southern and western part of Ethiopia. The local communities consider *Borde* as a meal replacement. Particularly, low-income local groups of the population may consume up to 3 L of *Borde* per day [47]. The nutritional value is high due to the high number of live cells present in freshly produced *Borde* [32].

The *Borde* making process starts with germinating barley grain by following the same procedure described for the *Tella* malt preparation process. This malt, a source of amylase enzymes, is ground to become a malt flour [33]. In parallel, maize grits are mixed with a proportional volume of water and fermented for about 44 to 48 h (Figure 1). The fermented blend is divided into three portions. Similar to Uganda's kwete [9], about 40% of the blend is roasted on a hot pan and a bread locally called *"Enkuro"* is produced. Then, the prepared *"Enkuro"* is mixed with malt flour and additional water and allowed to ferment for about 24 h in the same mixing tank [32]. The other 40% of the fermented maize grits are mixed with additional fresh maize flour and water. This mixture is shaped into a ball-like structure and cooked using steam to form *"Gafuma"* [29]. Subsequently, *"Gafuma"* is added to previously prepared *"Tinsis"* to become the thick brown mash called *"Difdif"* [13]. The remaining 20% of the fermented maize grits are mixed with additional flour and water and boiled to form thick porridge. Then, the prepared porridge, extra malt, and water are mixed into the earlier produced *"Difdif"*. Finally, the mixture is filtered and a small amount of water is added before serving to consumers as the final product *Borde* [33].

A good-quality *Borde* can be described as opaque, fizzy, of uniform turbidity, gray in color, with a thick consistency, a fairly smooth texture, and a flavor somewhere in the middle between sweet and sour [29]. The average pH values of *Borde* lie within the range of 3.6–4.1. The type of ingredients used and the processing conditions are the major causes for variation in the final product [32]. The conductivity, salinity and TDS values of *Borde* are 7139 µs/cm, 3.9%, and 3830 mg/L, respectively. As in Kenya's busaa [10], yeast and lactic acid bacteria are the dominant microorganisms in *Borde*. Around 10^9 CFU/mL counts have been recorded for both mesophilic bacteria and lactic acid bacteria [47]. In addition, a 10^5–10^7 CFU/mL yeast count has been reported for freshly prepared *Borde* (Table 3). Due to these high microorganism counts, *Borde* becomes unfit for consumption after 12 h of room temperature storage [29].

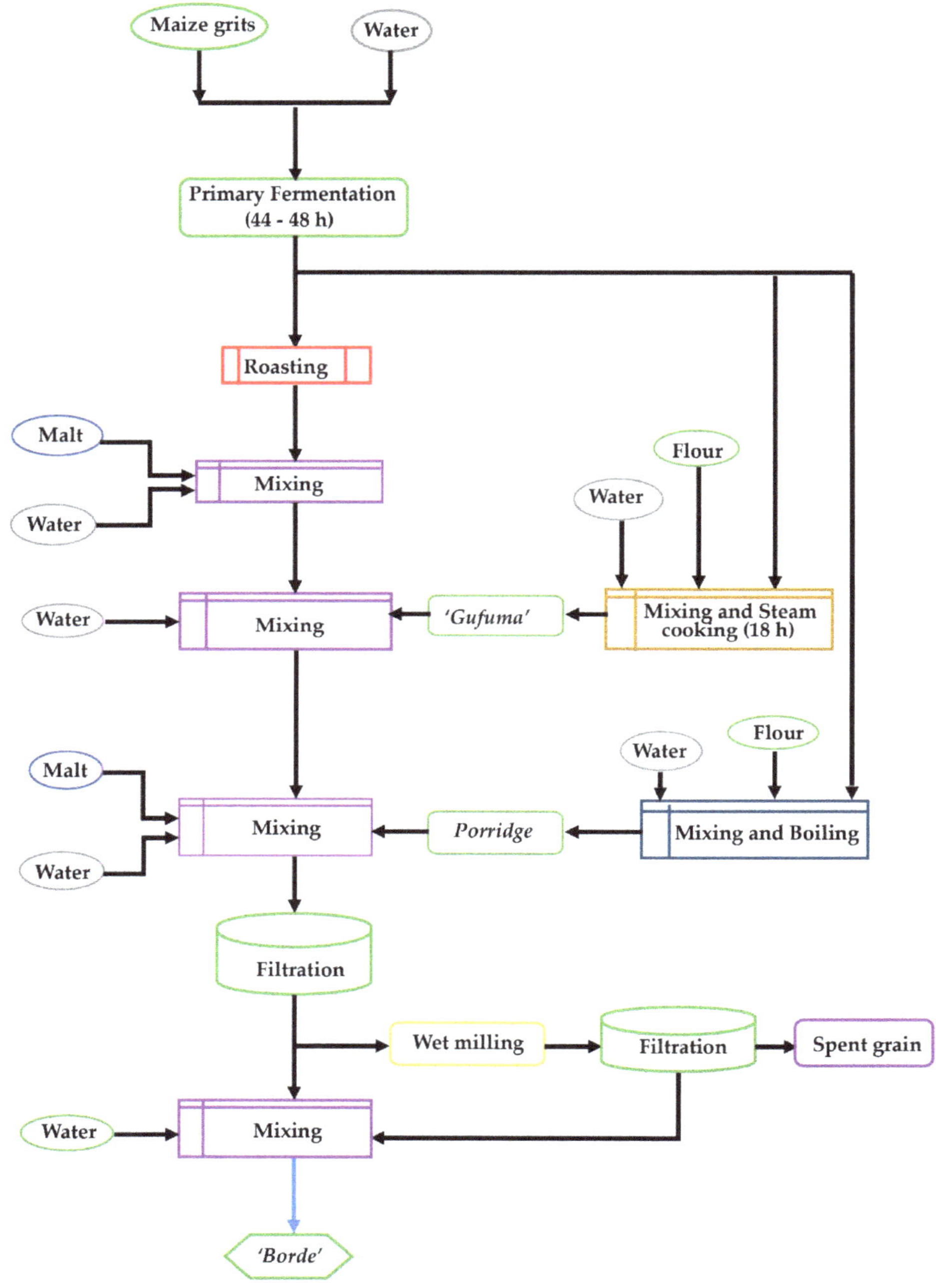

Figure 1. *Borde* processing flow chart [29].

Table 3. Processing methods and microbiological properties of *Borde*.

Area of Investigation	Microbial Load, Microbial Dynamics, and Processing Methods	Concluding Remarks	References
Isolation and characterization of lactic acid bacteria (LAB) involved in *Borde* fermentation	• Heterofermentative *lactobacillus* (79.4%) is the predominate microorganism in *Borde^*; • Dominant species are: *Weissella confusa* (30.9%), *Lactobacillus viridescens*, (26.5%), *Lactobacillus brevis* (10.3%) and *Pediococcus pentosaceus* (7.4%).	Dominant microorganisms have a potential to be used as a starter culture	[13]
Antagonist effect of lactic acid bacteria over pathogenic microorganisms	■ At the beginning of fermentation Esherichia coli (E. coli O157:H7), Staphylococcus aureus (S. aureus), Shigella flexneri (S. flexneri) and Salmonella species counts are greater than 10^7 CFU/mL; ■ After 16 h of fermentation the count of Salmonella species is less than log 2 CFU/mL, and all pathogenic microorganisms are below the detectable limit after 24 h of fermentation.	Secondary metabolites of LAB have a significant antimicrobial effect	[48]
Modified process technology for *Borde* production	• Maize flour is substituted by maze grits; • Remove wet milling from last stage of the process.	*Borde* making process can be simplified without compromising quality	[47]
Survey on local methods of processing and sensory analysis of *Borde*	■ Developed a traditional processing method with four-stage flow charts; ■ Maize, wheat, finger millet and sorghum used as raw materials; ■ Shelf life is no longer than 12 h at room temperature storage.	• *Borde* has short shelf life; • Production process is time-consuming.	[29]
Microbial dynamics of *Borde* fermentation	• *Enterobacteriaceae* and coliform decreased from 10^4 CFU/mL to below the detectable limit after 8 h of fermentation; • Lactic acid bacteria increased from 10^6 to 10^9 CFU/mL within 24 h of fermentation time; • Total fermentative yeast increased from 10^5 to 10^7 CFU/mL after 24 h fermentation time.	• Yeast biota is dominated by *Saccharomyces* species; • Keeping quality of *Borde* is very short.	[32]

2.3. Shamita

Shamita is another traditional low alcoholic beverage that is produced and consumed in different parts of Ethiopia. Roasted and ground barley is used as a major substrate during the fermentation stage [49]. This beverage also serves as a meal replacement for low income workers. Like other traditional Ethiopian fermented beverages (*Tella* and *Borde*), *Shamita* production does not require malt for the saccharification process [15].

To prepare *Shamita,* barley flour, salt, linseed flour, and a small amount of spice are mixed together with water to form a slurry liquid. As a starter culture, 1 to 2 L of previously produced slurry is added to the blend. The mixture is allowed to ferment overnight. Then, a small amount of bird's eye chili (*C. annuum*) is added and the beverage is ready to serve for consumption [34].

The first full-length article on *Shamita* was published by Ashenafi and Mehari [34], which focused on the enumeration of microorganisms in samples collected from different vendors. The report found that lactic acid bacteria and yeasts are the dominant microorganisms in *Shamita*. Four years later, Bacha et al. [14] studied *Shamita* fermentation microbial dynamics and the microbial load of raw materials. Their study showed that barley is the major source of fermentative microorganisms. The count of these fermentative microbes reached 10^9 CFU/mL after a 24 h fermentation period. Later Tadesse et al. [49] studied the antimicrobial effect of lactic acid bacteria isolated from *Shamita* on pathogenic microorganisms. The isolated lactic acid bacteria were found to inhibit the growth of the *Salmonella* species *S. flexneri*, and *S. aureus*. Similar inhibition was observed for lactic acid bacteria isolated from Nigeria's oti-oka [8]. Additionally, the pH, conductivity, salinity and TDS values of *Shamita* were 3.8, 8391 μs/cm, 4.6% and 4520 mg/L, respectively [44].

2.4. Korefe

Korefe is a foamy fermented low alcoholic beverage popular in the northern and northwestern parts of Ethiopia. Similar to other Ethiopian fermented beverages, the fermentation system is natural and spontaneous. Barley, malted barley, "gesho"' (*R. prinoides*), and water are the major ingredients used to prepare this indigenous beverage [50].

The process of making *Korefe* begins by mixing "gesho" (*R. prinoides*) and water to produce *"Tijit"* in a traditional container locally known as *"Gan"* (Figure 2). The blend is left for 72 h to extract flavor, aroma, bitterness and fermenting microorganisms [15]. While that is happening, non-malted barley powder is mixed with water to form a dough. The dough is then baked to make unleavened bread locally called *"Kitta"*. Then, *"Tijit"*, a small sized *"Kitta"* and an adequate amount of water are mixed together and left to ferment for about 48 h [39]. The semisolid mixture obtained at this stage is locally called *"Tenses"*. Subsequently, non-malted roasted barley powder, locally called *"Derekot"*, is added to the previously prepared *"Tenses"*. At this stage the blend is allowed to ferment for an additional 72 h. Finally, water is added to the mixture in a ratio of 1:3. After another 2 to 3 h of further fermentation the *Korefe* is ready to be served [15].

According to Getnet and Berhanu [15], the titratable acidity, ethanol, and crude fat content of *Korefe* are 32 g/L, 2.7% and 7.01%, respectively. In addition, the pH, conductivity, salinity and TDS values of *Korefe* are 3.7, 3199 μs/cm, 1.7% and 1610 mg/L, respectively [44]. After 72 h of fermentation, lactic acid bacteria and yeast counts were more than 10^9 CFU/mL, whereas the *enterobacteriaceae* count was below the detectable limit due to the antagonistic effect of lactic acid bacteria [15].

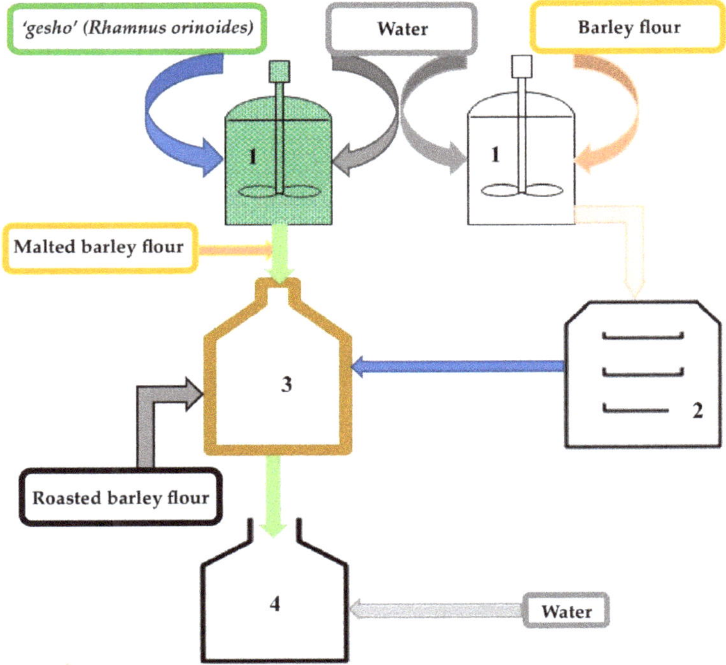

Figure 2. *Korefe* production process flow diagram: (**1**) mixer, (**2**) baking oven, (**3**) primary fermentation tank, (**4**) secondary fermentation tank.

2.5. Cheka

Cheka is a traditional low alcoholic fermented beverage commonly consumed in the southwestern parts of Ethiopia and particularly in Dirashe and the Konso district [36]. It is a cereal- and vegetable-based fermented low alcoholic beverage. Sorghum (*S. bicolor*), maize (*Z. mays*), finger millet (*E. coracana*), and vegetables such as leaf cabbage (*Brassica spp.*), moringa, (*Moringa stenopetala*), decne (*Leptadenia hastata*), and root of taro (*Colocasia esculenta*) are the main ingredients for *Cheka* preparation [17].

Worku et al. [35] reported a survey of raw materials and the production process of *Cheka*. According to their report, *Cheka* preparation starts by malting. The malt is prepared either from a single or a combination of the cereals listed above. Cabbage leaves and/or taro roots are cut into pieces and fermented anaerobically for about 4 to 6 d in a clean container. Then, a small amount of maize flour is added to the vegetable mixture and is fermented for an additional 2 to 3 d. The fermented vegetable mixture is then ground, filtered, and mixed with fresh maize flour. The fermentation continues for another 12 to 24 h. Then, water is added to the mixture and the mixture is allowed to ferment for one month. This fermented mixture is shaped into a dough ball, locally called *"Gafuma"*, and cooked at a temperature of 96 °C. After cooling, the cooked *"Gafuma"* is mixed with an adequate amount of previously prepared malt. The mixture is then allowed to ferment for an extra 12 h. This fermented mixture is locally called *"Sokatet"*. At this stage of the process a very thick porridge, locally called *"koldhumat"*, is prepared from maize flour. The prepared porridge is added to the vessel containing *"Sokatet"* with a sufficient amount of water. Finally, the mixture is left to ferment for another 4 to 12 h and served to consumers as *Cheka*.

Worku et al. [35] also published a paper that focused on the nutritional and alcohol content of *Cheka*. This report contained the physicochemical properties, ethanol, and methanol content of *Cheka* collected from *Cheka* producers. The average pH, ethanol, iron (Fe) and calcium (Ca) contents of *Cheka* samples are 3.76, 6%, 0.2 mg/g and 0.14 mg/g, respectively.

2.6. Keribo

Keribo is another alcoholic traditional beverage consumed by many Ethiopians, especially by those who prefer low alcoholic drinks. The production process is relatively less complicated [51].

Abawari [34] reported the raw materials and processing conditions of *Keribo*. According to the report, making *Keribo* begins by mixing roasted barley with hot water. Then, the mixture is boiled for about 20 min, after which the solid residue is removed by filtration. Subsequently, sugar and bakery yeast are added into the separated filtrate and left overnight to ferment. Finally, extra sugar is added to the mixture and the beverage is served to the consumer.

Abawari [16] published a second report that dealt with the microbial dynamics of *Keribo* fermentation. Based on the findings, average lactic acid bacteria, aerobic mesophilic bacteria, aerobic spore formers and yeasts counts were 2.70, 2.34, 4.96 and 2.01 log CFU/mL, respectively. However, the average *enterobacteriaceae*, *staphylococci*, and mold counts were below the detectable levels. Additionally, the shelf life of *Kerbio* is not more than two days at room temperature storage [40].

3. Fruit-Based Traditional Alcoholic Beverages

3.1. Tej

Tej is an Ethiopian wine that uses honey as a substrate and *"gesho"* (*Rhamnus prinoides*) as a source of bitterness. Previously, *Tej* was produced and consumed only for cultural festivities and for the royal families [52]. These days, *Tej* is a popular drink in rural, semi-urban, and urban areas of Ethiopia. It is produced and sold at the household level. The final product usually lacks consistency in quality due to differences in the manner of preparation and the ratio of ingredients used [21].

Ethiopia has the potential to produce 500,000 tons of bee honey annually. However, production has not surpassed 10% of that potential [53]. About 80% of the total honey produced in the country serves as raw material for producing *Tej* [54]. Traditionally, crude honey rather than refined honey is preferred for the production of *Tej* due to the distinct sensorial properties that local consumers prefer [18].

The *Tej* making process begins by cleaning and drying the traditional fermenting container. Then, honey and water are mixed in a ratio of 1:3 and allowed to ferment for 2 to 3 d. Afterwards, leaves and stems of *"gesho"* (*R. prinoides*) are boiled, cooled to room temperature and added to the previously fermented honey and water mixture. This mixture is allowed to ferment for 8 to 10 more days during the hot season or 20 d during the cold season [52]. After the intended period of fermentation, the product is ready to serve to the consumer in a special glass, locally known as *"Berele"*.

The microorganisms involved in the fermentation process originate from the raw materials, equipment and utensils. Because of this, *Tej* fermentation is lengthy, spontaneous, and uncontrolled. Thus, the final product have inconsistent physicochemical properties, microbiological profile, and sensory attributes [21].

Good quality *Tej* is yellow, sweet, fizzy, and cloudy due to the presence of active yeasts [43]. The flavor of *Tej* is highly dependent on the type of honey used and amount of "gesho" (*R. prinoides*) added. Additionally, the diversity and population of microorganisms also contribute to the distinctive flavor of *Tej* [55]. Like Mexican pulque [12], the Ethiopian *Tej's* microorganism community is dominated by Lactic acid bacteria (LAB) and yeasts (Table 4). The shelf life and keeping quality of *Tej* is very short [40].

Table 4. Physicochemical and microbiological properties of *Tej*, *Ogol* and *Booka*.

Area of Investigation	Microbial and Physicochemical Properties	Concluding Remarks	References
Flora of yeast and lactic acid bacteria of *Tej*	• *S.cerevisiae* (25%), *K. bulgaricus* (16%), *D. phaffi* (14%) and *K. veronae* (10%) are dominant yeast species; • Lactic biota is composed of *Lactobacillus*, *Streptococcus*, *Leuconostoc* and *Pediococcus* species.	Yeasts and LAB are among the dominant microbes in *Tej* fermentation	[40]
Physicochemical properties of *Tej*	• pH values of collected samples ranged between 3.07 and 4.90; • Titratable acidity of samples ranged between 1 g/L and 1.03 g/L; • Total alcohol content ranged between 2.7% and 21.7%; • Average total dissolved solids (TDS) is 387%; • Average electrical conductivity is 811 µs/cm; • Average Salinity is 0.4 mg/L.	Natural and spontaneous fermentation is a major source of physicochemical variation in collected *Tej* samples	[18,44]
Isolating fermentative yeast from *Ogol*	• *S. cerevisiae* is isolated from *Ogol* sample; • Isolated species produce *16.5%* (*v/v*) ethanol; • Titratable acidity and pH are 60 g/L and 3.8, respectively.	Isolated yeast from *Ogol* has the potential to be used for ethanol fermentation	[19]
Physicochemical properties of *Booka*	• pH value ranges from 2.90–3.12; • Moisture content—82.18%; • Ash content—0.82%; • Crude fat content—1.43%; • Total nitrogen—7.01%; • Total carbohydrate—8.56%; • Mean alcoholic content—1.53%.	*Booka* can be used as a meal replacement	[20]

3.2. Ogol

Ogol is another traditional fermented honey wine beverage commonly consumed in the western part of Ethiopia. The preparation process starts by pulverizing the bark of the native tree *"Mange"* (*B. unijungata*). The pulverized bark, wild honey, and water are mixed in a container and the mixture is allowed to ferment for about two weeks. After completing the intended period of fermentation a small amount of water is added and the mixture is allowed to ferment anaerobically in a hot place for additional 12 to 36 h. Finally, it is filtered through a clean cloth and served to consumers as *Ogol* [19].

3.3. Booka

Booka is a low alcoholic traditional beverage that is popular in southern Oromia, Ethiopia (Table 4). The preparation process is relatively simple and easily adaptable. First the bladder of a cow is carefully removed from a dressed carcass and cleaned properly to remove residue urine. Honey and water are added to the prepared cow bladder in a ratio of 1:4. After 2 to 3 d of fermentation, a small amount of honey is added to the mixture and it is left to ferment anaerobically for an additional 2 d [43]. After the fermentation process is completed, the filtrate is ready to be served to consumers as *Booka*. Good quality *Booka* is yellowish in color, sweet in taste, and attractive in odor [20].

4. Nutritional Value, Function Properties and Safety Issues of Ethiopian Alcoholic Beverages

The nutritional values of Ethiopian traditional alcoholic beverages can be seen in two ways. In low alcoholic beverages, the nutritional values are higher than their respective raw materials [29]. The main justification forwarded by authors is the live microorganisms present in these beverages [14,34,47]. In high alcoholic beverages, the nutritional values are lower than that of low alcoholic traditional beverages [1,18]. As shown in Table 5, *Borde*, *Shamita* and *Cheka* have a good nutritional value compared to that of high alcoholic beverages like *Tella* and *Tej*. As the fermentation continues, from the fermentation dynamics point of view, only limited microorganisms withstand the adverse environmental effect of the growth medium. Thus, the microorganisms that do not cope with the new environment will be lysed and become a source of protein for cell maintenance for the surviving species. This analysis works even better in natural, spontaneous and uncontrolled fermentation systems. Hence, this competition in return decreases the nutritional value of the beverages while increasing secondary metabolites like ethanol [56–58].

The functional properties of the beverage are manifested in the content of total polyphenols (TP) and antioxidant activity (AA) [59]. These polyphenols and antioxidants have a health-promoting effect by scavenging free radicals and regulating metabolism [60]. Many Ethiopian alcoholic beverages have good TP and AA values (Table 5). The phenolic content of *Tella* is greater than that of *Korefe* and *Tej* [61]. Even though there are many factors responsible for this difference, raw materials, and especially the amount of gesho added to the mixture, take the lion's share of the contribution [38].

The safety issues of Ethiopian traditional alcoholic beverages should be understood from the perspective of microorganism growth, higher alcohol and fluoride contents. Although the presence of a large amount of live fermentative microorganisms in low alcoholic traditional beverages contributes to their good nutritional value, there are major concerns related to food safety [17]. The microbiological safety issues were discussed in the previous section of this paper. This section focuses only on food safety issues related to higher alcohol and fluoride content. Higher alcohols contents of Isobutanol, 1-Butanol, 2-Butanol and 1-Propanol can be called collectively fusel oil or fuselol [62]. Fusel oil in a minute quantity contributes to the good flavor of the product. However, if it is consumed at a level above 1000 g/hL of pure alcohol, fusel oil is harmful for health [63]. The higher methanol content in traditional beverages also has a negative health impact [44]. Most of the time methanol is formed due to natural, spontaneous and uncontrolled fermentation [18]. As shown in Table 5, the methanol content of *Tella* and *Cheka* is very much lower than the maximum standard set by the European Union (EEC No 1576/89). Since Ethiopia is located in the region of the Great Rift Valley, fluoride ion concentration is another important food safety concern in traditional alcoholic beverages. A level of fluoride ions above 1.5 mg/L in the beverage creates dental and skeletal fluorosis [64]. Traditional beverages collected near Rift Valley localities showed a higher fluoride ion concentration (Table 5).

Table 5. Nutritional value, functional properties and safety issues of Ethiopian alcoholic beverages.

Beverages	Nutritional Value	Functional Properties (Average Values)	Higher Alcohol and Fluoride Ion (Average Values)	References
Tella	Total protein—0.4% Carbohydrate—1.98%	TP (μg mL$^{-1)}$—232.40 AA (μg mL$^{-1)}$—296.00 Folate (mgcg^{-1})—0.093	Fusel oil (ppm)—51 Methanol (ppm)—41.5 Fluoride ion (mg/L)—4.26	[1,46,61,65]
Borde	Total protein—9.55%, Crude fat—6.88%, Total ash—3.66%	TP (μg mL$^{-1)}$—9.50 AA (μg mL$^{-1)}$—198.5	Fluoride ion (mg/L)—4.95	[29,33,39,65]
Shamita	Total protein—10.37% Crude fat—6.85% Total ash—3.46%	-	Fluoride ion (mg/L)—5.21	[33,34,65]
Korefe	-	TP (μg mL$^{-1)}$—167.60 AA (μg mL$^{-1)}$—278.13	Fluoride ion (mg/L)—1.39	[63,65]
Cheka	Total protein—3.83% Crude fat—1.49% Carbohydrate—16.59% Total ash—0.79%	-	Methanol(ppm)—271.55	[17]
Keribo	-	TP (μg mL$^{-1)}$—12.65 AA (μg mL$^{-1)}$—64.66	-	[38]
Tej	Total protein—0.35% Crude fat—0.35% Carbohydrate—3.58% Total ash—0.04%	TP (μg mL$^{-1)}$—197.00 AA (μg mL$^{-1)}$—240.37	Fusel oil (ppm)—205.08 Fluoride ion (mg/L)—6.68	[18,21,38,61,65]
Bokaa	Moisture content—82.18% Ash content—0.82% Crude fat content—1.43% Total Nitrogen—7.01%	-	-	[20]

TP in gallic acid equivalent (GAE); AA in ascorbic acid equivalent (AAE); —values not available in the literatures.

5. Conclusions and Future Perspectives

The most commonly produced and consumed Ethiopian traditional alcoholic beverages are *Tella, Borde, Shamita, Korefe, Cheka, Keribo, Tej, Ogol* and *Booka*. The ingredients, ratios, procedures and equipment used to prepare these beverages vary from place to place, but they all are produced through natural and spontaneous fermentation processes. Low alcoholic Ethiopian beverages have a higher nutritional value. Thus, they can be used as a meal replacement. These traditional alcoholic beverages also contain a significant amount of total polyphenols and antioxidants. The alcohol content and pH values of these beverages range from 1.53–21.7% and 2.9–4.9, respectively. As the fermentation continues, counts of lactic acid bacteria and yeasts species flourish while mesophilic aerobic bacteria and coliform counts decrease significantly. The source of microorganisms responsible for fermentation is mainly from the ingredients and utensils. These traditional alcoholic beverages show inconsistent quality within and between productions, and have a short shelf life. This is due to the high number of live cells present in freshly produced beverages.

Until now, research on Ethiopian traditional fermented beverages has mainly focused on the identification of raw materials and traditional processing methods. Moreover, microbial characterization and microbial dynamics have been reported for the last two decades. All of the reports have used culture-dependent phenotypic characterization. Hence, the current findings lack the completeness needed to lead these traditional beverages, which hold equal local market share with commercial products, into large-scale production. Thus, we find that future research has to shift its gear to a higher level by studying microbial metagenomics, starter culture development, rheological study, shelf life extension, process modification, kinetics, modeling and optimization.

Author Contributions: Conceptualization, writing—original draft preparation, rewriting, E.G.F.; conceptualization, writing—review and editing, supervision, S.A.E.; writing—review and editing, H.D.D.;

writing—review and editing, D.W.D.; writing—review and editing, supervision, J.-H.S. All authors have read and agreed to the published version of the manuscript.

Funding: This work was supported by the Strategic Initiative for Microbiomes in Agriculture and Food (Grant No. 918010-4), Ministry of Agriculture, Food, and Rural Affairs, South Korea.

Acknowledgments: The Authors would like to acknowledge the Addis Ababa Science and Technology University, Addis Ababa University, and Kyungpook National University.

Conflicts of Interest: The authors declare no conflict of interest.

References

1. Tekle, B.; Jabasingh, S.; Fantaw, D.; Gebreslassie, T.; Rao, S.; Baraki, H.; Gebregziabher, K. An insight into the Ethiopian traditional alcoholic beverage: *Tella* processing, fermentation kinetics, microbial profiling and nutrient analysis. *LWT Food Sci. Technol.* **2019**, *107*, 9–15. [CrossRef]
2. Tamang, J. Diversity of fermented beverages and alcoholic drinks. In *Fermented Foods and Beverages of the World*; Tamang, J., Kailasapathy, K., Eds.; Taylor and Francis Group: New York, NY, USA, 2010; pp. 86–117.
3. Marshall, E.; Mejia, A. *Traditional Fermented Food and Beverages for Improved Livelihoods*; FAO: Rome, Italy, 2011.
4. CSA. *Agricultural Sample Survey 2015/2016 (2009 E.C.). Volume I. Report on Area and Production of Major Crops (Private Peasant Holdings, Meher Season)*; Addis: Ababa, Ethiopia, 2016.
5. WHO. *Global Status Report on Alcohol and Health 2018*; World Health Organization: Geneva, Switzerland, 2018.
6. Tamang, J.P.; Thapa, N.; Bhalla, T.C. Ethnic Fermented Foods and Beverages of India. In *Ethnic Fermented Foods and Alcoholic Beverages of Asia*, 1st ed.; Tamang, J., Ed.; Springer India: New Delhi, India, 2016; pp. 17–72.
7. Lyumugabe, F.; Kamaliza, G.; Bajyana, E.; Thonart, P. Microbiological and physico-chemical characteristic of Rwandese traditional beer Ikigage. *Afr. J. Biotechnol.* **2010**, *9*, 4241–4246.
8. Ogunbanwo, S.; Ogunsanya, B. Quality assessment of oti-oka like beverage produced from pearl millet. *J. Appl. Biosci.* **2012**, *51*, 3608–3617.
9. Namugumya, B.; Muyanja, C. Traditional processing, microbiological, physiochemical and sensory characteristics of kwete, a Ugandan fermented maize based beverage. *Afr. J. Food Agric. Nutr. Dev.* **2009**, *9*, 1046–1059.
10. Kirui, M.; Alakonya, A.; Talam, K.; Tohru, G.; Bii, C. Total aflatoxin, fumonisin and deoxynivalenol contamination of busaa in Bomet county, Kenya. *Afr. J. Biotechnol.* **2014**, *13*, 2675–2678.
11. Kim, E.; Chang, Y.; Ko, J.; Jeong, J. Physicochemical and microbial properties of the Korean traditional rice wine, *Makgeolli*, supplemented with banana during fermentation. *Prev. Nutr. Food Sci.* **2013**, *18*, 203–209. [CrossRef]
12. Valadez-Blanco, R.; Bravo-Villa, G.; Santos-Sánchez, N.; Velasco-Almendarez, S.; Montville, T. The Artisanal Production of Pulque, a Traditional Beverage of the Mexican Highlands. *Probiotics Antimicrob. Proteins* **2012**, *4*, 140–144. [CrossRef]
13. Abegaz, K. Isolation, characterization and identification of lactic acid bacteria involved in traditional fermentation of *Borde*, an Ethiopian cereal beverage. *Afr. J. Biotechnol.* **2007**, *6*, 1469–1478.
14. Bacha, K.; Mehari, T.; Ashenafi, M. Microbiology of the fermentation of 'shamita', a traditional Ethiopian fermented beverage. *Ethiop. J. Sci.* **1999**, *22*, 113–126. [CrossRef]
15. Getnet, B.; Berhanu, A. Microbial dynamics, roles and physico-chemical properties of '*Korefe*', a traditional fermented Ethiopian beverage. *Biotechnol. Int.* **2016**, *9*, 56–175.
16. Abawari, R. Microbiology of keribo fermentation; an Ethiopian tradtional fermented beverage. *Pak. J. Biol. Sci.* **2013**, *16*, 1113–1121. [PubMed]
17. Worku, B.; Gemede, H.; Woldegiorgis, A. Nutritional and alcoholic contents of cheka: A traditional fermented beverage in Southwestern Ethiopia. *Food Sci. Nutr.* **2018**, *6*, 2466–2472. [CrossRef] [PubMed]
18. Bahiru, B.; Mehari, T.; Ashenafi, M. Chemical and nutritional properties of 'tej', an indigenous Ethiopian honey wine: Variations within and between production units. *J. Food Technol. Afr.* **2001**, *6*, 104–108.
19. Teramoto, Y.; Sato, R.; Ueda, S. Characteristics of fermentation yeast isolated from traditional Ethiopian honey wine, ogol. *Afr. J. Biotechnol.* **2005**, *4*, 160–163.
20. Elema, T.; Olana, B.; Elema, A.; Gemeda, H. Processing methods, physical properties and proximate analysis of fermented beverage of honey wine Booka in Gujii, Ethiopia. *J. Nutr. Food Sci.* **2018**, *8*, 2. [CrossRef]
21. Nemo, R.; Bacha, K. Microbial, physicochemical and proximate analysis of selected Ethiopian traditional fermented beverages. *LWT—Food Sci. Technol.* **2020**, *131*, 109713. [CrossRef]

22. Herman, J.; Maynard, A. Wine. In *Microbial Technology: Fermentation Technology*; Peppler, H., Perlman, D., Eds.; Academic Press Inc.: London, UK, 1979; Volume 2, p. 132.
23. Pederson, S. *Microbiology of Food Fermentation*; AVI Publishing Co. Inc.: Westport, CT, USA, 1979.
24. Steinkraus, K. *Indigenous Fermented Foods in Which Ethanol Is a Major Product*, 2nd ed.; Taylor & Francis: New York, NY, USA, 1996; pp. 365–402.
25. Amabye, T. Evaluation of phytochemical, chemical composition, antioxidant and antimicrobial screening parameters of rhamnus prinoides (Gesho) available in the market of mekelle, tigray, Ethiopia. *Nat. Prod. Chem. Res.* **2015**, *3*, 6. [CrossRef]
26. Abegaz, B.; Kebede, T. Geshoidin: A bitter principle of Rhamus Prinoides and other constituents of the leaves. *Bull. Chem. Soc. Ethiop.* **1995**, *9*, 107–114.
27. Alemu, H.; Abegaz, B.; Bezabih, M. Electrochemical behaviour and voltammetric determination of geshoidin and its spectrophotometric and antioxidant properties in aqueous buffer solutions. *Bull. Chem. Soc. Ethiop.* **2007**, *21*, 89–204.
28. FAO. *Traditional Fermented Food and Beverage for Improved Livelihoods*; A Global Perspective; FAO: Rome, Italy, 2012.
29. Abegaz, K.; Beyene, F.; Langsrud, T.; Narvhus, J. Indigenous processing methods and raw materials of '*Borde*', an Ethiopian traditional fermented beverage. *J. Food Technol. Afr.* **2002**, *7*, 59–64. [CrossRef]
30. Andualem, B.; Shiferaw, M.; Berhane, N. Isolation and characterization of saccaromyces cervisiae yeasts isolates from '*Tella*' for beer production. *Annu. Res. Rev. Biol.* **2017**, *15*, 1–12. [CrossRef]
31. Tekluu, B.; Gebremariam, G.; Aregai, T.; Harikrishna, R. Determination of alcoholic content and other parameters of local alcoholic beverage (*Tella*) at different stages in Gondar, Ethiopia. *Int. J. IT Eng. Appl. Sci. Res.* **2015**, *4*, 37–40.
32. Bacha, K.; Mehari, T.; Ashenafi, M. The microbial dynamics of '*Borde*' fermentation, a traditional Ethiopian fermented beverage. *Ethiop. J. Sci.* **1998**, *21*, 195–205. [CrossRef]
33. Ashenafi, M.; Mehari, T. Some microbiological and nutritional properties of '*Borde*' and 'Shamita', traditional Ethiopian fermented beverages. *Ethiop. J. Health Dev.* **1995**, *9*, 105–110.
34. Abawari, R. Indigenous Processing Methodsand Raw Materials of Keribo: An Ethiopian Traditional Fermented Beverage. *J. Food Resour. Sci.* **2013**, *2*, 13–20.
35. Worku, B.; Woldegiorgis, A.; Gemeda, H. Indigenous processing methods of Cheka: A traditional fermented beverage in southwestern Ethiopia. *J. Food Process. Technol.* **2015**, *7*, 1.
36. Desta, B.; Melese, G. Determination of protein value and alcoholic content in locally prepared different types of Cheka at different stages using CHNS elemental analyzer and specific gravity methods. *Am. J. Appl. Chem.* **2019**, *7*, 168–174. [CrossRef]
37. Lee, M.; Regu, M.; Seleshe, S. Uniqueness of Ethiopian traditional alcoholic beverage of plant origin, *Tella*. mboxemphJ. Ethn. Foods. **2015**, 2, 110–114. [CrossRef]
38. Debebe, A.; Chandravanshi, B.; Redi-Abshiro, M. Total contents of phenolics, flavonoids, tannins and antioxidant capacity of selected traditional ethiopian alcoholic beverages. *Bull. Chem. Soc. Ethiop.* **2016**, *30*, 27–37. [CrossRef]
39. Tafere, G. A review on Traditional Fermented Beverages of Ethiopian. *J. Nat. Sci. Res.* **2015**, *5*, 94–103.
40. Bahiru, B.; Mehari, T.; Ashenafi, M. Yeast and lactic acid flora of tej, an indigenous Ethiopian honey wine: Variations within and between production units. *Food Microbiol.* **2006**, *23*, 277–282. [CrossRef] [PubMed]
41. Berza, B.; Wolde, A. Fermenter Technology Modification Changes Microbiological and Physicochemical Parameters, Improves Sensory Characteristics in the Fermentation of *Tella*: An Ethiopian Traditional Fermented Alcoholic Beverage. *J. Food Process. Technol.* **2014**, *5*, 1–8.
42. Berhanu, A. Microbial profile of *Tella* and the role of gesho (*Rhamnus prinoides*) as bittering and antimicrobial agent in traditional *Tella* (Beer) production. *Int. Food Res. J.* **2014**, *21*, 357–365.
43. Yohannes, T.; Melak, F.; Siraj, K. Preparation and physicochemical analysis of some Ethiopian traditional alcoholic beverages. *Afr. J. Food Sci.* **2013**, *7*, 399–403. [CrossRef]
44. Tadesse, S.; Chandravanshi, B.; Ele, E.; Zewge, F. Ethanol, methanol, acid content and other quality parameters of Ethiopian traditional fermented, distilled and factory produced Ethiopian traditional fermented, distilled and factory produced. *SINET Ethiop. J. Sci.* **2017**, *40*, 16–35.
45. Fite, A.; Tadesse, A.; Urga, K.; Seyoum, E. Methanol, fusel oil and ethanol contents of some Ethiopian traditional alcoholic beverages. *SINET Ethiop. J. Sci.* **1991**, *14*, 19–27.

46. Getaye, A.; Tesfaye, D.; Zerihun, A.; Melese, F. Production, optimization and characterization of Ethiopian traditional fermented beverage 'Tella' From Barley. *J. Emerg. Technol. Innov. Res.* **2018**, *5*, 797–799.
47. Abegaz, K.; Langsrud, T.; Beyene, F.; Narvhus, J. The effect of technological modifications on the fermentation of '*Borde*', an Ethiopian tradtional fermented cereal beverage. *J. Food Technol. Afr.* **2004**, *9*, 3–12.
48. Tadesse, G.; Ashenafi, M.; Ephraim, E. Survival of *E. coli* O157: H7 *Staphylococcus aureus*, *Shigella flexneri* and *Salmonella* spp. in fermenting '*Borde*', a traditional Ethiopian beverage. *Food Control* **2005**, *16*, 189–196. [CrossRef]
49. Tadesse, G.; Ephraim, E.; Ashenafi, M. Assessment of the antimicrobial activity of lactic acid bacteria isolated from *Borde* and Shamita, traditional Ethiopian fermented beverages, on some food-borne pathogens and effect of growth medium on the inhibitory activity. *Internet J. Food Saf.* **2005**, *4*, 13–20.
50. Gebreyohannes, G.; Welegergs, B. Determination of alcoholic content and other parameters of locally preparedalcoholic beverages (*korefe* and tej) at different stages in Gondar town. *Int. J. Integr. Sci. Technol.* **2015**, *4*, 1–4.
51. Dibaba, K.; Tilahun, L.; Satheesh, N.; Geremu, M. Acrylamide occurrence in Keribo: Ethiopian traditional fermented beverage. *Food Control* **2018**, *86*, 77–82. [CrossRef]
52. Vogel; Gobezie. Indigenous fermented foods in which ethanol is a major product. In *Handbook of Indigenous Fermented Foods*, 2nd ed.; Steinkraus, K., Ed.; Marcel Dekker Inc.: New York, NY, USA, 1996; pp. 367–369.
53. Gebretsadik, T.; Negash, D. Honeybee production system, challenges and opportunities in selected districts of gedeo zone, southern nation, nationalities and peoples regional state, Ethiopia. *Int. J. Res.* **2016**, *4*, 49–63.
54. Gebremedhin, G.; Tadesse, G.; Kebede, E. Physiochemical characteristics of honey obtained from traditional and modern hive production systems in Tigray region, northern Ethiopia. *Momona Ethiop. J. Sci.* **2013**, *5*, 115–128. [CrossRef]
55. Lemi, B. Microbiology of Ethiopian Traditionally Fermented Beverages and Condiments. *Int. J. Food Microbiol.* **2020**, *2020*, 1478536.
56. Lucke, F. Indigenous lactic acid bacteria of various food commodities and factor affecting their growth. In *Lactic Acid Bacteria: Current Advances in Metabolism, Gentics and Application*, 1st ed.; Bozoglu, T., Ray, B., Eds.; Springer: Berlin, Germany, 2015.
57. Djeni, T.; Karen, H.; Ake, K.; Laurent, S. Microbial diversity and metabolite profiles of palm wine produced from three different palm tree species in Côte d'Ivoire. *Sci. Rep.* **2020**, *10*, 1715. [CrossRef]
58. Cason, E.; Mahlomaholo, B.; Taole, M.; Abong, G.; Vermeulen, J.; Smidt, O.; Vermeulen, M.; Steyn, L. Bacterial and fungal dynamics during the fermentation process of sesotho, a traditional beer of Southern Africa. *Front. Microbiol.* **2020**, *11*, 1451. [CrossRef]
59. Cory, H.; Passarelli, S.; Szeto, J.; Tamez, M.; Mattei, J. The Role of polyphenols in human health and food systems: A mini-review. *Front. Nutr.* **2018**, *5*, 87. [CrossRef]
60. Ganesan, K.; Xu, B. A critical review on polyphenols and health benefits of black soybeans. *Nutrients* **2017**, *9*, 455. [CrossRef]
61. Shewakena, S.; Chandravanshi, B.; Debeb, A. Levels of total polyphenol, flavonoid, tannin and antioxidant activity of selected Ethiopian fermented traditional beverages. *Int. Food Res. J.* **2017**, *24*, 2033–2040.
62. Lachenmeier, D.; Haupt, S.; Schulz, K. Defining maximum levels of higher alcohols in alcoholic beverages and surrogate alcohol products. *Regul. Toxicol. Pharmacol.* **2008**, *50*, 313–321. [CrossRef] [PubMed]
63. Lee, S.; Shin, J.; Lee, K. Determination of fusel oil content in various types of liquor distributed in Korea Pre-treatment of samples GC analysis of fusel oil. *Korean Soc. Food Preserv.* **2017**, *24*, 510–516. [CrossRef]
64. Fawell, J.; Bailey, K.; Chilton, J.; Dahi, E.; Fewtrell, L.; Magara, Y. *Fluoride in Drinking Water*; World Health Organization and IWA Publishing, Inc.: London, UK, 2006.
65. Belete, Y.; Chandravanshi, B.S.; Zewgea, F. Levels of the fluoride ion in six traditional alcoholic fermented beverages commonly consumed in Ethiopia. *Fluoride* **2017**, *50*, 79.

Publisher's Note: MDPI stays neutral with regard to jurisdictional claims in published maps and institutional affiliations.

MDPI

Article

Potential Functional Snacks: Date Fruit Bars Supplemented by Different Species of *Lactobacillus* spp.

Maria Maisto [1,*], Giuseppe Annunziata [1], Elisabetta Schiano [1], Vincenzo Piccolo [1], Fortuna Iannuzzo [1], Rosaria Santangelo [2], Roberto Ciampaglia [1], Gian Carlo Tenore [1], Ettore Novellino [3] and Paolo Grieco [1]

1 Department of Pharmacy, University of Naples Federico II, Via Domenico Montesano 59, 80131 Naples, Italy; giuseppe.annunziata@unina.it (G.A.); elisabetta.schiano@unina.it (E.S.); vincenzo.piccolo3@unina.it (V.P.); fortuna.iannuzzo@unina.it (F.I.); roberto.ciampaglia@unina.it (R.C.); giancarlo.tenore@unina.it (G.C.T.); paolo.grieco@unina.it (P.G.)
2 Pharma Biomateck s.r.l., Frattamaggiore, 80027 Naples, Italy; santangelo@pharmabiomateck.it
3 NGN Healthcare—New Generation Nutraceuticals s.r.l., Torrette Via Nazionale 207, 83013 Mercogliano, Italy; ngnhealthcare@gmail.com
* Correspondence: maria.maisto@unina.it

Abstract: The influence of the addition of four different potential probiotic strains, *Lactiplantibacillus plantarum* subsp. *plantarum* (*L. plantarum*), *Lactobacillus delbruekii* subsp. *bulgaricus* (*L. bulgaricus*), *Lactobacillus acidophilus* (*L. acidophilus*) and *Lactinocaseibacillus rhamnosus* (*L. rhamnosus*), in date fruit-based products was investigated in order to evaluate the possibility of producing a functional snack. All bacterial strains tested were able to grow in date fruit palp, reaching probiotic concentrations ranging from 3.1×10^9 to 4.9×10^9 colony-forming units after 48 h of fermentation, and the pH was reduced to 3.5–3.7 or below. The viability of inoculated probiotic bacteria after 4 weeks of storage at 4 °C was slightly reduced. Some biochemical features of the fermented snacks, such as the total phenolic content (TPC), antioxidant activity and detailed polyphenolic profile, were also evaluated. After fermentation, changes in the polyphenol profile in terms of increased free phenolic compounds and related activity were observed. These results may be attributed to the enzymatic activity of *Lactobacillus* spp. in catalyzing both the release of bioactive components from the food matrix and the remodeling of polyphenolic composition in favor of more bioaccessible molecules. These positive effects were more evident when the snack were fermented with *L. rhamnosus*. Our results suggest the use of lactic acid fermentation as an approach to enhance the nutritional value of functional foods, resulting in the enhancement of their health-promoting potential.

Keywords: lactofermentation; probiotic; date fruit bars; functional snack; polyphenols

Citation: Maisto, M.; Annunziata, G.; Schiano, E.; Piccolo, V.; Iannuzzo, F.; Santangelo, R.; Ciampaglia, R.; Tenore, G.C.; Novellino, E.; Grieco, P. Potential Functional Snacks: Date Fruit Bars Supplemented by Different Species of *Lactobacillus* spp. *Foods* **2021**, *10*, 1760. https://doi.org/10.3390/foods10081760

Academic Editors: Valentina Bernini and Juliano De Dea Lindner

Received: 18 June 2021
Accepted: 26 July 2021
Published: 29 July 2021

1. Introduction

Palm date fruit is one of the oldest fruits consumed by man. It is well known in folklore beliefs that palm date fruit possesses extraordinary health-promoting effects. In ancient times, it was largely used for its extraordinary effects on fertility and sexual performance, to care for gastrointestinal disturbances, but also to treat respiratory disease such as bronchitis and asthma [1,2]. Today, the latter beneficial effects have been scientifically studied and documented, and pre-clinical and clinical studies have confirmed the wide latter spectrum of health benefits after treatment with palm date fruit extract [3].

From the chemical point of view, the strength of this fruit is not only its peculiar polyphenolic content, characterized mainly in phenolic acid, followed by flavonoids, procyanidins, carotenoids, and sterols [3], but also its relevant nutritional properties, and especially its energy boosters. Particularly, palm date fruits are a rich source of minerals, such as potassium (864 mg/100 g), calcium (70.7 mg/100 g), sodium (32.9 mg/100 g), iron (0.3–6.03 mg/100 g), zinc (0.5 mg/100 g) and magnesium (64.2 mg/100 g), that are vital for human physiological process such as respiration (Na+), performance of the immune response (Zn) and physical potency (Fe) [4]. Palm date fruits are also a precious fruit, especially for their fiber content, mainly insoluble fiber (11.5 g/100 g) [5], protein content (2.5–6.5 g/100 g) [6], and of essential amino acids such as arginine and histidine that are fundamental for human health [7–9]. Moreover, the high level of glucose and fructose, easily absorbable at the intestinal level, make this fruit one of the most ancient and diffuse energy sources.

Over the last few decades, western nutraceutical and food industries have placed an increasing interest in the formulation of fruit- or vegetable-based fermented foods. These products have found a rapid diffusion on the nutraceutical and food market due to several reasons, primarily nutrition–health approaches, food safety, advantageous sensorial changing, shelf-life prolongation, the facility of preparation, the valorization of unused raw vegetal material, and sustainable development [10–12].

The main microorganism species employed for the formulation of such products are bacteria (manly *Bacillus subtilis, Bacillus thuringiensis, Aspergilus niger* and *Aspergilus oryzae*), yeast (mainly represented by Saccharomyces cerevisiae) and acid lactic bacteria (LAB) (mainly belonging to the species lactobacillus: *Lactiplantibacillus plantarum* subsp. *plantarum, Levilactobacillus brevis, Lactinocaseibacillus rhamnosus* and *Lactobacillus acidophilus*) [13,14]. Particularly, lactofermentation can be a useful tool for remodeling the polyphenolic composition of vegetables and fruit, enhancing their functional potential [15–17].

It is also well known that only some polyphenols occurring in foods are easily bioaccessible at the intestinal level, since a major part of these molecules are bound by various types of interactions to a food matrix, mainly represented by soluble and insoluble fiber and cell wall polysaccharides (PCWs) [18]. The enzymatic activity of microorganisms has been widely documented to be able to split the same types of bound molecules and/or degrade complex polyphenols in smaller ones, which in most cases are more bioaccessible at the intestinal level [13].

In light of the above statements, it can be hypothesized that lactofermentation may play a major role in improving the potential functional features of this fruit. Thus, the aim of the present study was the formulation of lactofermented palm date fruit bars (LDBs) as a potential functional food. In order to reach this aim, we have evaluated (i) the capability of the growth of most diffuse lactobacillus strains in palm date pulp and their survival after 4 weeks of storage at 4 °C, (ii) the effects of lactic acid fermentation on free polyphenolic compounds levels in palm date fruit bars, (iii) the desirable enhancement of LDB antioxidant potential, and (iv) the remodeling of the polyphenolic composition of LDB.

2. Materials and Methods

2.1. Inoculum and LDB Preparation

Lactiplantibacillus plantarum subsp. *plantarum* ATCC14917, *Lactobacillus delbruekii* subsp. *bulgaricus* ATCC 11842, *L. acidophilus* ATCC4356 and *Lactinocaseibacillus rhamnosus* ATCC 7469 (Farmalabor, Canosa, Italy) were reactivated by culturing twice in 25 mL of MRS broth (meat peptone 10.0 gL^{-1}; dextrose 20.0 gL^{-1}; yeast extract 5.0 gL^{-1}; beef extract 10.0 gL^{-1}; disodium phosphate 2.0 gL^{-1}; sodium acetate 5.0 gL^{-1}; ammonium citrate 2.0 gL^{-1}; magnesium sulfate 0.1 gL^{-1}; manganese sulphate 0.05 gL^{-1}; Tween 80 1.0 gL^{-1}) (Thermo scientific, Waltham, MA, USA) at 37 °C for 18 h to obtain 10^8 cells/mL. To prepare the inoculum, bacterial cultures were centrifuged and washed in sterile physiological solution (NaCl 8.5 gL^{-1}) and resuspended in 5 mL of the same solution. Fresh palm date fruits were purchased from a local supermarket, were boned, and 200 g of milled date fruit pulp, 100 g of cereals, 18 h cultures (final concentration > 10^6 CFU/mL) and sterile water were mixed in a food processor mixer for 2 min. For the fermentation, the mixtures obtained were placed in plastic bags and incubated at 37 °C for 48 h. After this time, the pH was measured followed by sample drying up to 15% moisture content. Extrusion was accomplished using a laboratory single screw extruder S-45 (Metalchem Gliwice) after the fermentation time.

to release polyphenolic compounds from the food matrix, but also a tool to remodel the polyphenolic composition of palm date fruit in favor of a smaller and less polar one, thus reasonably creating more absorbable bioactive molecules. HPLC-DAD analysis of the LDB polyphenolic extracts indicated a drastic change in terms of polyphenolic composition. The fermented version of palm date bars showed a radical reduction in chlorogenic acid in favor of its molecular component: caffeic acid [31]. This molecule may exert more health-promoting effects than its precursor and, due to its lower polarity and dimension, it is better absorbed in the small intestine than its precursors [32]. The observed increase in cinnamic acids (i.e., ferulic and caffeic acids) is also related to feruloyl esterase enzymes (E.C. 3.1.1.73), produced by bacteria, responsible for catalyzing the resolution of polyphenolic esterified forms to the vegetable cell wall, that turn into the release of free phenolic acids, available to be absorbed at the intestinal level [33]. Furthermore, the release of gallic acid in LDB was observed. This compound may be released via tannase activity. Tannase or tannin acyl hydrolase (EC 3.1.1.20) catalyzes the hydrolysis of ester bonds that occur in hydrolysable tannins and gallic acid esters, and releases glucose and gallic acid [34]. The fermented version, moreover, was also enriched in quercetin, a not polar polyphenol largely absorbed at the intestinal level. Microbial α-rhamnoside (widely produced by *L. rhamnosus*) can hydrolyze quercitrin (quercetin 3-O- rhamnoside) in its aglycone form (quercetin). Similarly, rutin may be hydrolyzed by a-rhamnosidase to produce quercetin-3-O-glucoside, and then further hydrolyzed by α-glucosidase (bacteria enzyme) to release quercetin [35,36]. In light of these considerations, the polyphenolic richness of these products was not only fortified by the deep remodeling operated by bacteria activities, but the general polyphenolic stability was improved by pH change (from neutral to mild acid), enhancing the intestinal bioaccessibility and bioavailability of these bioactive compounds [37].

5. Conclusions

On the basis of functional ingredients occurring in date fruit, LDB products, and especially *L. rhamnosus* LDB, may be proposed as a prototype of functional food, mainly indicated for athletics nutrition and supplementation. LDB products may be considered as bi-functional, because after the biotransformation, LDB increases the rate of free and simple phenolic compounds, which are more absorbable at the intestinal level, and, at the same time, acts as a carrier of probiotics. The higher amounts of free phenolic compounds may be a precious support able to contrast the strong oxidative stress to which athletes are constantly subjected. Moreover, the high glucose content is a good source of energy, and the high mineral levels may rebalance the loss of minerals during sportive activities. Finally, the active probiotics may aid athletes with secondary health benefits that could positively influence physical performance through improved recovery from fatigue, enhanced immune function, and the maintenance of healthy gastrointestinal and upper respiratory tract function [38]. Accordingly, a recent clinical trial has proven the importance of probiotic supplementation to contrast the high oxidative stress level [39] and to increase muscle strength and resistance [40]. Undoubtedly, further in vivo studies are necessary to confirm such promising results.

Author Contributions: Conceptualization, P.G., R.S., E.N., G.C.T. and M.M.; methodology, M.M., G.C.T., E.N.; validation, G.C.T., P.G.; formal analysis, M.M.; investigation, M.M., R.C., E.S., V.P., G.A. and F.I.; data curation, M.M.; writing—original draft preparation, M.M. and G.A.; writing—review and editing, M.M., G.A., G.C.T.; visualization, M.M., P.G., G.C.T. and E.N.; supervision, P.G., G.C.T. and E.N. funding acquisition, P.G., G.C.T. and E.N. All authors have read and agreed to the published version of the manuscript.

Funding: This work was financed by NGN Healthcare—New Generation Nutraceuticals s.r.l. and Pharma Biomateck s.r.l.

Institutional Review Board Statement: Not applicable.

Informed Consent Statement: Not applicable.

Data Availability Statement: The data used to support the findings of this study are included within the article.

Acknowledgments: The assistance of the staff is gratefully appreciated.

Conflicts of Interest: The authors declare no conflict of interest.

References

1. Amir, J.; Kumari, A.; Khan, M.N.; Medam, S.K. Evaluation of the combinational antimicrobial effect of Annona squamosa and Phoenix dactylifera seeds methanolic extract on standard microbial strains. *Int. J. Biol. Sci.* **2013**, *2*, 68–73.
2. Ammar, N.M.; Lamia, T.; Abou, E.; Nabil, H.S.; Lalita, M.C.; Tom, J.M. Flavonoid constituents and antimicrobial activity of date (*Phoenix dactylifera* L.) seeds growing in Egypt. *MedAromPl Sci. Biotech.* **2009**, *3*, 1–5.
3. Baliga, M.S.; Baliga, B.R.V.; Kandathil, S.M.; Bhat, H.P.; Vayalil, P.K. A review of the chemistry and pharmacology of the date fruits (*Phoenix dactylifera* L.). *Food Res.* **2010**, *44*, 1812–1822. [CrossRef]
4. Taleb, H.; Maddocks, S.E.; Morris, R.K.; Kanekanian, A.D. Chemical characterisation and the anti-inflammatory, anti-angiogenic and antibacterial properties of date fruit (*Phoenix dactylifera* L.). *J. Ethnopharmacol.* **2016**, *194*, 457–468. [CrossRef]
5. Al-Shahib, W.; Marshall, R.J. The fruit of the date palm: Its possible use as the best food for the future? *Int. J. Food Sci. Nutr.* **2009**, *54*, 247–259. [CrossRef]
6. Chaira, N.; Mrabet, A.; Ferchichi, A. Evaluation of antioxidant activity, phenolics, sugar and mineral contents in date palm fruits. *J. Food Biochem.* **2009**, *33*, 390–403. [CrossRef]
7. Al-Aswad, M.B. The amino acids content of some Iraqi dates. *J. Food Sci.* **1971**, *36*, 1019–1020. [CrossRef]
8. Auda, H.; Al-Wandawi, H.; Al-Adhami, L. Protein and amino acid composition of three varieties of Iraqi dates at different stages of development. *J. Agric. Food Chem.* **1976**, *24*, 365–367. [CrossRef]
9. Auda, H.; Al-Wandawi, H. Effect of gamma irradiation and storage conditions on amino acid composition of some Iraqi dates. *J. Agric. Food Chem.* **1980**, *28*, 516–518. [CrossRef]
10. Motarjemi, Y. Impact of small scale fermentation technology on food safety in developing countries. *Int. J. Food Microbiol.* **2002**, *75*, 213–229. [CrossRef]
11. Ray, R.C.; Ward, O.P. *Microbial Biotechnology in Horticulture*; CRC Press: Plymouth, UK, 2006; Volume 1.
12. Sabater, C.; Ruiz, L.; Delgado, S.; Ruas-Madiedo, P.; Margolles, A. Valorization of Vegetable Food Waste and By-Products Through Fermentation Processes. *Front. Microbiol.* **2020**, *11*, 2604. [CrossRef]
13. Septembre-Malaterre, A.; Remize, F.; Poucheret, P. Fruits and vegetables, as a source of nutritional compounds and phytochemicals: Changes in bioactive compounds during lactic fermentation. *Food Res.* **2017**, *104*, 86–99. [CrossRef] [PubMed]
14. Hur, S.J.; Lee, S.Y.; Kim, Y.C.; Choi, I.; Kim, G.B. Effect of fermentation on the antioxidant activity in plant-based foods. *Food Chem.* **2014**, *160*, 346–356. [CrossRef] [PubMed]
15. Fessard, A.; Kapoor, A.; Patche, J.; Assemat, S.; Hoarau, M.; Bourdon, E.; Bahorun, T.; Remize, F. Lactic Fermentation as an Efficient Tool to Enhance the Antioxidant Activity of Tropical Fruit Juices and Teas. *Microorganisms* **2017**, *5*, 23. [CrossRef] [PubMed]
16. Curiel, J.A.; Pinto, D.; Marzani, B.; Filannino, P.; Farris, G.A.; Gobbetti, M.; Rizzello, C.G. Lactic acid fermentation as a tool to enhance the antioxidant properties of Myrtuscommunis berries. *Microb. Cell Fact.* **2015**, *14*, 67. [CrossRef] [PubMed]
17. Saubade, F.; Hemery, Y.M.; Guyot, J.P.; Humblot, C. Lactic Acid fermentation as a tool for increasing the folate content of foods. *Crit. Rev. Food Sci. Nutr.* **2017**, *57*, 3894–3910. [CrossRef] [PubMed]
18. Padayachee, A.; Netzel, G.; Netzel, M.; Day, L.; Zabaras, D.; Mikkelsen, D.; Gidley, M.J. Binding of polyphenols to plant cell wall analogues—Part 1: Anthocyanins. *Food Chem.* **2012**, *134*, 155–161. [CrossRef]
19. Annunziata, G.; Maisto, M.; Schisano, C.; Ciampaglia, R.; Daliu, P.; Narciso, V.; Tenore, G.C.; Novellino, E. Colon Bioaccessibility and Antioxidant Activity of White, Green and Black Tea Polyphenols Extract after In Vitro Simulated Gastrointestinal Digestion. *Nutrients* **2018**, *10*, 1711. [CrossRef]
20. Nazzaro, F.; Fratianni, F.; Coppola, R.; Sada, S.; Orlando, P. Synbiotic potential of carrot juice supplemented with Lactobacillus spp. and inulin or fructooligosaccharides. *J. Sci. Food Agric.* **2008**, *88*, 2271–2276. [CrossRef]
21. Marco, M.L.; Heeney, D.; Binda, S.; Cifelli, C.J.; Cotter, P.D.; Foligné, B.; Gänzle, M.; Kort, R.; Pasin, G.; Pihlanto, A.; et al. Health benefits of fermented foods: Microbiota and beyond. *Curr. Opin. Biotech.* **2017**, *44*, 94–102. [CrossRef]
22. Harima-Mizusawa, N.; Kamachi, K.; Kano, M.; Nozaki, D.; Uetake, T.; Yokomizo, Y.; Nagino, T.; Tanaka, A.; Miyazaki, K.; Nakamura, S. Beneficial effects of citrus juice fermented with Lactobacillus plantarum YIT 0132 on atopic dermatitis: Results of daily intake by adult patients in two open trials. *Biosci. Microbiota Food Health* **2015**, *35*, 29–39. [CrossRef]
23. Lee, C.H. Lactic acid fermented foods and their benefits in Asia. *Food control* **1997**, *8*, 259–269. [CrossRef]
24. Nielsen, E.S.; Garnås, E.; Jensen, K.J.; Hansen, L.H.; Olsen, P.S.; Nielsen, D.S. Lacto-fermented sauerkraut improves symptoms in IBS patients independent of product pasteurization—A pilot study. *Food Funct.* **2018**, *9*, 5323–5335. [CrossRef] [PubMed]
25. Bartkiene, E.; Bartkevics, V.; Starkute, V.; Zadeike, D.; Juodeikiene, G. The Nutritional and Safety Challenges Associated with Lupin Lacto-Fermentation. *Front. Plant Sci.* **2016**, *7*, 951. [CrossRef]
26. Shah, N.P. Functional foods from probiotics and prebiotics. *Food Technol.* **2001**, *55*, 46–53.

27. Granito, M.; Frias, J.; Doblado, R.; Guerra, M.; Champ, M.; Vidal-Valverde, C. Nutritional improvement of beans (*Phaseolus vulgaris*) by natural fermentation. *Eur. Food Res. Technol.* **2002**, *214*, 226–231. [CrossRef]
28. Martin-Cabrejas, M.A.; Sanfiz, B.; Vidal, A.; Molla, E.; Esteban, R.; Lopez-Andreu, F.J. Effect of fermentation and autoclaving on dietary fiber fractions and anti-nutritional factors of beans (*Phaseolus vulgaris* L.). *J. Agric. Food Chem.* **2004**, *52*, 261–266. [CrossRef] [PubMed]
29. Siebenhandl, S.; Lestario, L.N.; Trimmel, D.; Berghofer, E. Studies on tape ketan-an Indonesian fermented rice food. *Int. J. Food Sci. Nutr.* **2001**, *52*, 347–357. [CrossRef] [PubMed]
30. Crozier, A.; Jaganath, I.B.; Clifford, M.N. Dietary phenolics: Chemistry, bioavailability and effects on health. *Nat. Prod. Rep.* **2009**, *26*, 1001–1043. [CrossRef] [PubMed]
31. Silva, F.A.; Borges, F.; Guimarães, C.; Lima, J.L.; Matos, C.; Reis, S. Phenolic acids and derivatives: Studies on the relationship among structure, radical scavenging activity, and physicochemical parameters. *J. Agric. Food. Chem.* **2000**, *48*, 2122–2126. [CrossRef]
32. Olthof, M.R.; Hollman, P.C.; Katan, M.B. Chlorogenic acid and caffeic acid are absorbed in humans. *J. Nutr.* **2001**, *131*, 66–71. [CrossRef]
33. Scalbert, A.; Manach, C.; Morand, C.; Rémésy, C.; Jiménez, L. Dietary polyphenolsand the prevention of diseases. *Crit. Rev. Food Sci. Nutr.* **2005**, *45*, 287–306. [CrossRef] [PubMed]
34. Vaquero, I.; Marcobal, Á.; Muñoz, R.; Marcobal, A.; Muñoz, R. Tannaseactivityby lactic acid bacteria isolated from grape must and wine. *J. Food Microbiol.* **2004**, *96*, 199–204. [CrossRef]
35. Francesca, N.; Barbera, M.; Martorana, A.; Saiano, F.; Gaglio, R.; Aponte, A.; Moschetti, G.C.; Settanni, L. Optimised method for the analysis of phenolic compounds from caper(*Capparisspinosa*, L.) berries and monitoring of their changes during fermentation. *Food Chem.* **2016**, *196*, 1172–1179. [CrossRef] [PubMed]
36. Lin, S.; Zhu, Q.; Wen, L.; Yang, B.; Jiang, G.; Gao, H.; Chen, F.; Jiang, Y. Production ofquercetin, kaempferol and their glycosidic derivatives from the aqueousorganic extracted residue of litchi pericarp with Aspergillus awamori. *Food Chem.* **2014**, *145*, 220–227. [CrossRef] [PubMed]
37. Zhao, D.; Shah, N.P. Changes in antioxidant capacity, isoflavone profile, phenolic and vitamin contents in soymilk during extended fermentation. *LWT-Food Sci. Technol.* **2014**, *58*, 454–462. [CrossRef]
38. Leite, G.S.; Resende, A.M.S.; West, N.P.; Lancha, J.A. Probiotics and sports: A new magic bullet. *Nutrition* **2018**, *60*, 152–160. [CrossRef] [PubMed]
39. Huang, W.C.; Wei, C.C.; Huang, C.C.; Chen, W.L.; Huang, H.Y. The Beneficial Effects of Lactobacillus plantarum PS128 on High-Intensity, Exercise-Induced Oxidative Stress, Inflammation, and Performance in Triathletes. *Nutrients* **2019**, *11*, 353. [CrossRef] [PubMed]
40. Ibrahim, N.S.; Muhamad, A.S.; Ooi, F.K.; Meor-Osman, J.; Chen, C.K. The effects of combined probiotic ingestion and circuit training on muscular strength and power and cytokine responses in young males. *Appl. Physiol. Nutr. Metab.* **2017**, *43*, 180–186. [CrossRef] [PubMed]

Article

Rice Bran Fermentation Using *Lactiplantibacillus plantarum* EM as a Starter and the Potential of the Fermented Rice Bran as a Functional Food

Song-Hee Moon and Hae-Choon Chang *

Kimchi Research Center, Department of Food and Nutrition, Chosun University, 309 Pilmun-daero, Dong-gu, Gwangju 61452, Korea; dal_moon@chosun.ac.kr
* Correspondence: hcchang@chosun.ac.kr

Abstract: Rice bran was fermented using a functional starter culture of *Lactiplantibacillus plantarum* EM, which exhibited high cholesterol removal and strong antimicrobial activity. Highest viable cell counts (9.78 log CFU/mL) and strong antimicrobial activity were obtained by fermenting 20% rice bran supplemented with 1% glucose and 3% corn steep liquor (pH 6.0) at 30 °C for 48 h. The fermented rice bran slurry was hot air-dried (55 °C, 16 h) and ground (HFRB). HFRB obtained showed effective cholesterol removal (45–68%) and antimicrobial activities (100–400 AU/mL) against foodborne pathogenic bacteria and food spoilage fungi. Phytate levels were significantly reduced during fermentation by 53% due to the phytase activity of *L. plantarum* EM, indicating HFRB does not present nutrient deficiency issues. In addition, fermentation significantly improved overall organoleptic quality. Our results indicate that HFRB is a promising functional food candidate. Furthermore, HFRB appears to satisfy consumer demands for a health-promoting food and environmental and legal requirements concerning the re-utilization of biological byproducts.

Keywords: *L. plantarum* EM; rice bran fermentation; cholesterol removal; antimicrobial activity; sensory quality

Citation: Moon, S.-H.; Chang, H.-C. Rice Bran Fermentation Using *Lactiplantibacillus plantarum* EM as a Starter and the Potential of the Fermented Rice Bran as a Functional Food. *Foods* **2021**, *10*, 978. https://doi.org/10.3390/foods10050978

Academic Editors: Valentina Bernini and Juliano De Dea Lindner

Received: 3 March 2021
Accepted: 27 April 2021
Published: 29 April 2021

1. Introduction

Rice bran is produced during rice milling and is one of the most abundant agricultural byproducts. Most of the rice bran produced is used as an animal feed ingredient and to produce fertilizers due to its poor taste and smell [1]. However, rice bran is rich in dietary fiber, protein, minerals, and phytochemicals, which are important health-promoting food ingredients [1,2]. Thus, to enhance the nutritional qualities of rice bran, many food processing techniques such as fermentation using fungi or extraction with different solvents have been used [3–6]. In particular, fermentation has been used to improve sensory qualities, enhance nutritional qualities, and reduce undesirable compound levels [7]. Specifically, for rice bran fermentation, most investigators have adopted solid-state fermentation using non-toxigenic filamentous fungi, such a *Rhizopus* and/or *Aspergillus* strains, as starter cultures to develop prebiotic, antioxidant or preservative, or to enhance cosmeceutical properties [2–5]. Most bacterial rice bran fermentations described have been carried out using lactic acid bacteria (LAB) to produce useful compounds (lactic acid or ornithine), to improve sensory characteristics, or to enhance LAB viabilities or phytochemical levels [8–13]. Generally, fungi grow well in rice bran cultures without additional nutrients, but LAB are difficult to grow in rice bran, as they have complex nutrient requirements, which include amino acids, vitamins, fatty acids, purines, and pyrimidines [12,14]. Thus, other investigations on rice bran fermentation using LAB used enzyme hydrolyzed rice bran, rice bran extract, or rice bran supplemented with other nutrients such as milk, yeast extract, soybean hydrolysate, arginine, or whey [9,12–14]. However, the utilization of rice bran as a functional food is limited despite its potential as a rich source of valuable health-promoting compounds.

Hypercholesterolemia is strongly associated with coronary heart disease, which is an important cause of death [7]. Several LAB strains have been reported to have cholesterol-lowering effects in vitro or in vivo [7]. We also reported on the cholesterol-lowering effect of *Lactobacillus plantarum* EM isolated from kimchi [15]. *Lactobacillus plantarum* was renamed *Lactiplantibacillus plantarum* according to a reclassification of the genus *Lactobacillus* in 2020 [16]; thus, we refer to it throughout as *Lactiplantibacillus plantarum*. *Lactiplantibacillus plantarum* EM showed high cholesterol removal by growing cells and even dead bacterial cells. Cholesterol removal mechanisms by *L. plantarum* EM were verified to enzymatic assimilation including bile salt hydrolase assimilation and cell surface-binding. *L. plantarum* EM also appeared to meet the functional criteria required for health-promoting probiotics, including acid and bile tolerance and antibiotic susceptibility [15]. Moreover, *L. plantarum* EM exhibited strong antimicrobial activities against different foodborne pathogenic bacteria and food spoilage fungi, and the active compounds involved were identified as 3-hydroxy-5-dodecenoic acid and lactic acid [17]. Cabbage-apple juice was also fermented using *L. plantarum* EM as a functional starter culture, and the fermented juice obtained exhibited significant hypocholesterolemic and anti-obesity effects in rats [7,18].

Recently, environmental and legal pressures have forced the food and agricultural industries to find means of re-utilizing biological byproducts [2]. As consumer's interests in the health benefits of food continue to increase, functional LAB starter culture-based fermentation has attracted attention as a means of producing food materials containing a wide range of health-promoting compounds. We considered rice bran fermentation using a functional starter culture such as *L. plantarum* EM might satisfy the demands of consumers and meet environmental and legal requirements.

In this study, rice bran was fermented using *L. plantarum* EM as a functional starter culture. We describe the optimization of fermentation conditions with respect to *L. plantarum* EM growth and antimicrobial activity, and the fermented bran produced was hot air-dried to produce a fermented rice bran product, which was evaluated for its functional and organoleptic properties to assess its potential as a functional food candidate.

2. Materials and Methods

2.1. Microbial Cultures and Media

Microorganisms and culture media used in this study are listed in Table S1. LAB were cultivated in de Man Rogosa Sharpe (MRS; Difco, Sparks, MD, USA) broth. Non-LAB bacteria were cultivated in Luria-Bertani (LB, Difco) agar. Molds were grown on malt extract agar (MEA, Difco) or potato dextrose agar (PDA, Difco). Strains were purchased from the American Type Culture Collection (ATCC; Manassas, VA, USA). The LAB *L. plantarum* EM and *Weissella koreensis* DB1 were isolated from kimchi, as we previously described [13,15].

2.2. Optimization of Rice Bran Fermentation Conditions

To prepare a starter culture, *L. plantarum* EM was cultivated in MRS broth at 30 °C for 24 h, centrifuged (9950× *g*, 5 min, 4 °C), washed with sterile distilled water, and resuspended in sterile distilled water containing the same volume of culture. Thereafter, the prepared starter culture was inoculated (1%; equivalent to ~7.0 log CFU/mL) in rice bran slurry as described in Sections 2.2.1–2.2.3 and fermented at 30 °C for 48 h.

To prepare rice bran filtrate, fermented rice bran slurry was filtered through four layers of sterile thin cloth, and filtrates were tested for pH, viable LAB cell count, and antimicrobial activities [13].

2.2.1. Rice Bran Concentration in Rice Bran Slurry

Rice bran slurries containing 5, 10, 15, 20, 25, or 30% (*w*/*v*) of rice bran powder (Henanum Co., Yangpyeong, Korea) in distilled water (Table S2) were autoclaved (121 °C, 15 min), immediately cooled (20 °C), and viscosities were determined using different spindle speeds of 12–60 rpm using a viscometer (AMETEK Brookfield, Middleboro, MA, USA) equipped with a No. 4 spindle.

2.2.2. Rice Bran Slurry Supplementation

To investigate the effects of different nutrients on rice bran fermentation, different nutrients including carbon (glucose, maltose, sucrose, or fructose), nitrogen (peptone, soytone, beef extract, or yeast extract), and/or complex compound sources (corn steep liquor; CSL) at different concentrations (1–5%) were added to rice bran slurry containing 20% rice bran powder (Table S2).

2.2.3. pH and Temperature

To investigate the effect of temperature, *L. plantarum* EM was cultivated at 24, 30, or 37 °C for 24 h or 48 h in pH non-adjusted rice bran slurry. To investigate the effect of pH, *L. plantarum* EM was cultivated in rice bran slurry consisting of 20% rice bran with 3% CSL and 1% glucose at pH values ranging from 4.0 to 8.0 for 24 h or 48 h at 30 °C. Thereafter, viable cells [13] and antimicrobial (antibacterial and antifungal) activities [17] were determined.

2.2.4. Viable Cells and pH

To investigate the effects of rice bran concentration and other nutrients on the cell growth and antimicrobial activity of *L. plantarum* EM, *L. plantarum* EM was inoculated (7.0 log CFU/mL) into each prepared rice bran slurry and cultivated at 30 °C for 48 h, after which viable cells were counted. Viable cell counts in fermented rice bran were counted by plating on MRS agar [13]. pH values were determined using a pH meter (Fisher, Hanover Park, IL, USA).

2.2.5. Antimicrobial Activities

Antimicrobial activities were determined using spot-on-the-lawn assays [17]. To prepare MRS cultures, *L. plantarum* EM was cultivated in MRS broth for 24 h at 30 °C. Prepared MRS broths of *L. plantarum* EM and rice bran filtrates were centrifuged (9950× *g*, 15 min, 4 °C) and filtered (0.45 μm; Millipore, Beverly, MA, USA). Prepared cell-free MRS filtrates and cell-free rice bran filtrates were analyzed for antimicrobial activities. Plates were prepared by adding mold spores (6.0 log spores/20 mL of MEA or PDA) to 1.5% bacto agar for the antifungal assay or by spreading bacterial cells (6.0 log CFU/mL) onto LB agar for the antibacterial assay. Thereafter, 10 μL aliquots of the sample were spotted onto prepared plates. Antimicrobial activity, expressed as arbitrary units (AU) per milliliter, was defined as the reciprocal of the highest dilution that produced an inhibitory zone towards sensitive microorganisms. Activities were calculated using (1000/d) D, where D is the dilution factor and d is dose (10 μL of prepared antimicrobial sample).

2.3. *Preparation of Fermented Rice Bran*

Under optimized fermentation conditions, rice bran was fermented at 30 °C for 48 h using *L. plantarum* EM as a starter culture, and hot air-dried at 55 °C for 16 h (Temperature & Humidity Chamber, HB-105SP, Hanbaek, Bucheon, Korea) to produce hot air-dried fermented rice bran (HFRB). Thereafter, it was ground using a blender (BW-3000, Buwon, Daegu, Korea) and tested for cholesterol removal, antimicrobial activity, phytic acid content, and organoleptic qualities. Raw rice bran (RRB) and hot air-dried non-fermented rice bran (HNRB) were used as controls.

2.4. *Cholesterol Removal*

Cholesterol removal by fermented rice bran was determined using the method devised by Rude and Morris [19]. To assay cholesterol removal, ~2 g of HFRB (obtained from 10 mL of fermented rice bran slurry) was resuspended in MRS broth containing 0.5% oxgall and 0.1 g/L of water-soluble cholesterol as well as 0.5% TDCA and 0.1 g/L of water-soluble cholesterol in 10 mL. Separately, RRB or HNRB slurries were also prepared. To prepare dead LAB cells, *L. plantarum* EM was incubated at 30 °C in MRS broth (10 mL) containing 0.5% oxgall or 0.5% TDCA and harvested (9950× *g*, 5 min, 4 °C). Cell pellets were suspended in saline and autoclaved at 121 °C for 15 min. Dead cells were harvested and then resuspended in MRS broth (10 mL) containing 0.5% oxgall and 0.1 g/L water-soluble cholesterol as well as 0.5% TDCA and 0.1 g/L water-soluble cholesterol [15].

Prepared resuspensions (10 mL) were incubated at 37 °C for 24 h in a GasPak EZ system (Becton Dickinson, Sparks, MD, USA) and harvested (9950× *g*, 4 °C, 5 min), and cholesterol concentrations in supernatants were determined as previously described [15]. Briefly, 1 mL of supernatant was added to 2 mL of 33% (*w*/*v*) potassium hydroxide and 3 mL of 95% (*v*/*v*) ethanol, heated for 10 min in a 60 °C water bath, and then cooled with tap water. Hexane (5 mL) was then added, mixed, and 1 mL of distilled water was added. After standing for 10 min to allow phase separation, the hexane phase was evaporated under a nitrogen stream. The concentrated aqueous fraction was then added to 4 mL of freshly prepared *o*-phthalaldehyde (Sigma-Aldrich, St. Louis, MO, USA, 0.5 mg in 1 mL acetic acid), mixed, allowed to stand for 10 min, and treated with 2 mL of concentrated sulfuric acid for 10 min. Thereafter, absorbances were measured at 550 nm using an Ultrospec 2100 pro (Biochrom, Cambridge, UK). Cholesterol removal (%) was expressed as {0.1 g/L of cholesterol—concentration of remaining cholesterol in the cultures}/0.1 g/L of cholesterol ×100.

2.5. Phytase Activity and Phytic Acid

Phytase activity of *L. plantarum* EM was determined using phytase-specific medium as previously described [20]. Overnight grown *L. plantarum* EM was harvested (9950× *g*, 5 min, 4 °C), and suspended in the same volume of distilled water. *W. koreensis* DB1, which was used as a starter culture for rice bran fermentation in our previous study [13], and *L. plantarum* ATCC 14,917 were used as controls. Paper discs (diameter 8 mm; Advantec, Tokyo, Japan) were placed on phytase-specific medium (1.5% glucose, 0.5% calcium phytate, 0.5% NH_4NO_3, 0.05% KCl, 0.05% $MgSO_4 \cdot 7H_2O$, 0.001% $FeSO_4 \cdot 7H_2O$, 0.001% $MnSO_4$, and 1.5% micro agar, pH 5.5), and 100 μL of prepared LAB suspensions was spotted onto the paper discs. Plates were incubated at 30 °C for 24–48 h, and the diameters of haloes formed on plates were measured using a caliper (Mitutoyo, Tokyo, Japan).

Total phytic acid contents of the rice bran products were determined using a phytic acid assay kit K-PHYT (Megazyme; Wicklow, Ireland) according to the manufacturer's instructions. Briefly, the rice bran products (1 g) were extracted using 0.66 M hydrochloric acid and then neutralized with 0.75 M sodium hydroxide. Extracted samples were tested for total phosphorus and free phosphorus produced by phytase and alkaline phosphatase using the assay kit. Phytate contents were calculated using the following Equation (1):

$$\text{Phytate content } (\text{g}/100\text{ g}) = \frac{\text{Phosphorus content } (\text{g}/100\text{ g})}{0.282} \quad (1)$$

where, phosphorous content (g/100 g) = mean M × 0.1112 × ΔA phosphorous, ΔA phosphorous = Absorbance of total phosphorous − Absorbance of free phosphorous, and mean M (μg/ΔA phosphorous) = mean of phosphorous standard values (standard solutions 1–4 in the Megazyme K-PHYT kit).

2.6. Sensory Evaluations

Sensory evaluations were performed after obtaining approval from the Institutional Review Board of Chosun University (Gwangju, Korea; IRB#2-1041055-AB-N-01-2019-33). Nine trained students who had performed more than 30 evaluations per year at the Department of Food and Nutrition, Chosun University, constituted the sensory evaluation panel. Four samples of RRB, HNRB, HFRB, or HFRB containing 0.07% stevia (Viomix, Seoul, Korea; HFRB-S) were presented to the panel. Prior to sensory evaluations, panelists rinsed their mouths with water and waited for 1–2 min before performing each evaluation.

Sensory attributes were evaluated using a 5-point scale for sourness, bitterness, sweetness, hay smell, and pleasant flavor (1 = very weak, 3 = moderate, and 5 = very strong) and mouthfeel texture and overall acceptability (1 = very bad, 3 = moderate, and 5 = very good).

2.7. Statistical Analysis

Data were expressed as the mean and standard deviation (mean ± SD) of triplicate determinations. Experimental data were statistically analyzed by Duncan's multiple range test for one-way ANOVA in SPSS 26.0 software, and $p < 0.05$ was considered statistically significant.

3. Results and Discussion

3.1. Rice Bran Fermentation

3.1.1. Rice Bran as a Nutrient Source for LAB Cell Growth

High *L. plantarum* EM cell growth during rice bran fermentation is most important to obtain in terms of developing a functional rice bran product, as *L. plantarum* EM, whether alive or dead, has shown high cholesterol removal ability [15]. *L. plantarum* EM was inoculated (7.00–7.04 log CFU/mL) in rice bran slurries (5–30% rice bran powder in distilled water) and cultivated at 30 °C for 24–48 h. Highest viable cell counts were obtained at 24 h (8.30–8.67 log CFU/mL), and these were maintained at 48 h. Viable cell counts were not affected by rice bran concentration (5–30%) (Table 1). The viscosities of 25–30% rice bran slurries could not be determined because they were almost solid. However, 20% of slurry viscosities ranged from 5733 to 17,514 cp at 12–60 rpm (data not shown); thus, we used this concentration for further experiments. The majority of previous studies have been conducted using solid-state fungal fermentations of 30–50% (*w*/*v*) rice bran slurries [2–5], whereas liquid-state rice bran fermentations of bacteria (LAB) have been performed using lower concentrations (<10% of rice bran or rice bran extract) [9,12,14].

The effects of other nutrients added to rice bran on LAB cell growth were also determined (Table 1). In carbon source supplementation to 20% rice bran, viable LAB cells of the cultures were 8.60–8.69 log CFU/mL. Whereas supplementation with 3% beef extract as a nitrogen source or 3% CSL as a complex compound source resulted in significantly better cell growth. Based on the above results, we decided that nutrients should be added to rice bran fermentations at 1% for glucose (as a primary carbon source), 3% CSL, and/or 3% beef extract. Two or three of these nutrients were used in combination, and then viable cell counts were determined. The highest viability (9.60–9.78 log CFU/mL) was obtained from 20% rice bran slurries containing 3% CSL supplemented with other nutrients (Table 1; 20% RB + combined nutrients); this cell growth was almost the same as that obtained (9.59 log CFU/mL) by MRS cultivation. In the present study, neither beef extract, peptone, soytone, nor yeast extract significantly affected the growth of *L. plantarum* EM, whereas CSL obviously increased *L. plantarum* EM growth. These results indicated that CSL addition is important to enhance LAB cell growth in rice bran culture. The cost of fermentation medium has been reported to account for almost 30% of the total cost of microbial fermentation [21]. CSL is a byproduct of corn wet-milling and contains carbon and nitrogen sources as well as minerals and vitamins. Thus, CSL has been proposed as an alternative to expensive nutrients such as beef extract, yeast extract, or peptone [21]. Behr suggested CSL be used as a nutrient for microorganisms as long ago as 1909 [22]. The usage of CSL as a component of nutrient media can be summarized as follows: Production of penicillin by mold, *Penicillium* sp. [22], production of ethanol, succinic acid, or arabinase by yeasts, *Zymomonas* sp. [23], *Pichia* sp. [24], *Anaerobiospirillum* sp. [25], or *Fusarium* sp. [26], and production of lactic acid by LAB [21,27–29]. However, its use in microbiology was limited to yeast fermentation until recently. Specifically, in studies on the fermentation using LAB, CSL has been used as an alternative for yeast extract to produce lactic acid [21,27–29]. However, results were less than satisfactory as CSL did not produce as much lactic acid as yeast extract [30]. In this study *L. plantarum* EM cell growth in rice bran culture was successfully enhanced byadding the biological byproduct CSL instead of expensive nutrients such as beef extract or yeast extract, which satisfies environmental and legal requirements concerning the re-utilization of biological byproducts.

Table 1. Effect of rice bran and other nutrients on the cell growth of *L. plantarum* EM.

Culture		Viable Cells (log CFU/mL)
Nutrient	**Concentration (%)**	
MRS (control)		9.59 ± 0.13 [A]
Rice bran (RB)	5	8.58 ± 0.13 [a]
	10	8.39 ± 0.41 [a]
	15	8.64 ± 0.39 [a]
	20	8.67 ± 0.38 [aB]
	25	8.36 ± 0.01 [a]
	30	8.30 ± 0.01 [a]
20% RB + Carbon-source		
Glucose	1	8.69 ± 0.16 [aB]
	3	8.66 ± 0.10 [a]
Maltose	1	8.62 ± 0.02 [a]
	3	8.63 ± 0.05 [a]
Sucrose	1	8.65 ± 0.12 [a]
	3	8.60 ± 0.04 [a]
Fructose	1	8.63 ± 0.05 [a]
	3	8.62 ± 0.12 [a]
20% RB + Nitrogen-source		
Peptone	1	8.86 ± 0.19 [a]
	3	8.94 ± 0.06 [ab]
Beef extract	1	8.79 ± 0.11 [a]
	3	9.05 ± 0.03 [aB]
Yeast extract	1	8.84 ± 0.12 [a]
	3	8.76 ± 0.04 [a]
Soytone	1	8.81 ± 0.04 [a]
	3	8.89 ± 0.04 [ab]
20% RB + Complex compound source		
Corn steep liquor (CSL)	1	9.29 ± 0.18 [b]
	2	9.33 ± 0.11 [b]
	3	9.70 ± 0.11 [aA]
	4	9.38 ± 0.09 [b]
	5	9.40 ± 0.09 [b]
20% RB + Combined nutrients		
20% RB + 1% Glucose + 3% Beef extract		9.03 ± 0.28 [b]
20% RB + 1% Glucose + 3% CSL		9.78 ± 0.06 [aA]
20% RB +3% Beef extract + 3% CSL		9.60 ± 0.05 [aA]
20% RB + 1% Glucose + 3% CSL + 3% Beef extract		9.65 ± 0.12 [aA]

L. plantarum EM was cultivated in rice bran slurry at 30 °C for 48 h when viable cells were counted. Values are the mean ± SD of three independent cultivations. Different lowercase letters indicate significant differences ($p < 0.05$) between supplemented nutrients in the same column. Different uppercase letters indicate significant differences ($p < 0.05$) between cultures (indicated by shading).

3.1.2. Antimicrobial Activity of Fermented Rice Brans

MRS culture of *L. plantarum* EM (9.59 log CFU/mL) showed strong antifungal activity against *A. fumigatus* ATCC 96,918 (600 AU/mL) and antibacterial activity against *B. cereus* ATCC 14,579 (300 AU/mL) (Table 2). Rice bran cultures of *L. plantarum* EM (8.30–8.67 log CFU/mL) prepared from 5 to 30% rice bran/distilled water slurries did not exhibit antibacterial or antifungal activity. Rice bran cultures in rice bran slurries composed of 20% rice bran supplemented with different carbon or nitrogen sources showed antibacterial activity (100 AU/mL) against *B. cereus* ATCC 14,579 but no antifungal activity against *A. fumigatus* ATCC 96918. However, supplementation (>2% by weight) of CSL in rice bran slurry resulted in antibacterial (200 AU/mL) and antifungal (100 AU/mL) activities, and these antimicrobial activities were markedly higher than those of rice bran cultures supplemented with carbon or nitrogen but lower than those obtained by MRS culture. The addition of 3% CSL with glucose or/and beef extract increased the antifungal activity of fermented rice bran to 400 AU/mL, and the effect of beef extract supplementation on antimicrobial activity was similar to that of glucose. Supplementation with 3% CSL or 3% CSL + 1% glucose to rice bran culture both resulted in viable LAB cell counts of 9.70–9.78 CFU/mL, but the antifungal activity of rice bran culture supplemented with 3% CSL + 1% glucose was higher than that supplemented with 3% CSL (400 vs. 100 AU/mL) (Table 1). The results in Tables 1 and 2 show supplementation with 3% CSL and 1% glucose achieved high *L. plantarum* EM cell growth (9.78 log CFU/mL) and strong antibacterial activity (200 AU/mL) against *B. cereus* ATCC 14,579 and antifungal activity (400 AU/mL) against *A. fumigatus* ATCC 96918. Notably, these results demonstrate that replacement of beef extract, peptone, soytone, or yeast extract with CSL resulted in higher LAB cell growth and antimicrobial activities. The antimicrobial mechanism was verified as *L. plantarum* EM with antimicrobial activity induced dimples on the surface of vegetative *B. cereus* cells, which resulted in cell death. *L. plantarum* EM showing antimicrobial activity also affected the cell membranes of *A. fumigatus* conidium and *B. cereus* endospore, which led to collapsed and shrunken morphologies resulting in cell death [17].

Based on these results, *L. plantarum* EM was cultivated in 20% rice bran slurry supplemented with 3% CSL + 1% glucose in subsequent experiments.

3.1.3. Effects of pH and Temperature

Prior to adjustment, the pH of the prepared rice bran slurry composed of 20% rice bran + 3% CSL + 1% glucose was pH 5.6 ± 0.3. To investigate the effects of pH and temperature on the fermentation, pH values and temperatures of rice bran slurry were adjusted in ranges pH 4.0–8.0 and 25–37 °C, respectively, and then viable LAB counts and antibacterial and antifungal activity were measured (Table 3). Highest cell growth and antimicrobial activity were obtained at pH from 6.0 to 7.0 and at 30 °C. Based on these results, rice bran at pH 6.0 was fermented at 30 °C in further experiments.

3.2. *Characterization of Rice Bran Products*

We fermented rice bran using a functional starter culture *L. plantarum* EM, which exhibited effective cholesterol removal regardless of LAB viability coupled with strong antimicrobial activities against foodborne pathogenic bacteria and food spoilage fungi. To address the unsuitability of rice bran as a growth medium for LAB, others have used enzyme hydrolyzed rice bran, rice bran extract, or rice bran with other nutrient supplementations [8–13]. However, in this study highest LAB cell growth (9.78 log CFU/mL) and strong antimicrobial activity were obtained by simply fermenting 20% rice bran in the presence of 3% CSL and 1% glucose (pH 6.0) at 30 °C for 48 h.

For customer convenience and to extend shelf-life, fermented slurries must be further processed by a drying process such as freeze-drying or hot air-drying. We adopted a hot air-drying method, which is more cost-effective than freeze-drying. After the fermented rice bran had been hot air-dried, no viable LAB cells were detected. Advantages of the hot air-drying method used in this study are that the used starter culture *L. plantarum* EM can represent its bioactivities (cholesterol removal and antimicrobial activity) even in dead cells [15,17], and that the used substrate in this fermentation is a heat-stable cereal-based material, rice bran.

Table 2. Effect of rice bran and other nutrients on the antimicrobial activities of *L. plantarum* EM.

Culture.		Antimicrobial Activity (AU/mL)	
Nutrient	**Concentration (%)**	**Antibacterial**	**Antifungal**
MRS (control)		300	600
Rice bran (RB)	5	0	0
	10	0	0
	15	0	0
	20	0	0
	25	0	0
	30	0	0
20% RB + Carbon-source			
Glucose	1	100	0
	3	100	0
Maltose	1	100	0
	3	100	0
Sucrose	1	100	0
	3	100	0
Fructose	1	100	0
	3	100	0
20% RB + Nitrogen-source			
Peptone	1	100	0
	3	100	0
Beef extract	1	100	0
	3	100	0
Yeast extract	1	100	0
	3	100	0
Soytone	1	100	0
	3	100	0
20% RB + Complex compound source			
Corn steep liquor (CSL)	1	100	100
	2	200	100
	3	200	100
	4	200	100
	5	200	100
20% RB + Combined nutrients			
20% RB + 1% Glucose + 3% Beef extract		100	0
20% RB + 1% Glucose + 3% CSL		200	400
20% RB + 3% Beef extract + 3% CSL		200	400
20% RB + 1% Glucose + 3% CSL + 3% Beef extract		200	400

Rice bran slurry was fermented at 30 °C for 48 h, after which antimicrobial activities were assayed on *B. cereus* ATCC 14,579 and *A. fumigatus* ATCC 96,918 lawn plates as described in Materials and Methods.

3.2.1. Cholesterol Removal by Fermented Rice Bran Products

The effect of the hot air-dried fermented rice bran product (HFRB) on reducing cholesterol levels in vitro was examined (Table 4). In this assay, raw rice bran (RRB), hot air-dried non-fermented rice bran product (HNRB), and dead cells of *L. plantarum* EM were used as controls. In our previous study, *L. plantarum* EM showed high cholesterol removal regardless of cell viability, but cholesterol removal by growing cells (88.12%) was greater than that by dead cells (39.02%) in oxgall assay [15]. Meanwhile, it has been reported that bioactive compounds, including tocotrienols, γ-oryzanol, and dietary fiber, in rice bran have cholesterol-lowering effects [31]. As shown in Table 4, RRB and HNRB achieved 7.77–9.75% cholesterol removal, and dead cells of *L. plantarum* EM achieved 34.67–39.58% cholesterol removal, while HFRB attained 44.93–67.58% in oxgall and TDCA assays. Cholesterol removal by HFRB was slightly lower than that achieved by *L. plantarum* EM live cells (47.66–88.12% in oxgall and TDCA assays [15]). These results demonstrated that the cholesterol removal efficacy of rice bran was significantly enhanced by fermentation. We suppose these results were due to the synergistic effects of bioactive compounds in rice bran, the strong cholesterol-binding ability of *L. plantarum* EM cell walls, and the effects of unidentified compounds produced during fermentation.

Table 3. Effect of pH and temperature on cell growth and the antimicrobial activities of *L. plantarum* EM.

Factor	Viable Cells (CFU/mL)		Antimicrobial Activity (AU/mL)			
			Antibacterial		Antifungal	
	24 h	48 h	24 h	48 h	24 h	48 h
Temp. (°C) *						
25	9.49 ± 0.08 [b]	9.28 ± 0.25 [a]	100	200	100	200
30	9.78 ± 0.11 [a]	9.33 ± 0.21 [a]	200	200	200	400
37	9.47 ± 0.06 [b]	8.69 ± 0.09 [b]	100	200	100	200
pH **						
4.0	9.39 ± 0.10 [c]	8.90 ± 0.30 [cd]	100	100	100	100
5.0	9.44 ± 0.17 [c]	9.22 ± 0.15 [bc]	100	200	100	100
6.0	9.81 ± 0.10 [a]	9.59 ± 0.07 [a]	200	200	200	400
7.0	9.76 ± 0.13 [ab]	9.49 ± 0.16 [ab]	200	200	200	400
8.0	9.53 ± 0.14 [bc]	8.70 ± 0.14 [d]	100	200	100	200

* Rice bran slurries were incubated at 25, 30, or 37 °C without pH adjustment. ** pH values of rice bran slurries were adjusted to 4.0–8.0 and then incubated at 30 °C. Different letters (a–d) represent significant differences ($p < 0.05$) on the same factor at 24 h and 48 h, respectively.

Table 4. Cholesterol removal by the fermented rice bran products.

Sample	Cholesterol Removal (%)	
	0.5% Oxgall	0.5% TDCA
Dead cells of *L. plantarum* EM	39.58 ± 0.49 [b]	34.67 ± 1.34 [b]
RRB	8.87 ± 3.77 [c]	7.77 ± 1.33 [c]
HNRB	9.75 ± 1.06 [c]	9.29 ± 3.10 [c]
HFRB	67.58 ± 3.34 [a]	44.93 ± 1.21 [a]

Rice bran slurry consisting of 20% rice bran powder (RRB) + 3% CSL + 1% glucose in distilled water (pH 6.0) was fermented using *L. plantarum* EM for 0 h (HNRB) or 48 h (HFRB) at 30 °C, hot air-dried at 55 °C, and then ground. Separately, *L. plantarum* EM was cultivated in MRS containing 0.5% TDCA or 0.5% oxgall at 37 °C for 24 h. Cell pellets were harvested and suspended in saline and heat-killed at 121 °C for 15 min. Prepared samples of RRB, HNRB, HFRB, and of the dead cells of *L. plantarum* EM were incubated at 37 °C for 48 h in 0.5% oxgall or 0.5% TDCA containing cholesterol media as described in Materials and Methods. Values in the same column with different letters indicate significant difference ($p < 0.05$).

3.2.2. Antimicrobial Activity

As we previously reported, *L. plantarum* EM showed antimicrobial activities against various foodborne pathogenic bacteria and food spoilage molds [17]. The compounds showing antibacterial and antifungal activities were identified as 3-hydroxy-5-dodecenoic acid and lactic acid. Furthermore, the antimicrobial effects of these active compounds were found to act synergistically [17].

Cholesterol removal and antimicrobial (both antibacterial and antifungal) activity are distinctive and important bioactivities of *L. plantarum* EM. Fermented rice bran products intended to be used as functional food or ingredients should possess these bioactivities. Thus, in this study, we examined whether the fermented rice bran product produced by hot air-drying (at 55 °C for 16 h) retained antimicrobial activity. To determine the antimicrobial activity of HFRB, 2 g (equivalent to 10 mL of rice bran slurry) was resuspended in 10 mL of distilled water, allowed to stand for 4–5 h at 4 °C, and filtered. The antimicrobial activities of the filtrates obtained were then evaluated against various foodborne pathogens and food spoilage microorganisms (Table 5). The antimicrobial activities of RRB (pH 6.3), HNRB (pH 5.6), HFRB (pH 3.8), and MRS culture filtrate (pH 3.8) of *L. plantarum* EM were measured. HFRB showed the same or slightly less antimicrobial activity against foodborne pathogenic bacteria (200–400 AU/mL) and food spoilage fungi (100–400 AU/mL) than *L. plantarum* EM MRS filtrate (200–400 AU/mL antibacterial and 100–600 AU/mL antifungal), but no antimicrobial activity was observed for RRB or HNRB (Table 5). Fermented rice bran product and MRS culture of *L. plantarum* EM had similar cell counts (Table 1) and pH values, and lactic acid amounts detected in HFRB and MRS culture filtrates were similar (data not shown). The reason why the antimicrobial activities of MRS culture filtrate against *A. fumigatus*, *P. expansum*, and *B. cereus* were slightly higher than those of HFRB (in Table 5) is believed to be due to the sodium acetate present in MRS medium. Stiles et al. reported that sodium acetate, a basic component of commercial MRS medium, has microorganism-dependent antifungal effects [32].

Table 5. Antimicrobial activity of the fermented rice bran products.

Indicator Strains		Antimicrobial Activity (AU/mL)			
		RRB	**HNRB**	**HFRB**	**MRS Culture Filtrate ***
Molds	*Aspergillus flavus* ATCC 22546	0	0	200	200
	Aspergillus fumigatus ATCC 96918	0	0	400	600
	Penicillium roqueforti ATCC 10110	0	0	100	100
	Penicillium expansum ATCC 7861	0	0	0	100
Bacteria	*Bacillus cereus* ATCC 14579	0	0	200	300
	Escherichia coli O157:H7 ATCC 43895	0	0	200	200
	Pseudomonas aeruginosa ATCC 29853	0	0	400	400
	Salmonella enterica servoar. Typhi ATCC 14028	0	0	200	200

Rice bran slurry consisting of 20% rice bran powder (RRB) + 3% CSL + 1% glucose in distilled water (pH 6.0) was fermented using *L. plantarum* EM for 0 h (HNRB) or 48 h (HFRB) at 30 °C, hot air-dried (55 °C), and ground. Products were resuspended in distilled water, and cell-free filtrates were prepared to assay antimicrobial activities, as described in Materials and Methods. * To prepare MRS culture filtrate, *L. plantarum* EM was cultivated in MRS broth at 30 °C for 48 h, centrifuged, and filtered.

These results showed that the fermented rice bran product even after hot air-drying (HFRB) retained strong antimicrobial activities against different foodborne pathogenic bacteria and food spoilage molds. The results suggest that the natural antimicrobial agent, HFRB, could be used as a functional food.

3.2.3. Phytic Acid

As shown in Table S3, *L. plantarum* EM showed obvious phytase activity (18 mm clear zone on phytase-specific medium), while other LAB strains showed lower activities (9–12 mm clear zone).

Phytic acid (*myo*-inositol hexaphophoric acid) is found in the seeds of plants and has some beneficial properties, which include antioxidant, anticancer, cholesterol-lowering, and blood-sugar-lowering activities [33]. However, phytic acid also inhibits the absorptions of essential minerals such as iron, zinc, calcium, and magnesium in the digestive tract due to its ability to bind polycations and incorporate them in insoluble complexes. Thus, phytic acid is considered an antinutrient, especially in children and the elderly, in whom calcium absorption is important [33].

As shown in Table 6, phytic acid contents were 3.99 g per 100 g for HFRB, 7.93 g per 100 g for HNRB, and 8.48 g per 100 g for RRB; that is, *L. plantarum* EM phytase reduced phytic acid content by 53% during fermentation. It has been reported that the amount of phytic acid present in foods depends on the preparation methods used, such as milling, soaking, germinating, or fermentation [33]. In healthy people eating a balanced diet, the inhibitory effect of phytic acid on mineral absorption is minimal, and phytic acid consumption at 1000–2000 mg per day has not been associated with nutrient deficiencies [34]. Therefore, consumption of HFRB is unlikely to have any antinutritional effect.

Table 6. Phytic acid content of fermented rice bran products.

Sample	Phytic Acid (g/100 g)
RRB	8.48 ± 0.47 [a]
HNRB	7.93 ± 0.44 [a]
HFRB	3.99 ± 0.26 [b]

Values displayed with different letters are significantly different ($p < 0.05$).

3.2.4. Sensory Evaluation

As shown in Table 7, HFRB had a pleasant flavor, a soft mouthfeel, and a strong sour taste. The strong hay smell and bitterness of RRB and HNRB were notably reduced by fermentation. The sourness of HFRB was attributed to low pH (3.80 due to 83,726 mg/kg of lactic acid; data not shown), but this was improved by adding only 0.07% by weight of stevia without affecting sweetness. The consumption of rice bran has been limited despite its health-promoting compound contents because of its smell, taste, and coarse texture [1]. This study shows that the organoleptic qualities of rice bran, except sourness, were significantly improved by *L. plantarum* EM fermentation, and that this sourness was easily managed by adding a calorie-free sweetener, such as stevia (HFRB-S).

Table 7. Results of sensory evaluations of fermented rice bran products.

Indicator Strains	RRB	HNRB	HFRB	HFRB-S
Sourness	1.11 ± 0.33 [d]	2.22 ± 0.44 [c]	4.67 ± 0.50 [a]	3.89 ± 0.78 [b]
Bitterness	3.33 ± 0.71 [b]	4.11 ± 0.33 [a]	2.00 ± 0.00 [c]	2.00 ± 0.00 [c]
Sweetness	2.00 ± 0.00 [a]	2.00 ± 0.00 [a]	2.00 ± 0.00 [a]	2.00 ± 0.00 [a]
Hay smell	5.00 ± 0.00 [a]	4.00 ± 0.00 [b]	1.67 ± 0.71 [c]	1.67 ± 0.71 [c]
Pleasant flavor	1.78 ± 0.83 [c]	2.78 ± 0.83 [b]	4.89 ± 0.33 [a]	4.89 ± 0.33 [a]
Mouthfeel texture	1.56 ± 0.53 [c]	2.67 ± 0.50 [b]	4.11 ± 0.60 [a]	4.11 ± 0.33 [a]
Overall acceptability	1.67 ± 0.50 [c]	2.00 ± 0.00 [c]	3.00 ± 0.50 [b]	4.11 ± 0.60 [a]

Rice bran slurry was fermented at 30 °C for 0 h (HNRB) or 48 h (HFRB), hot air-dried at 55 °C for 14 h, and then ground. Sensory evaluations were carried out on raw rice bran (RRB), fermented rice bran product (HFRB), non-fermented rice bran product (HNRB), and HFRB containing 0.07% by weight of stevia (HFRB-S). Evaluations were performed using a 5-point scale; 1, 3, and 5 corresponding to "very bad, moderate, and very good" for mouthfeel texture and overall acceptability, and using "very weak, moderate, and very strong" for the other items. Results are expressed as the mean ± SD. Different letters for sensory items indicate significant differences ($p < 0.05$).

4. Conclusions

In this study, the fermented rice bran product (HFRB) showed high cholesterol removal (45–68%) and antimicrobial activities (100–400 AU/mL) against foodborne pathogenic bacteria and food spoilage fungi. Levels of phytic acid, which has a combination of beneficial and antinutrient properties, were significantly reduced (from 8480 mg/100 g to 3990 mg/100 g) during fermentation, which showed the antinutritional aspects of HFRB were not a cause for concern. HFRB showed much better organoleptic qualities than RRB and HNRB, except sourness, though this was easily improved by adding a small amount (0.07% by weight) of stevia. Commercially successful functional food products must satisfy consumer requirements for health-promoting properties and good sensory properties, and the results of this study show that HFRB well meets these requirements. Thus, we conclude that HFRB produced using *L. plantarum* EM is a promising functional food candidate or ingredient that can be incorporated into foods as a natural antimicrobial agent and health-promoting material without adversely affecting flavor or texture. The most important aspect of this study is that it resulted in the low-cost production of a valuable functional food candidate or functional food ingredient from an inexpensive food byproduct. We are currently evaluating HFRB in animals to confirm its beneficial health effects in vivo.

Supplementary Materials: The following are available online at https://www.mdpi.com/article/10.3390/foods10050978/s1, Table S1: Microorganisms used in this study, Table S2: Rice bran slurry supplemented with different nutrients, Table S3: Phytase activity of *L. plantarum* EM.

Author Contributions: Conceptualization, H.-C.C.; methodology, H.-C.C.; validation, S.-H.M.; formal analysis, H.-C.C.; investigation, S.-H.M.; data curation, S.-H.M.; writing—original draft preparation, H.-C.C.; writing—editing and review, H.-C.C.; visualization, S.-H.M.; supervision, H.-C.C.; project administration, H.-C.C.; funding acquisition, H.-C.C. All authors have read and agreed to the published version of the manuscript.

Funding: This research was funded by the Ministry of Agriculture, Food and Rural Affairs, Republic of Korea, grant number 918005-4.

Institutional Review Board Statement: The study was conducted according to the guidelines of the Declaration of Helsinki, and approved by the Institutional Review Board of Chosun University (IRB#2-1041055-AB-N-01-2019-33, 2020.09.02).

Informed Consent Statement: Informed consent was obtained from all subjects involved in the study.

Data Availability Statement: Data is contained within the article or supplementary material.

Acknowledgments: This research was supported by the Strategic Initiative for Microbiomes in Agriculture and Food, Ministry of Agriculture, Food and Rural Affairs, Republic of Korea (as part of the (multi-ministerial) Genome Technology to Business Translation Program), grant number 918005-4.

Conflicts of Interest: The authors declare no conflict of interest.

References

1. Demirci, T.; Aktaş, K.; Sözeri, D.; Öztürk, H.İ.; Akın, N. Rice bran improve probiotic viability in yoghurt and provide added antioxidative benefits. *J. Funct. Foods* **2017**, *36*, 396–403. [CrossRef]
2. Yang, M.; Ashraf, J.; Tong, L.; Wang, L.; Zhang, L.; Li, N.; Zhou, S.; Liu, L. Effects of *Rhizopus oryzae* and *Aspergillus oryzae* on prebiotic potentials of rice bran pretreated with superheated steam in an in vitro fermentation system. *LWT* **2021**, *139*, 110482. [CrossRef]
3. Denardi-Souza, T.; Luz, C.; Mañes, J.; Badiale-Furlong, E.; Meca, G. Antifungal effect of phenolic extract of fermented rice bran with *Rhizopus oryzae* and its potential use in loaf bread shelf life extension. *J. Sci. Food Agric.* **2018**, *98*, 5011–5018. [CrossRef]
4. Massarolo, K.C.; Denardi de Souza, T.; Collazzo, C.C.; Badiale Furlong, E.; Souza Soares, L.A. The impact of *Rhizopus oryzae* cultivation on rice bran: Gamma-oryzanol recovery and its antioxidant properties. *Food Chem.* **2017**, *228*, 43–49. [CrossRef] [PubMed]
5. Ritthibut, N.; Oh, S.J.; Lim, S.T. Enhancement of bioactivity of rice bran by solid-state fermentation with *Aspergillus* strains. *LWT* **2021**, *135*, 110273. [CrossRef]
6. Chen, Y.; Ma, Y.; Dong, L.; Jia, X.; Liu, L.; Huang, F.; Chi, J.; Xiao, J.; Zhang, M.; Zhang, R. Extrusion and fungal fermentation change the profile and antioxidant activity of free and bound phenolics in rice bran together with the phenolic bioaccessibility. *LWT Food Sci. Technol.* **2019**, *115*, 108461. [CrossRef]
7. Jeon, Y.B.; Lee, J.J.; Chang, H.C. Characterization of juice fermented with *Lactobacillus plantarum* EM and its cholesterol-lowering effects on rats fed a high-fat and high-cholesterol diet. *Food Sci. Nutr.* **2019**, *7*, 3622–3634. [CrossRef]
8. Wang, M.; Lei, M.; Samina, N.; Chen, L.; Liu, C.; Yin, T.; Yan, X.; Wu, C.; He, H.; Yi, C. Impact of *Lactobacillus plantarum* 423 fermentation on the antioxidant activity and flavor properties of rice bran and wheat bran. *Food Chem.* **2020**, *330*, 127156. [CrossRef]
9. Hasani, S.; Khodadadi, I.; Heshmati, A. Viability of *Lactobacillus acidophilus* in rice bran-enriched stirred yoghurt and the physicochemical and sensory characteristics of product during refrigerated storage. *Int. Food Sci. Technol.* **2016**, *51*, 2485–2492. [CrossRef]
10. Liu, L.; Zhang, R.; Deng, Y.; Zhang, Y.; Xiao, J.; Huang, F.; Wen, W.; Zhang, M. Fermentation and complex enzyme hydrolysis enhance total phenolics and antioxidant activity of aqueous solution from rice bran pretreated by steaming with α-amylase. *Food Chem.* **2017**, *221*, 636–643. [CrossRef]
11. Yun, J.S.; Wee, Y.J.; Kim, J.N.; Ryu, H.W. Fermentative production of DL-lactic acid from amylase-treated rice and wheat brans hydrolysate by a novel lactic acid bacterium, *Lactobacillus* sp. *Biotechnol. Lett.* **2004**, *26*, 1613–1616. [CrossRef]
12. Gao, M.T.; Kaneko, M.; Hirata, M.; Toorisaka, E.; Hano, T. Utilization of rice bran as nutrient source for fermentative lactic acid production. *Bioresour. Technol.* **2008**, *99*, 3659–3664. [CrossRef] [PubMed]
13. Moon, S.Y.; Moon, S.H.; Chang, H.C. Characterization of high-ornithine-producing *Weissella koreensis* DB1 isolated from kimchi and its application in rice bran fermentation as a starter culture. *Foods* **2020**, *9*, 1545.
14. Kwon, S.; Lee, P.C.; Lee, E.G.; Chang, Y.K.; Chang, N. Production of lactic acid by *Lactobacillus rhamnosus* with vitamin-supplemented soybean hydrolysate. *Enzym. Microb. Technol.* **2000**, *26*, 209–215. [CrossRef]

15. Choi, E.A.; Chang, H.C. Cholesterol-lowering effects of a putative probiotic strain *Lactobacillus plantarum* EM isolated from kimchi. *LWT Food Sci. Technol.* **2015**, *62*, 210–217. [CrossRef]
16. Zheng, J.; Wittouck, S.; Salvetti, E.; Franz, C.M.A.P.; Harris, H.M.B.; Mattarelli, P.; O'Toole, P.W.; Pot, B.; Vandamme, P.; Walter, J.; et al. A taxonomic note on the genus *Lactobacillus*: Description of 23 novel genera, emended description of the genus *Lactobacillus* Beijerinck 1901, and union of *Lactobacillaceae* and *Leuconostocaceae*. *Int. J. Syst. Evol. Microbiol.* **2020**, *70*, 2782–2858. [CrossRef]
17. Mun, S.Y.; Kim, S.K.; Woo, E.R.; Chang, H.C. Purification and characterization of an antimicrobial compound produced by *Lactobacillus plantarum* EM showing both antifungal and antibacterial activities. *LWT Food Sci. Technol.* **2019**, *114*, 108403. [CrossRef]
18. Park, S.; Son, H.K.; Chang, H.C.; Lee, J.J. Effects of cabbage-apple juice fermented by *Lactobacillus plantarum* EM on liqid profile improvement and obesity amelioration in rats. *Nutrients* **2020**, *12*, 1135. [CrossRef] [PubMed]
19. Rudel, L.L.; Morris, M.D. Determination of cholesterol using o-phthalaldehyde. *J. Lipid Res.* **1973**, *14*, 364–366. [CrossRef]
20. Quan, C.; Zhang, L.; Wang, Y.; Ohta, Y. Production of phytase in a low phosphate medium by a novel yeast *Candida krusei*. *J. Biosci. Bioeng.* **2001**, *92*, 154–160. [CrossRef]
21. Rivas, B.; Moldes, B.; Domínguez, J.M.; Parajó, J.C. Development of culture media containing spent yeast cells of *Debaryomyces hansenii* and corn steep liquor for lactic acid production with *Lactobacillus rhamnosus*. *Int. J. Food. Microbiol.* **2004**, *97*, 93–98. [CrossRef]
22. Liggett, R.W.; Koffler, H. Corn steep liquor in microbiology. *Bacteriol. Rev.* **1948**, *12*, 297–311. [CrossRef] [PubMed]
23. Silveira, M.M.; Wisbeck, E.; Hoch, I.; Jonas, R. Production of glucose-fructose oxidoreductase and ethanol by *Zymomonas mobilis* ATCC 29191 in medium containing corn steep liquor as a source of vitamins. *Appl. Microbiol. Biotechnol.* **2001**, *55*, 442–445. [CrossRef] [PubMed]
24. Amartey, S.; Jeffries, T.W. Comparison of corn steep liquor with other nutrients in the fermentation of D-xylose by *Pichia stipites* CBS 6054. *Biotechnol. Lett.* **1994**, *16*, 211–214. [CrossRef]
25. Lee, P.C.; Lee, W.G.; Lee, S.Y.; Chang, H.N.; Chang, Y.K. Fermentative production of succinic acid from glucose and corn steep liquor by *Anaerobiospirillum succiniciproducens*. *Biotechnol. Bioprocess Eng.* **2000**, *5*, 379–381. [CrossRef]
26. Cheilas, T.; Stoupis, T.; Christakopoulos, P.; Katapodis, P.; Mamma, D.; Hatzinikolaou, D.G.; Kekos, D.; Macris, B.J. Hemicellulolytic activity of *Fusarium oxysporum* grown on sugar beet pulp. *Process Biochem.* **2000**, *35*, 557–561. [CrossRef]
27. Demirci, A.; Pometto, A.L.; Lee, B.; Hinz, P.N. Media evaluation of lactic acid repeated-batch fermentation with *Lactobacillus plantarum* and *Lactobacillus casei* subsp. *rhamnosus*. *J. Agric. Food Chem.* **1998**, *46*, 4771–4774. [CrossRef]
28. Hujanen, M.; Linko, Y.Y. Effect of temperature and various nitrogen sources on L (+)-lactic acid production by *Lactobacillus casei*. *Appl. Microbiol. Biotechnol.* **1996**, *45*, 307–313. [CrossRef]
29. Arasaratnam, V.; Senthuran, A.; Balasubramaniam, K. Supplementation of whey with glucose and different nitrogen sources for lactic acid production by *Lactobacillus delbrueckii*. *Enzym. Microb. Technol.* **1996**, *19*, 482–486. [CrossRef]
30. Nancib, N.; Nancib, A.; Boudjelal, A.; Benslimane, D.; Blanchard, F.; Boudrant, J. The effect of supplementation by different nitrogen sources on the production of lactic acid from data juice by *Lactobacillus casei* subsp. *rhamnosus*. *Bioresour. Technol.* **2001**, *78*, 149–153. [CrossRef]
31. Rohrer, C.A.; Siebenmorgen, T.J. Nutraceutical concentrations within the bran of various rice kernel thickness fractions. *Biosyst. Eng.* **2004**, *88*, 453–460. [CrossRef]
32. Stiles, J.; Penkar, S.; Plocková, M.; Chumchalová, J.; Bullerman, L.B. Antifungal activity of sodium acetate and *Lactobacillus rhamnosus*. *J. Food Prot.* **2002**, *65*, 1188–1191. [CrossRef] [PubMed]
33. Nissar, J.; Ahad, T.; Naik, H.R.; Hussain, S.Z. A review phytic acid: As antinutrient or nutraceutical. *J. Pharm. Phytochem.* **2017**, *6*, 1554–1560.
34. Schlemmer, U.; Frølich, W.; Prieto, R.M.; Grases, F. Phytate in foods and significance for humans: Food sources, intake, processing, bioavailability, protective role and analysis. *Mol. Nutr. Food Res.* **2009**, *53*, S330–S375. [CrossRef] [PubMed]

 MDPI

Article

Effect of Fermentation Parameters on Natto and Its Thrombolytic Property

Yun Yang [1,2], Guangqun Lan [1,2], Xueyi Tian [1,2], Laping He [1,2,*], Cuiqin Li [1,3], Xuefeng Zeng [1,2] and Xiao Wang [1,2]

1 Key Laboratory of Agricultural and Animal Products Store & Processing of Guizhou Province, Guizhou University, Guiyang 550025, China; 18984280534@163.com (Y.Y.); langq@ssjf49.com (G.L.); 14785719587@163.com (X.T.); licuiqin2345@163.com (C.L.); heiniuzxf@163.com (X.Z.); wangzi8903@126.com (X.W.)
2 College of Liquor and Food Engineering, Guizhou University, Guiyang 550025, China
3 School of Chemistry and Chemical Engineering, Guizhou University, Guiyang 550025, China
* Correspondence: lphe@gzu.edu.cn; Tel./Fax: +86-0851-88236702

Abstract: Natto is a popular food because it contains a variety of active compounds, including nattokinase. Previously, we discovered that fermenting natto with the combination of *Bacillus subtilis* GUTU09 and *Bifidobacterium animalis subsp. lactis* BZ25 resulted in a dramatically better sensory and functional quality of natto. The current study further explored the effects of different fermentation parameters on the quality of natto fermented with *Bacillus subtilis* GUTU09 and *Bifidobacterium* BZ25, using Plackett–Burman design and response surface methodology. Fermentation temperature, time, and inoculation amount significantly affected the sensory and functional qualities of natto fermented with mixed bacteria. The optimal conditions were obtained as follows: soybean 50 g/250 mL, triangle container, 1% sucrose, *Bacillus subtilis* GUTU09 to *Bifidobacterium* BZ25 ratio of 1:1, inoculation 7%, fermentation temperature 35.5 °C, and fermentation time 24 h. Under these conditions, nattokinase activity, free amino nitrogen content, and sensory score were increased compared to those before optimization. They were 144.83 ± 2.66 FU/g, 7.02 ± 0.69 mg/Kg and 82.43 ± 5.40, respectively. The plate thrombolytic area and nattokinase activity both increased significantly as fermentation time was increased, indicating that the natto exhibited strong thrombolytic action. Hence, mixed-bacteria fermentation improves the taste, flavor, nattokinase activity, and thrombolysis of natto. This research set the groundwork for the ultimate manufacturing of natto with high nattokinase activity and free amino nitrogen content, as well as good sensory and thrombolytic properties.

Keywords: natto; nattokinase; combination fermentation; thrombolytic property

Citation: Yang, Y.; Lan, G.; Tian, X.; He, L.; Li, C.; Zeng, X.; Wang, X. Effect of Fermentation Parameters on Natto and Its Thrombolytic Property. *Foods* **2021**, *10*, 2547. https://doi.org/10.3390/foods10112547

Academic Editors: Valentina Bernini and Juliano De Dea Lindner

Received: 15 September 2021
Accepted: 19 October 2021
Published: 22 October 2021

1. Introduction

People's living standards have gradually risen in recent years, and the incidence of thrombosis in blood arteries has increased year after year [1]. Acute myocardial infarction, hypertension, and stroke, for example, are commonly linked to excessive fibrin deposition in blood arteries [2]. Meanwhile, the prevalence of cardiovascular and cerebrovascular diseases has skyrocketed. As a result, the importance of healthy products and foods in the prevention of cardio-cerebrovascular diseases has grown. In recent years, a natural fermentation enzyme known as nattokinase has been a prominent study issue among those nutritious foods. Natto is a soybean that has a distinct flavor and texture, which is made by fermenting soybeans with *Bacillus subtilis*. It originated in ancient China and evolved into a new form of fermented food after being exported to Japan during the Tang Dynasty. The fermentation of natto produces a number of bioactive components, including nattokinase, daidzein, phytosterols, superoxide dismutase, and several biologically active peptides, in addition to preserving the nutritional value of soybeans [3]. Nattokinase is a fibrinolytic enzyme released by the *Bacillus subtilis* natto bacteria during the fermentation process [4]. Nattokinase has better thrombolytic activity and selectivity when compared to other fibrinolytic enzymes [5]. Furthermore, after oral administration, nattokinase is rapidly absorbed across the gastrointestinal tract [6], causing fibrinolysis. Thus, it is evaluated as a possible clot-dissolving agent for the treatment and prevention of cardiovascular disease [7].

Natto has limited public awareness due to its distinct flavor produced by *Bacillus subtilis*. The dispute continues to focus on natto's pungent ammonia smell and flavor, as well as a lack of convincing proof that it reduces the risk of thrombotic disorders. Overall, the higher the nattokinase activity, free amino nitrogen content and sensory score, the better the natto's health advantages and acceptance. As a result, increasing the content of multiple biological activities, and improving the sensory score and flavor of natto for the general population are demanded. *Bifidobacterium* may produce lactic acid and acetic acid, which can cover the ammonia and its distinct taste, as well as other functional compounds; thus, adding additional *Bifidobacterium* to the natto may improve the flavor and quality.

We previously isolated nattokinase-producing *Bacillus subtilis* GUTU09 (CCTCC M 2021641) which may be used to produce natto and *Bifidobacterium* BZ25 (CGMCC NO.10225) [8]. However, the two bacteria strains' fermentation parameters for producing natto were not investigated. As a result, the goal of the current study was to apply Plackett–Burman design (PB) in conjunction with Box–Behnken design (BBD) to investigate the effect of fermentation parameters to produce high-quality natto. In addition, the natto's thrombolytic properties were examined.

2. Materials and Methods

2.1. Materials

The *Bacillus subtilis* GUTU09 (B9) and *Bifidobacterium animalis subsp. lactis* BZ25 were strains with excellent performance screened by our laboratory from Guizhou local characteristic food [9,10]. Thrombin (1000U) and bovine fibrinogen were purchased from Solarbio Science and Technology Co. (Beijing, China). Other culture media and analytical chemicals were purchased from Sinopharm Chemical Reagent Co., Ltd. (Shanghai, China).

2.2. Seed Preparation

BZ25 was inoculated in MRS medium (natural pH) and anaerobically cultured at 37 °C for 48 h [11]. B9 was cultivated in the liquid seed medium containing 10 g/L glucose, 5 g/L yeast extract, 10 g/L beef extract, and 5 g/L NaCl with pH of 7.0–7.5 at 37 °C and 180 r/min for 18 h. After that, the cells were extracted by centrifugation at $4000\times g$ (TGL20M High-speed freezing centrifuge, Maijiassen instrument, and equipment Co., Ltd., Changsha, China) for 10 min and resuspended in a small amount of sterile physiological saline. The concentrations of BZ25 and B9 were then determined by anaerobic or aerobic culture in either MRS agar plates (MRS agar medium was prepared by adding 20 g of agar per liter MRS broth medium) or B9 liquid medium (B9 agar medium was prepared by adding 20 g of agar per liter B9 liquid medium) at 37 °C for 48 h. Subsequently, its cell concentrations were diluted to 1×10^8 CFU/mL with physiological saline [12].

2.3. Natto Preparation

Soybeans from Heilongjiang, China, were soaked overnight in 20 °C water. Wet beans (50 g) were sterilized at 121 °C for 20 min, then cooled to 25 °C in a 250 mL conical flask. Soybeans were inoculated by 4% of BZ25(1×10^8 CFU/mL) and 4% of B9 (1×10^8 CFU/mL) and fermented under static conditions (250 mL flask is sealed with eight layers of gauze) for 24 h at 37 °C. The fermented soybeans were then ripened for 24 h at 4 °C before being used to make natto.

2.4. One-Factor-at-a-Time Experiments

Using pH (initial pH was 7.5), nattokinase activity, free amino nitrogen, and sensory scores as indicators, the effects of NaCl, sucrose, the ratio of B9 to BZ25 strains, inoculation amount, fermentation temperature, fermentation time, and after-ripening time on natto were investigated by one-factor-at-a-time experiments.

2.5. PB Design

NaCl (A), sucrose (B), fermentation temperature (D), fermentation time (E), strain ratio (G), inoculation amount (J), and after-ripening time (L) were chosen for PB design based on the findings of the single-factor experiment. Seven independent and four dummy variables (C, F, H, and K) were examined at two levels in total (Table 1).

Table 1. Minimum and maximum range of the parameters selected in the PB design.

Variables	Units	Levels	
		1	**−1**
A—NaCl content	%	0	1
B—Sucrose addition	%	1	3
D—Fermentation temperature	°C	37	34
E—Fermentation time	h	30	24
G—strain proportion		2:1	1:1
J—inoculation volume	%	8	6
L—after-ripening time	d	3	1
C, F, H, K		1	−1

2.6. BBD

The results of the PB design demonstrate that the most critical parameters influencing soybean fermentation were fermentation time, fermentation temperature, and inoculation amount. We applied BBD and used nattokinase activity, free amino nitrogen, and sensory scores as response values. Each of the parameters is listed in Table 2.

Table 2. Coded and real values of variables in the BBD.

Variables	Units	Levels		
		−1	**0**	**1**
D—Fermentation temperature	°C	31	34	37
E—Fermentation time	h	24	30	36
J—Inoculation volume	%	4	6	8

2.7. pH Measurement and Extraction of Crude Enzyme Solution

Natto (10 g) was dissolved in 90 mL deionized water, homogenized for 30 s, and extracted for 24 h at 4 °C. A pH meter was used to determine the pH of the suspension. The supernatant was kept at 4 °C after centrifugation at 16,000× g (TGL20M High-speed freezing centrifuge, Maijiassen instrument, and equipment Co., Ltd., Changsha, China) for 10 min for nattokinase activity determination.

2.8. Free Amino Nitrogen Contents (FANs)

The natto suspension was centrifuged at 1800× g for 10 min after being extracted at 4 °C for 24 h, and the supernatant was kept at 4 °C for later use. The method described by Lu et al. [13] was used to determine the free amino nitrogen.

2.9. Nattokinase (NK) Activity Assay

The fibrinolytic activity of nattokinase was measured using the V. Deepak et al. [14] technique.

2.10. Sensory Properties

The sensory properties of natto were evaluated by the method of Feng et al. [15] with some modifications. A sensory evaluation team of 15 teachers and graduate students with sensory evaluation experience in food-related courses was assembled. The sensory evaluation of natto was carried out in a well-ventilated food lab with plenty of light and room. Appearance, viscosity, flavor, mouthfeel, and chewiness are the key determinants of sensory properties. The appearance of the product was evaluated by color intensity, color consistency, brightness, and gloss. The length, density, and adhesion of natto to the chopsticks were used to determine stickiness. The flavor of natto was evaluated by fragrance, ammonia, and beany flavor. It was expected that the flavor would be mild or somewhat sour. The overpowering bitterness was unappealing. Teeth's feedback to the softness, hardness, stickiness, and smoothness of natto comprised chewiness. Rating values of 1 to 5 were utilized for independent evaluation of these five sensory properties (5 = like very much, 4 = like more, 3 = average, 2 = dislike very much, 1 = dislike very much). Finally, each sensory rating was multiplied by a factor of four to provide an index score. The greater the index score, the higher the natto's quality [16].

2.11. Thrombolytic Effect

The area and diameter of the dissolution circle in the fibrin plate were used to evaluate the thrombolytic action of crude enzyme solution using the agarose fibrinogen plate method as described by Gao [10].

2.12. Anticoagulant Activity

The anticoagulant activity was determined by measuring the inhibition of fibrinogen to fibrin conversion, using Wei's [17] approach with minor changes. The details are as follows: First, 0.3 mL of fibrinogen solution (7.2 mg/mL) was combined with 1.2 mL of Tris–HCl (50 mM, pH 7.8) buffer in a colorimetric cuvette with 1.0 mL of the diluted sample solution. After that, 0.1 mL of thrombin solution (20 U/mL) was added, the mixture was incubated at 37 °C for a while, and the absorbance was measured at 405 nm (UV-Vis Spectrophotometer TU-1810PC Pullout General Instrument Co., Ltd., Beijing, China.). As a control, no sample solution was used. The anticoagulant activity of natto is estimated by formula (1) below:

$$\text{Activity of coagulation inhibitory \%} = \left(1 - OD_{\text{sample}}\right)/OD_{\text{Control}} \times 100\% \quad (1)$$

2.13. Statistical Analysis

All of the tests were performed in triplicate. The data are presented as mean ± standard deviation (SD). One-way ANOVA was performed using SPSS 19.0 (SPSS Inc., Chicago, IL, USA), and differences were determined using Tukey's HSD test, with p values < 0.05 considered statistically significant. Design-Expert 8.0.6 software (Stat-Ease, Inc., Minneapolis, MN, USA) was used for the PB test and BBD and for their data analysis, and Origin 2018 software (OriginLab, Northampton, MA, USA) was used for drawing the graph.

3. Results

3.1. Effect of NaCl on Natto

Figure 1(a_1,b_1,c_1) show the effect of NaCl on natto. The sensory qualities of natto were significantly affected by different NaCl concentrations. The highest sensory score was earned by 1% NaCl. The sensory scores, on the other hand, showed no significant difference ($p > 0.05$) between 0% and 1%. However, as the salt concentration increased above 1%, the sensory score rapidly decreased, and the difference was extremely significant ($p < 0.01$). Figure 1(b_1,c_1) demonstrate that as NaCl levels rise, pH, nattokinase activity, and free amino nitrogen levels fall, with the highest indices appearing in the control group (NaCl equal to 0).

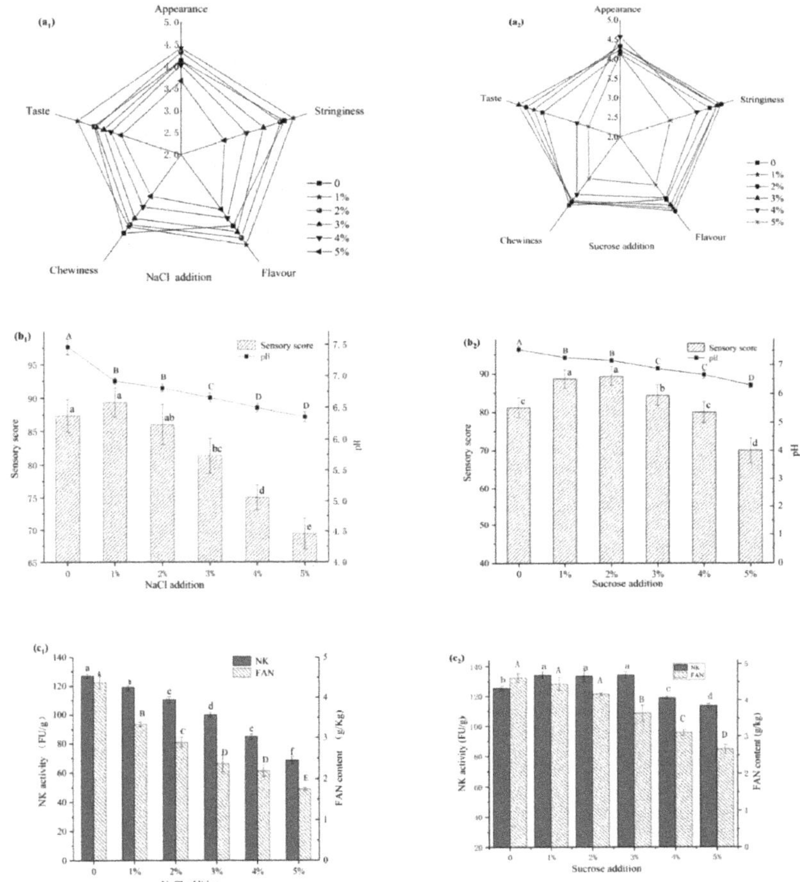

Figure 1. Effects of NaCl addition (**a**1,**b**1,**c**1) and sucrose addition (**a**2,**b**2,**c**2) on natto fermentation. Different letters above each bar means significant differences.

3.2. Effect of Sucrose on Natto

Figure 1(a2,b2,c2) reveals that the amount of sucrose has a highly significant ($p < 0.01$ obtained by one-way ANOVA analysis) impact on the sensory score of natto. The sensory score was highest when the sugar concentration was 1–2%. When the sugar level in the flask fermentation environment reached over 2%, the higher sugar content made the mixed-bacteria fermentation produce more acid. This is mainly due to the fact that O_2 would become almost depleted in the solid fermentation of soybean, and this favored the growth of the *Bifidobacterium* and inhibited the strictly aerobic *Bacillus* strain. The extra acid made the natto taste and flavor worse, as well as making the bean difficult to chew and increasing the stringiness viscosity. When the sugar content approached 3%, the sensory score, nattokinase activity, and free amino nitrogen levels all decreased significantly ($p < 0.05$).

3.3. Effect of Strain Proportion on Natto

Figure 2(a1,b1,c1) depict the influence of strain ratio on natto fermentation. The ratio of strains did not affect free amino nitrogen or sensorial properties ($p > 0.05$). The pH, on the other hand, rose initially and subsequently fell as the strain ratio rose. When the strain ratios were 2:1, 1:1, and 1:2, there was no significant change in pH ($p > 0.05$). Furthermore, when the ratio increased, nattokinase activity reduced progressively but not significantly.

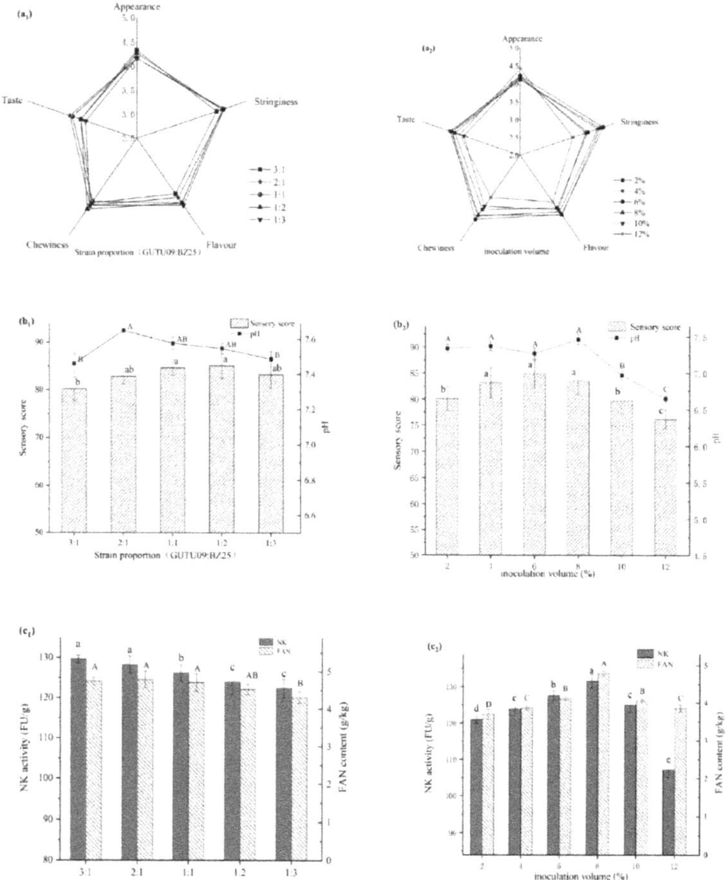

Figure 2. Effects of strain proportion (**a$_1$**,**b$_1$**,**c$_1$**) and inoculation volume (**a$_2$**,**b$_2$**,**c$_2$**) on natto fermentation. Different letters above each bar means significant differences.

3.4. *Effect of Inoculation Amount on Natto*

Figure 2(a$_2$,b$_2$,c$_2$) show the effect of inoculation amount on fermented natto. The activity of nattokinase and free amino nitrogen concentration increased and then reduced when the inoculation amount was increased, which was consistent with earlier studies [18]. The highest levels of nattokinase activity, sensory score, and free amino nitrogen were attained when the inoculation amount reached 6–8%. When the inoculation concentration was less than 6% or higher than 8%, nattokinase activity, free amino nitrogen, and sensory score decreased.

3.5. *Effect of Fermentation Temperature on Natto*

As demonstrated in Figure 3(a$_1$,b$_1$,c$_1$), the nattokinase activity was highest at 40 °C; however, the sensory score was not the best at this temperature. In terms of flavor, texture, stringiness, and appearance, the temperature has a bigger impact on these sensory characteristics. When the temperature was between 31–34 °C, the bacteria grew slower and had an impact on various biochemical reactions throughout the fermentation. When the temperature was higher than 37 °C, the sensory score gradually decreased, and the activity of nattokinase first increased and eventually decreased. In addition, as the temperature rose, the pH increased as well (Figure 3(b_1)), and the amount of free amino nitrogen also increased.

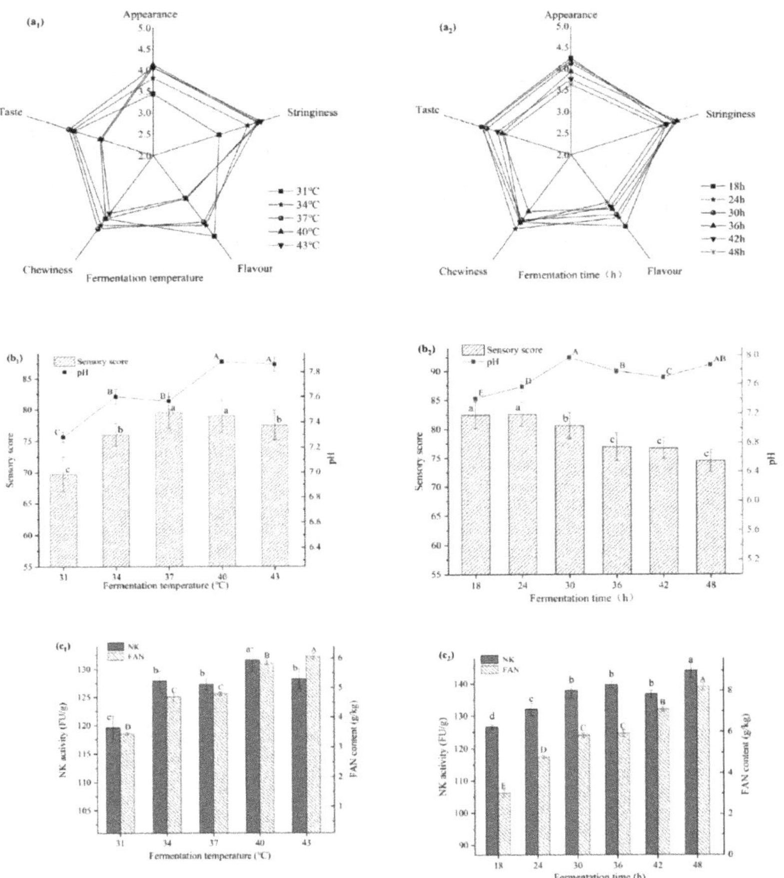

Figure 3. Effects of fermentation temperature (a_1–c_1) and fermentation time (a_2–c_2) on natto fermentation. Different letters above each bar means significant differences.

3.6. *Effect of Fermentation Time on Natto*

Figure 3(a_2,b_2,c_2) show the effect of fermentation time on fermented natto. Natto hade a high sensory score when fermented for 18 or 24 h. The color of the natto darkened as the fermentation time progressed, and the ammonia smell became stronger. Excessive fermentation softened the bean grains, lowering the sensory score. However, the activity of nattokinase and the quantity of free amino nitrogen increased as the fermentation duration increased.

3.7. *Effect of After-Ripening Time on Natto*

Figure 4a–c depicts the influence of after-ripening time on natto. Natto scored a higher sensory score of 2–3 days after ripening time. However, extending the after-ripening time reduces the product quality and freshness, lowering the sensory score. The pH changes slightly during the after-ripening process, suggesting the dynamics of the composition of natto. The free amino nitrogen steadily rose as the after-ripening period increased, demonstrating that the protease was still hydrolyzing soybean protein even at low temperatures. Despite the fact that the overall trend in nattokinase was increasing, it seemed that there was a downward trend in the end, which is consistent with prior observations [19].

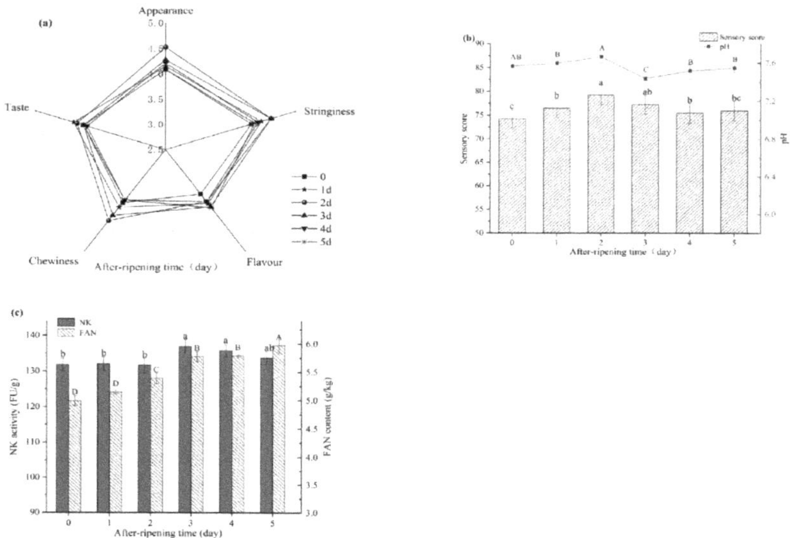

Figure 4. Effects of after-ripening time on natto (**a**–**c**). Different letters above each bar means significant differences.

3.8. Plackett–Burman (PB) Design

According to the analysis of the single factor test, the Design-Expert software was used to carry out the PB design (Table 1, Supplementary Materials Table S1), and the key factors that significantly affected the fermentation of natto were screened out.

The analysis of variance response values with nattokinase activity, free amino nitrogen content, and sensory scores are shown in Table 3. The *p*-values of the three ANOVA models are all less than 0.05, indicating that the three ANOVA models are significant and have a good degree of fit. We discovered that NaCl, sucrose, fermentation temperature, and fermentation duration all had significant ($p < 0.05$) influences on nattokinase activity. The following is the sequence in which factors impact nattokinase: NaCl > sucrose > fermentation temperature > fermentation time. The amount of NaCl and sucrose, fermentation temperature, and fermentation time had an extremely significant ($p < 0.01$) effect on the content of free amino nitrogen. The following is the sequence in which factors impact the production of free amino nitrogen: NaCl > sucrose > fermentation time > fermentation temperature. Fermentation temperature, fermentation period, and inoculum amount all had an impact on the sensory score ($p < 0.05$). The following is the sequence in which factors affected sensory score: inoculum amount > fermentation time > fermentation temperature.

Table 3. Experimental analysis of variance-response values was based on nattokinase activity, amino acid nitrogen, and sensory score.

Source	Nattokinase Activity						Amino Acid Nitrogen						Sensory Scores					
	Sum of Squares	df	Meansquare	F-Value	p	Significance	Sum of Squares	df	Meansquare	F-Value	p	Significance	Sum of Squares	df	Meansquare	F-Value	p	Significance
Model	492.89	7	70.41	8.92	0.0258	*	29.87	4	7.47	31.9	0.0001	**	280.06	3	93.35	13.10	0.0019	**
A—NaCl content	118.94	1	118.9	15.07	0.0178	*	17.74	1	17.74	75.77	<0.0001	**						
B—Sucrose addition	131.87	1	131.9	16.71	0.015	*	5.45	1	5.45	23.30	0.0019	**						
D—Fermentation temperature	88.35	1	88.35	11.19	0.0287	*	3.19	1	3.19	13.64	0.0077	**	55.04	1	55.04	7.72	0.024	*
E—Fermentation time	62.38	1	62.38	7.9	0.0482	*	3.49	1	3.49	14.9	0.0062	**	101.5	1	101.5	14.24	0.0054	**
G—Strain proportion	28.34	1	28.34	3.59	0.131													
J—Inoculation volume	9.08	1	9.08	1.15	0.3438								123.52	1	123.52	17.33	0.0032	**
L—After-ripening time	53.93	1	53.93	6.83	0.0592													
Residual	31.57	4	7.89	8.92	0.0258		1.64	7	0.23				57.02	8	7.13			
sum	524.46	11		15.07			31.51	11					337.08	11				

Note: ** indicates $p < 0.01$, * indicates $p < 0.05$. Statistical analysis was performed using test design experts.

In order to obtain products with good sensory characteristics, as a result, three factors were chosen for the next response surface analysis: fermentation time (E), fermentation temperature (D), and inoculation amount (J).

3.9. Response Surface Analysis by Box–Behnken Design (BBD)

The result of the BBD is shown in Table S2 (Supplementary Materials). All data are fitted to a second-order polynomial model (Y_1, Y_2, Y_3). The significance of these models is expressed as a low *p* value (Supplementary Materials Table S3, nattokinase activity $p = 0.0174$; free amino nitrogen content $p = 0.0041$; sensory score $p = 0.0004$), and the lack of fit of these models was not significant, suggesting the models as being credible. When the R^2 value is greater than 0.8, the response surface model is generally considered appropriate [20]. The higher R^2 ($R^2 > 0.8$) of the three responses indicates a high correlation between the predicted value and the actual value.

$$Y_1 = 132.58 + 5.08 \times A - 1.16 \times B - 3.04 \times C + 1.79 \times AB - 1.29 \times AC + 0.78 \times BC + 7.27 \times A^2 - 3.28 \times B^2 + 2.7 \times C^2 \quad (2)$$

$$Y_2 = 5.51 + 1.71 \times A + 0.65 \times B - 0.28 \times C + 0.89 \times AB + 0.12 \times AC - 0.24 \times BC + 0.94 \times A^2 - 0.73 \times B^2 + 0.59 \times C^2 \quad (3)$$

$$Y_3 = 85.28 - 3.15 \times A - 2.18 \times B + 1.83 \times C - 3.90 \times AB - 1.30 \times AC + 2.75 \times BC - 8.76 \times A^2 + 0.085 \times B^2 - 6.22 \times C^2 \quad (4)$$

According to the model equation Y_1, the linear term A, cross-product terms AB and BC, and quadratic terms A^2–C^2 had a positive effect on the nattokinase activity, while the linear terms B and C, cross-product term AC, and quadratic term B^2 showed a negative effect. Table S3 (Supplementary Materials) shows that only the linear terms of fermentation time (A), inoculation volume (C), and quadratic terms of A^2 ($p < 0.05$) were substantially linked with NK production.

The 3D surfaces of the interactive effects of fermentation time (A), fermentation temperature (B), and inoculation volume (C) on the NK are illustrated in Figure 5. Figure 5(A_1) shows with the increase in temperature and time, the activity of nattokinase decreased at first and then increased. As shown in Figure 5(B_1), with the extension of the fermentation time, the activity of nattokinase showed an increasing trend. This is because the longer the fermentation time, the more time the bacteria will have to grow and ferment, resulting in more nattokinase being produced by the bacteria. The combined effect of inoculum amount and fermentation temperature revealed that nattokinase activity was affected by fermentation temperatures between 31 and 37 °C, but the effect was minor (Figure 5(C_1)).

Similar to nattokinase activity, Table S3 (Supplementary Materials) shows that the linear terms A-B, cross-product terms AB and AC, and quadratic terms A^2-C^2 had a significantly ($p < 0.05$) positive influence on the free amino nitrogen (FAN)content, while the linear terms C, cross-product terms BC and quadratic terms B^2 presented a significantly ($p < 0.05$) negative effect (equation Y_2).

The level of FAN increased as the temperature and fermentation time increased (Figure 5(A_2)). The fermentation period influences the accumulation of products. The free amino nitrogen concentration first declined and subsequently increased as the amount of inoculation and fermentation time increased (Figure 5(B_2)).

Table S3 (Supplementary Materials) shows that the linear term C, cross-product term BC, and quadratic term B^2 had a significantly ($p < 0.05$) positive influence on the sensory score, while the linear terms A-B, cross-product terms AB and AC, and quadratic terms A^2-C^2 presented a significantly ($p < 0.05$) negative effect (equation Y_3). The sensory score of natto climbed as fermentation time, fermentation temperature, and inoculum amount increased, reaching a maximum. Then, it began to decline as these three elements continued to rise (Figure 6).

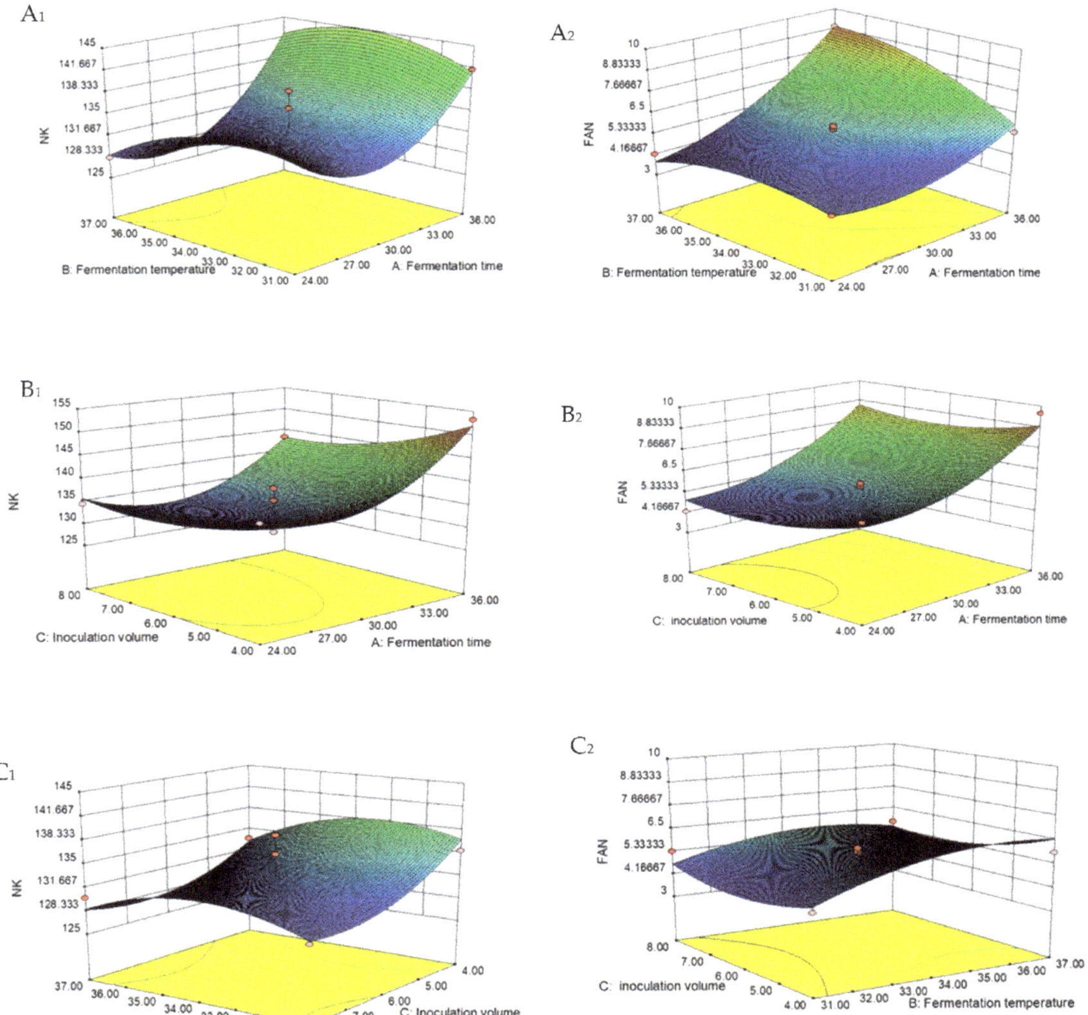

Figure 5. Response surface diagram of NK activity and FAN content as a response to the interaction of: (A_1,A_2) fermentation temperature and fermentation time (inoculation amount = 6%), (B_1,B_2) inoculation amount and fermentation time (temperature = 34 °C), (C_1,C_2) inoculation amount and fermentation temperature (time = 30 h).

A
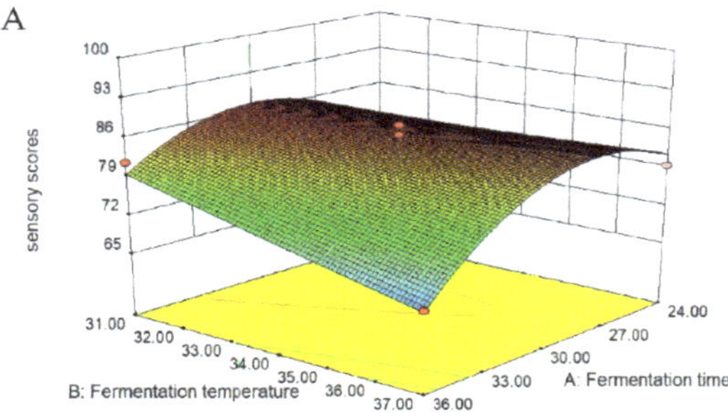

B
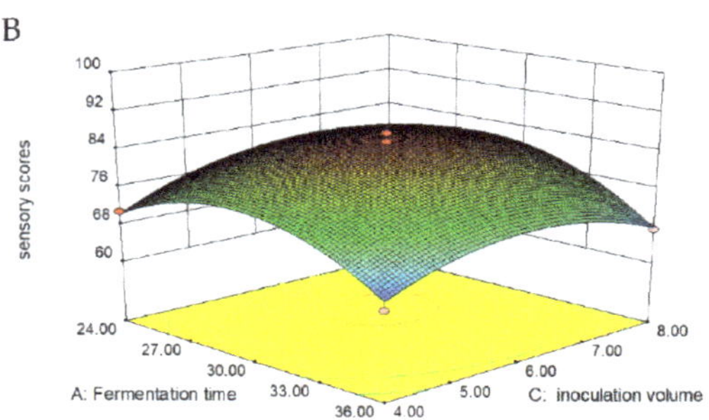

C
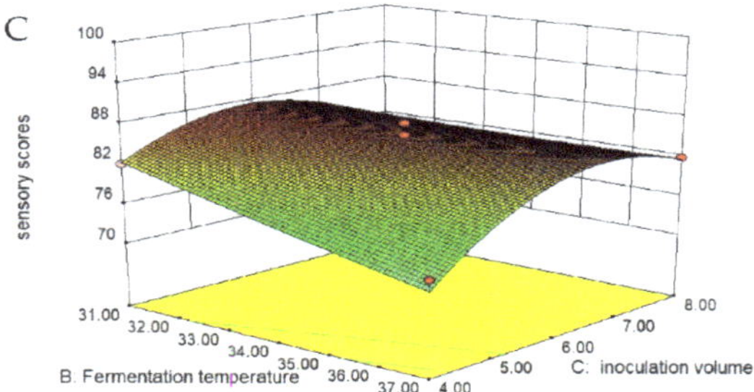

Figure 6. Response surface diagram of sensory score as a response to the interaction of: (**A**) fermentation temperature and fermentation time (inoculation amount = 6%), (**B**) inoculation amount and fermentation time (temperature = 34 °C), (**C**) inoculation amount and fermentation temperature (time = 30 h).

3.10. Determination and Verification of Natto Process

After response surface optimization, according to the second-order model formula, the theoretical optimal conditions were: sucrose addition was 1%, inoculation amount was 6.96%, fermentation temperature was 35.44 °C, and fermentation time was 24 h. Under these conditions, the predicted value of nattokinase activity was 132.56 FU/g, the sensory score predictive value was 81.21, and the free amino nitrogen content was 4.34 mg/kg. The preceding parameters were slightly modified as follows: sucrose at 1%, inoculation amount at 7%, fermentation temperature at 35.5 °C, and fermentation period at 24 h, taking into account the feasibility of the actual operation.

In order to verify the effectiveness and feasibility of the model, three replicate tests were performed under the optimal culture conditions. The results show that the activity of nattokinase was 144.83 ± 2.66 FU/g, the sensory score was 82.43 ± 5.40, and the free amino nitrogen content was 7.02 ± 0.69 mg/Kg. All of the above values were in line with the expected outcomes. As a result, we came to the conclusion that the model could accurately predict the real fermentation data for natto.

3.11. Thrombolytic Activity, Fibrin Degradation Ability, and Anticoagulant Activity

The thrombolytic activities of natto were investigated utilizing the agarose–fibrinogen plate method under the abovementioned ideal circumstances. The fibrinolysis action of natto extract is shown in Figure 7A. The diameter of the dissolution circle given in Table 4 was measured under aseptic conditions. The area of the dissolution circle was calculated according to the diameter (π = 3.14). The area of thrombolysis is the area of the dissolution circle minus the area of the central pore (1.98 ± 0.39 mm ^{2}).

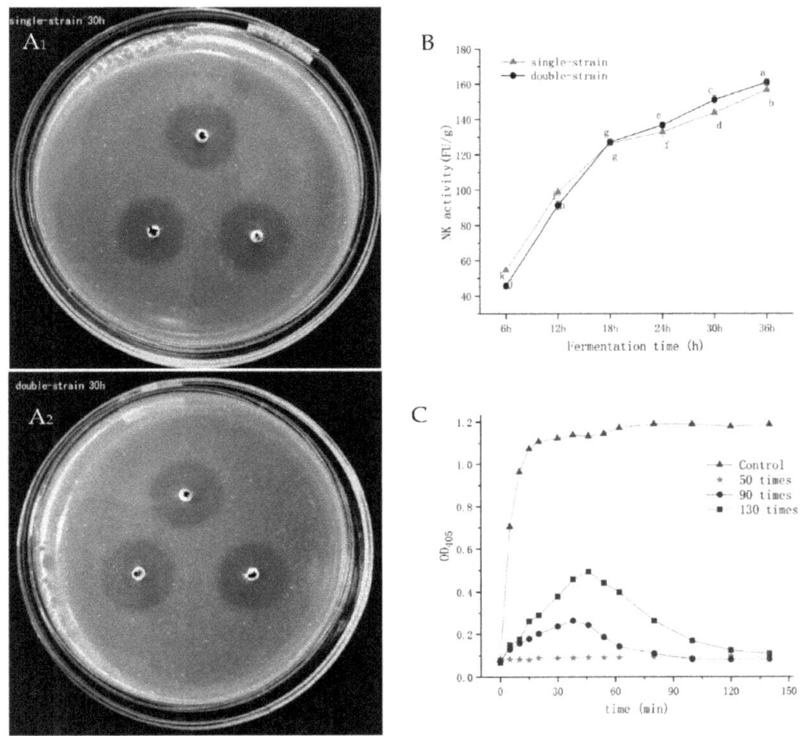

Figure 7. (A_1,A_2) Comparison of the effect of single-bacteria and mixed-bacteria fermented natto on fibrin plate dissolution circle; (**B**) changes in nattokinase activity during fermentation; (**C**) prothrombin condensation inhibition ability.

Table 4. Changes in fibrinolytic activity during fermentation.

Center Hole Area 1.98 ± 0.39 mm^2	GUTU09 Single-Strain		Double-Strain Fermentation	
	Dissolving Circle Diameter (mm)	Thrombolytic Area (mm^2)	Dissolving Circle Diameter (mm)	Thrombolytic Area (mm^2)
6 h	5.46 ± 0.45 ^{g}	21.53 ± 3.4 ^{G}	4.77 ± 0.69 ^{g}	16.16 ± 4.75 ^{G}
12 h	14.17 ± 0.31 ^{f}	155.76 ± 6.55 ^{F}	13.75 ± 0.51 ^{f}	146.64 ± 10.66 ^{F}
18 h	16.13 ± 0.48 ^{e}	202.46 ± 11.86 ^{E}	16.12 ± 0.45 ^{e}	202.11 ± 11.02 ^{E}
24 h	17.1 ± 0.20 ^{d}	232.64 ± 5.08 ^{D}	17.77 ± 0.2 ^{c}	245.83 ± 5.18 ^{C}
30 h	18.28 ± 0.38 ^{b}	272.8 ± 10.66 ^{B}	19.70 ± 0.35 ^{a}	302.73 ± 10.37 ^{A}

Note: Data are mean values of three independent experiments ± standard deviation. Mean values displaying different letters within each row are significantly different according to the Duncan test at 95% confidence level.

As shown in Table 4, as the fermentation time rose, the corresponding diameter of the dissolving circle increased in proportion to the thrombolytic area, and the thrombolysis became more obvious.

Figure 7B depicts the variations in nattokinase activity throughout single-strain and mixed-strain fermentation. The activity of nattokinase gradually increased as the fermentation period was extended. The maximum activity of nattokinase was found in a 36 h fermentation of mixed bacteria, with an enzyme activity of 161 ± 1.96 FU/g recorded. When comparing the two fermentation methods, single-bacteria fermentation had higher natto enzyme activity before 18 h, and equal the enzyme activity at 18 h. After 18 h, the mixed-bacteria fermentation had higher enzyme activity than single-bacteria, which was similar to the fibrinolytic activity indicated before. Finally, the mixed-bacteria fermentation method produced higher nattokinase activity, more free amino nitrogen, and a superior sensory evaluation. As a result, it is a nutritious food that is readily available to the general public.

The antithrombotic process was recorded when the crude enzyme extract of natto (151.06 ± 2.34 FU/g) was diluted 50 times, 90 times, and 130 times with no added sample as a control (Figure 7C). The control group reached the maximum fibrin concentration (OD_{405} = 1.107) at 20 min. When the dilution ratio was 130 times, the formation of fibrin was inhibited and the OD_{405} was 0.494 at 46 min, and fibrin was completely degraded within 100 min. In contrast, the formation of fibrin in a 90-fold diluted sample solution was inhibited at 38 min, and fibrin was almost completely degraded within 60 min. The inhibitory effect, on the other hand, was better and the time was shorter. When the dilution ratio was 50 times, there was no significant change in the curve, indicating that it had strong anticoagulant activity.

4. Discussion

The formation of a blood clot in a blood vessel is one of the main causes of cardiovascular diseases [5]. Blood clots are formed from fibrinogen via proteolysis by thrombin, and can be hydrolyzed by plasmin to avoid thrombosis in blood vessels [21]. Natural products with anticoagulant activity can hinder the conversion of fibrinogen into fibrin and play a role in the final stage of the anticoagulation process [22]. Natto offers several health benefits, with nattokinase being the most essential functional component. Nattokinase, a serine protease secreted by *Nattobacillus*, has high fibrinolytic activity and has been proved to be an effective thrombolytic in vitro and in vivo [23]. The health benefits of natto have always attracted people's interest [24], and the improvement in its quality has always been a goal that people pursue, including improving the flavor and taste of natto, and increasing the content of functional ingredients. The combination of *Bacillus subtilis* GUTU09 and *Bifidobacterium* BZ25 (mixed bacteria) was utilized to ferment natto in this study. Sensory score, pH, nattokinase, and free amino nitrogen concentration were used as indexes to investigate the impact of various variables on the sensory and functional quality of fermented natto.

The indexes of natto decreased with the addition of NaCl (Figure 1(b_1,c_1)). It may be due to the inhibition of some enzyme activities of the two strains under high- and medium-salt conditions. Yue [25] et al. found that histidine decarboxylase was completely inactivated in high- and medium-salt conditions, but remained active when the salt concentration decreased. Meanwhile, because of health concerns associated with high salt intakes, the demand for low-salt foods has surged. High salt intake is linked to diseases such as hypertension and renal impairment [26]. As a result, excessive salt should not be used in the fermentation process. Hypertension can be prevented and controlled by eating low-salt foods [27]. It is very beneficial to people with coronary heart disease and hypertension, and it meets the nutritional needs of modern society.

The effect of raising the sucrose content differs from that of raising the salt concentration (Figure 1). Adding about 1–2% sugar will improve the sensory characteristics. The stringiness, flavor, taste, and chewiness of natto are all affected by the sugar concentration. Despite the presence of oligosaccharides and polysaccharides in the steamed soybean, it lacks a carbon source that BZ25 can utilize directly. Increasing the carbon source sugar that the microbes can consume directly results in accelerated growth and metabolism. Our findings are in line with a previous study by Wu et al. [28]. On the one hand, more sucrose provides a better environment for BZ25 to grow, and acid production as a by-product of the microbe increases rapidly, lowering the pH throughout the fermentation process (Figure 1(b_2)) and improving the texture and aroma of natto. Adding sucrose for B9, on the other hand, causes rapid fermentation and the production of numerous enzymes to hydrolyze the macromolecular components in soybeans. This not only promotes BZ25 proliferation but also accumulates a significant number of fermentation products.

In the same fermentation environment, a high strain ratio not only prevents the synergy between the strains from working, but also lowers the quality of the fermented product due to nutrient competition between the two strains [29]. A high strain ratio of BZ25 to GUTU09 also produced much acid (Figure 2(b_1)), which led to the decline in sensory quality. The reason may be that the internal oxygen content in the solid fermentation was generally low, and increasing the ratio of *Bifidobacterium* BZ25 to *Bacillus subtilis* GUTU09 can enhance the resistance of *Bifidobacterium* to a small amount of oxygen and make it easy to propagate rapidly and produce more acid in the fermentation. Therefore, it is not beneficial to natto fermentation if the ratio of strains GUTU09 to BZ25 is too large or too small. As a result, the suitable strain ratio was approximately 2:1 or 1:1 (Figure 2(a_1,b_1,c_1)).

The number of initial bacteria in the fermentation process is determined by the amount of inoculation, and the appropriate starting bacteria concentration can shorten the strain's growth and proliferation period, allowing it to enter the fermentation stage sooner. An inoculation quantity ranging from 6 to 8% (Figure 2(a_2,b_2,c_2)) is more appropriate; too high or too low inoculation may impair the natto quality. If the initial bacteria concentration is lower, the time needed for reproduction will be longer and the fermentation speed will be slower, which is not conducive to the accumulation of fermentation products. Higher inoculation concentrations, on the other hand, resulted in a strong and speedy bacterial metabolism, requiring the cell to consume the majority of the nutrients in order to continue its rapid growth and hinder the accumulation of fermentation products. When a large amount of metabolic waste is produced, the bacteria's cell senescence may be accelerated, and its functionality may be lowered [30]. Ultimately, the quality of natto could be compromised.

Fermentation temperature is one of the important parameters affecting fermentation. The suitable growth temperature for *Bacillus subtilis* and *Bifidobacterium* is around 37 °C. The activity of nattokinase and the sensory score decreased as the fermentation temperature of natto increased (Figure 3(a_1,b_1,c_1)). The increasing temperature, on the one hand, causes the water in the soybeans to evaporate more quickly. The surface layer of the beans becomes dry, affecting mucus formation and, as a result, the taste. On the other hand, we used a 250 mL flask, which may have led to fermentation, as the heat could not be properly dissipated over time, and an accumulation of heat formed in the core. The high temperature impacts strain development and enzyme reactions, causing uneven natto fermentation and hence affecting nattokinase synthesis [31]. pH increased with the increasing temperature (Figure 3(b_1)). The amine bases produced by the breakdown of soybeans and the organic acids produced by BZ25 are the primary sources of the pH changes. The pH of mixed-bacteria fermented natto was practically the same as that of single-bacteria fermented soybean when the temperature was greater than 37 °C, indicating that the effect of BZ25 is not noticeable after 37 °C. It is possible that the high temperature prevents BZ25 from growing. BZ25's ability to generate acid may be hampered by the higher temperature. The alterations in free amino nitrogen and nattokinase were varied at different temperatures, as shown in Figure 3(c_1). The generation of nattokinase was inhibited by high temperatures, while free amino nitrogen levels increased dramatically. The breakdown of soybean protein by excess protease was the predominant source of free amino nitrogen. As a result, high temperatures suppress nattokinase expression while having little to no influence on protease activity.

One of the determining factors of product quality was fermentation time. As the fermentation duration increased (Figure 3(a_2,b_2,c_3)), sensory score decreased, and nattokinase activity and free amino nitrogen increased. It could be because the bacteria in the fermentation reached a steady state of productivity. The catalytic biological reaction did not cease and the fermentation products continued to increase. Nevertheless, longer fermentation times will cause bacteria to enter senescence, resulting in the accumulation of hazardous compounds [32]. As a result, choosing the right fermentation period can help reduce nutrient waste and the formation of toxic metabolites.

After-ripening time can help to minimize natto's unpleasant smell and make it taste richer and fuller. A longer after ripening time will degrade the product's quality and freshness, resulting in a lower sensory score (Figure 4). As a result, selecting the most appropriate after-ripening time is critical.

Through the PB test (Supplementary Materials Table S1 and Table 3) and response surface test (Supplementary materials Tables S2 and S3), it was determined that fermentation time, fermentation temperature, and inoculation amount were significant factors affecting natto fermentation. PB test and response surface test can effectively be used to estimate the effect of fermentation time, fermentation temperature, and inoculation amount, and their interactions on the natto. According to the findings of the study, the optimal fermentation conditions for achieving the finest natto quality were as follows: soybean 50 g/bottle, NaCl was 0%, sucrose 1%, ratio of GUTU09 to BZ25 1:1, inoculum 7%, fermentation temperature 35.5 °C, and fermentation time 24 h. The nattokinase activity in the above conditions was 144.83 ± 2.66 FU/g, which was higher than the nattokinase activity in the single-bacteria fermentation procedure. The level of free amino nitrogen was 7.02 ± 0.69 mg/Kg. Furthermore, these optimal conditions resulted in an overwhelming sensory score, with the mixed bacteria fermentation natto scoring much higher on appearance, stringiness, flavor, taste, and chewiness than the non-optimized natto.

Under the optimal circumstances outlined above, the thrombolytic action of natto was investigated (Table 4, Figure 7(A_1,A_2)). It indicates that as fermentation progressed, strains produced more nattokinase and its thrombolytic activity was improved. The antithrombotic effect includes two aspects: one is the fibrinolysis of the formed thrombus, which is evaluated by the fibrin degradation ability (nattokinase activity); the other is the anticoagulant activity during the formation of fibrin, that is, the inhibition ability of fibrin on the original coagulation [17]. The fibrinolytic activity and anticoagulant activity may reflect the thrombolytic ability to some extent [33].

Single-bacteria fermentation had higher enzyme activity for the first 18 h, but after that, mixed-bacteria fermentation had higher enzyme activity than single-bacteria fermentation (Figure 7B). When both *Bifidobacterium* BZ25 and *Bacillus subtilis* GUTU09 are inoculated at the same time, BZ25 growth is impeded due to the presence of oxygen in the soybean medium at the start of fermentation [34]. According to Hosoi et al. [35], the viability of Lactobacilli in the presence of Bacillus was greatly improved, and they speculated that the production of subtilisin and catalase may play a part in this improvement. Another explanation for the improvement was that the growth of *B. subtilis* GUTU09 in solid-state substrates used dissolved oxygen, allowing *Bifidobacterium* BZ25 to flourish [36]. Simultaneously, the released protease assisted in the hydrolysis of soy protein, providing a nitrogen source for BZ25. Its amylase and other enzymes degraded the polysaccharide in soybeans to give BZ25 a carbon source. These factors promoted the growth of BZ25 [37]. Furthermore, *Bifidobacterium* growth may produce certain compounds that aid Bacillus metabolism. As a result, BZ25 and GUTU09 boosted each other more than single-bacteria fermentation, which had more advantages and was more favorable for bioactive ingredient accumulation. Our experiments revealed that adding a diluent to a natto extract prevented fibrin production, and that the anticoagulant action increased as the enzyme concentration increased (Figure 7C). Thus, mixed-bacteria fermentation methods result in greater nattokinase activity, free amino nitrogen concentration, and sensory assessment. As a result, it is a nutritious food that is readily available to the general public.

5. Conclusions

Mixed fermentation increased nattokinase activity, free amino nitrogen and the sensory score of natto. Substantial thrombolytic and anticoagulant effects were also observed. A healthy fermented natto with good sensory characteristics was studied and developed.

Supplementary Materials: The following are available online at https://www.mdpi.com/article/10.3390/foods10112547/s1, Table S1 PB experiment resµlts; Table S2 response surface experimental results; Table S3 analysis of experimental variance with Nattokinase Activity, amino acid nitrogen and sensory scores as response value

Author Contributions: Y.Y.: Conceptualization, writing—original draft. G.L.: data curation, formal analysis, writing—original draft. X.T.: data curation, methodology. L.H.: supervising the experiment, writing—review and editing. C.L.: methodology. X.Z.: writing—review and editing. X.W.: writing—review and editing. All authors have read and agreed to the published version of the manuscript.

Funding: This research was supported by the Key Agricultural Project of Guizhou Province (QKHZC—[2021]278, QKHZC—[2021]184, QKHZC—[2021]142, QKHZC—[2019]2382, and QKHZC—[2016]2580), and the Natural Science Foundation of China (31870002 and 31660010), Qiankehe talents project ([2018]5781 and [2017]5788-11).

Conflicts of Interest: The authors have no competing interest to disclose.

References

1. Kotb, E. The biotechnological potential of fibrinolytic enzymes in the dissolution of endogenous blood thrombi. *Biotechnol. Prog.* **2014**, *30*, 656–672. [CrossRef] [PubMed]
2. Chang, C.-T.; Wang, P.-M.; Hung, Y.-F.; Chung, Y.-C. Purification and biochemical properties of a fibrinolytic enzyme from *Bacillus subtilis*-fermented red bean. *Food Chem.* **2012**, *133*, 1611–1617. [CrossRef]
3. Ahmad, A.; Hayat, I.; Arif, S.; Masud, T.; Khalid, N.; Ahmed, A. Mechanisms involved in the therapeutic effects of soybean (*Glycine Max*). *Int. J. Food Prop.* **2014**, *17*, 1332–1354. [CrossRef]
4. Liu, Z.; Zheng, W.; Ge, C.; Cui, W.; Zhou, L.; Zhou, Z. High-level extracellular production of recombinant nattokinase in *Bacillus subtilis* WB800 by multiple tandem promoters. *BMC Microbiol.* **2019**, *19*, 14. [CrossRef] [PubMed]
5. Wei, X.; Luo, M.; Xie, Y.; Yang, L.; Li, H.; Xu, L.; Liu, H. Strain screening, fermentation, separation, and encapsulation for production of nattokinase functional food. *Appl. Biochem. Biotech.* **2012**, *168*, 1753–1764. [CrossRef] [PubMed]
6. Weng, Y.; Yao, J.; Sparks, S.; Wang, K.Y. Nattokinase: An oral antithrombotic agent for the prevention of cardiovascular disease. *Int. J. Mol. Sci.* **2017**, *18*, 523. [CrossRef] [PubMed]
7. Li, C.; Du, Z.; Qi, S.; Zhang, X.; Wang, M.; Zhou, Y.; Lu, H.; Gu, X.; Tian, H. Food-grade expression of nattokinase in *Lactobacillus delbrueckii* subsp. *bulgaricus* and its thrombolytic activity in vitro. *Biotechnol. Lett.* **2020**, *42*, 2179–2187. [CrossRef] [PubMed]
8. Lan, G.; Li, C.; He, L.; Zeng, X.; Zhu, Q. Effects of different strains and fermentation method on nattokinase activity, biogenic amines, and sensory characteristics of natto. *J. Food. Sci. Technol.* **2020**, *57*, 4414–4423. [CrossRef] [PubMed]
9. Zhang, L. The Medium Optimization and Microcapsule Preparation of *Bifidobacterium* with Cholesterol-Reducing Ability from Guizhou Xiang Pigs. Master's Thesis, Guizhou University, Guiyang, China, 2015. (In Chinese)
10. Gao, Z. Screening of High-Yield Nattokinase Strains and Study of Its Enzymatic Stability. Master's Thesis, Guizhou University, Guiyang, China, 2018. (In Chinese)
11. Zhang, S.; Shi, Y.; Zhang, S.; Shang, W.; Gao, X.; Wang, H. Whole soybean as probiotic lactic acid bacteria carrier food in solid-state fermentation. *Food Control* **2014**, *41*, 1–6. [CrossRef]
12. Chen, T.; Wang, Y.; Yu, R. A supplement to the details of the experiment of "isolation of microorganisms decomposing cellulose". *Biol. Teach.* **2019**, *44*, 44–45. (In Chinese)
13. Lu, Y.; Chen, X.; Jiang, M.; Lv, X.; Rahman, N.; Dong, M.; Yan, G. Biogenic amines in Chinese soy sauce. *Food Control* **2009**, *20*, 593–597. [CrossRef]
14. Deepak, V.; Kalishwaralal, K.; Ramkumarpandian, S.; Babu, S.V.; Senthilkumar, S.R.; Sangiliyandi, G. Optimization of media composition for nattokinase production by *Bacillus subtilis* using response surface methodology. *Bioresour. Technol.* **2008**, *99*, 8170–8174. [CrossRef] [PubMed]
15. Feng, C.; Jin, S.; Luo, M.; Wang, W.; Xia, X.; Zu, Y.; Li, L.; Fu, Y. Optimization of production parameters for preparation of natto-pigeon pea with immobilized *Bacillus* natto and sensory evaluations of the product. *Innov. Food Sci. Emerg. Technol.* **2015**, *31*, 160–169. [CrossRef]
16. Yoshikawa, Y.; Chen, P.; Zhang, B.; Scaboo, A.; Orazaly, M. Evaluation of seed chemical quality traits and sensory properties of natto soybean. *Food Chem.* **2014**, *153*, 186–192. [CrossRef]
17. Wei, X.; Luo, M.; Liu, H. Preparation of the antithrombotic and antimicrobial coating through layer-by-layer self-assembly of nattokinase-nanosilver complex and polyethylenimine. *Colloids Surf. B* **2014**, *116*, 418–423. [CrossRef] [PubMed]
18. Ku, T.-W.; Tsai, R.-L.; Pan, T.-M. A simple and cost-saving approach to optimize the production of subtilisin NAT by submerged cultivation of *Bacillus subtilis* natto. *J. Agric. Food Chem.* **2009**, *57*, 292–296. [CrossRef]
19. Niu, H.; Miao, X.; Zheng, L.; Chi, Y.; Liu, J.; Gao, Y.; Wang, J.; Li, D. Changes of nutrients and bioactive substances in natto production and in cold storage. *Food Res. Dev.* **2021**, *42*, 48–54. (In Chinese) [CrossRef]
20. Yang, S.-Y.; Kim, S.-W.; Kim, Y.; Lee, S.-H.; Jeon, H.; Lee, K.-W. Optimization of maillard reaction with ribose for enhancing anti-allergy effect of fish protein hydrolysates using response surface methodology. *Food Chem.* **2015**, *176*, 420–425. [CrossRef]
21. Peng, Y.; Yang, X.; Zhang, Y. Microbial fibrinolytic enzymes: An overview of source, production, properties, and thrombolytic activity in vivo. *Appl. Microbiol. Biotechnol.* **2005**, *69*, 126–132. [CrossRef]
22. Gu, J.; Kang, N.; Li, D.; Zhou, C.; Du, L.; Fu, Q. Analysis of synergistic anticoagulant effect of acid auricularia auricula polysaccharide. *Mod. Food Sci. Technol.* **2017**, *33*, 45–52. (In Chinese) [CrossRef]

23. Xu, J.; Du, M.; Yang, X.; Chen, Q.; Chen, H.; Lin, D. Thrombolytic effects in vivo of nattokinase in a carrageenan-induced rat model of thrombosis. *Acta Haematol.* **2014**, *132*, 247–253. [CrossRef]
24. Sumi, H.; Hamada, H.; Tsushima, H.; Mihara, H.; Muraki, H.J.E. A novel fibrinolytic enzyme (nattokinase) in the vegetable cheese natto; a typical and popular soybean food in the Japanese diet. *Experientia* **1987**, *43*, 1110–1111. [CrossRef]
25. Yang, Y.; Niu, C.; Shan, W.; Zheng, F.; Liu, C.; Wang, J.; Li, Q. Physicochemical, flavor and microbial dynamic changes during low-salt *doubanjiang* (broad bean paste) fermentation. *Food Chem.* **2021**, *351*. [CrossRef]
26. Devanthi, P.V.P.; Gkatzionis, K. Soy sauce fermentation: Microorganisms, aroma formation, and process modification. *Food Res. Int.* **2019**, *120*, 364–374. [CrossRef] [PubMed]
27. He, F.; Tan, M.; Ma, Y.; MacGregor, G.A. Salt reduction to prevent hypertension and cardiovascular disease *JACC* state-of-the-art review. *J. Am. Coll. Cardiol.* **2020**, *75*, 632–647. [CrossRef] [PubMed]
28. Wu, R.; Chen, G.; Pan, S.; Zeng, J.; Liang, Z. Cost-effective fibrinolytic enzyme production by *Bacillus subtilis* WR350 using medium supplemented with corn steep powder and sucrose. *Sci. Rep.* **2019**, *9*. [CrossRef]
29. Wang, R.; Cai, T.; Xu, Y.; Ren, L.; Lei, H.; Xu, H. Study of improving nattokinase activity through solid-state fermentation of cold-pressing walnut dregs by compound strains. *Sci. Technol. Food Ind.* **2017**, *38*, 112–117. (In Chinese) [CrossRef]
30. Lee, B.; Lai, Y.; Wu, S. Antioxidation, angiotensin converting enzyme inhibition activity, nattokinase, and antihypertension of *Bacillus subtilis* (natto)-fermented pigeon pea. *J. Food Drug Anal.* **2015**, *23*, 750–757. [CrossRef]
31. Yang, Y.; Li, J.; Zhang, M.; Yang, Z.; Liang, P.; Mou, J. Study on the technology of natto fermentation by mixed bacteria. *China Brew* **2019**, *38*, 49–53. CNKI:SUN:ZNGZ.0.2019-06-009(In Chinese)
32. Jhan, J.-K.; Chang, W.-F.; Wang, P.-M.; Chou, S.-T.; Chung, Y.-C. Production of fermented red beans with multiple bioactivities using co-cultures of *Bacillus subtilis* and *Lactobacillus delbrueckii* subsp. *bulgaricus*. *LWT-Food Sci. Technol.* **2015**, *63*, 1281–1287. [CrossRef]
33. Luo, S.; Liu, H.; Yang, S.; Yan, Q.; Wu, S.; Jiang, Z. Nutrient composition, thrombolysis and antioxidant activity of three kinds of tempeh fermented by *Bacillus amyloliticus*. *J. Chin. Inst. Food Sci. Technol.* **2021**, *21*, 37–44. (In Chinese) [CrossRef]
34. Mozzetti, V.; Grattepanche, F.; Moine, D.; Berger, B.; Rezzonico, E.; Meile, L.; Arigoni, F.; Lacroix, C. New method for selection of hydrogen peroxide adapted *bifidobacteria* cells using continuous culture and immobilized cell technology. *Microb. Cell Factories* **2010**, *9*. [CrossRef] [PubMed]
35. Hosoi, T.; Ametani, A.; Kiuchi, K.; Kaminogawa, S. Improved growth and viability of lactobacilli in the presence of *Bacillus subtilis* (natto), catalase, or subtilisin. *Can. J. Microbiol.* **2000**, *46*, 892–897. [CrossRef] [PubMed]
36. Zhang, Y.-R.; Xiong, H.-R.; Guo, X.-H. Enhanced viability of *Lactobacillus reuteri* for probiotics production in mixed solid-state fermentation in the presence of *Bacillus subtilis*. *Folia Microbiol.* **2014**, *59*, 31–36. [CrossRef] [PubMed]
37. Wang, H.; Ng, Y.K.; Koh, E.; Yao, L.; Chien, A.S.; Lin, H.X.; Lee, Y.K. RNA-Seq reveals transcriptomic interactions of *Bacillus subtilis* natto and *Bifidobacterium animalis* subsp. *lactis* in whole soybean solid-state co-fermentation. *Food Microbiol.* **2015**, *51*, 25–32. [CrossRef] [PubMed]

Article

Metagenomic Study on Chinese Homemade Paocai: The Effects of Raw Materials and Fermentation Periods on the Microbial Ecology and Volatile Components

Linjun Jiang †, Shuang Xian †, Xingyan Liu, Guanghui Shen, Zhiqing Zhang, Xiaoyan Hou and Anjun Chen * 

College of Food Science, Sichuan Agricultural University, Ya'an 625014, China; jianglinjun2020@163.com (L.J.); xianshuang@stu.sicau.edu.cn (S.X.); lxy05@126.com (X.L.); Shenghuishen@163.com (G.S.); zqzhang721@163.com (Z.Z.); houxiaoyan106@163.com (X.H.)

* Correspondence: chen_anjun@sicau.edu.cn; Tel.: +86-0835-2882187

† These authors contributed equally to this work.

Abstract: "Chinese paocai" is typically made by fermenting red radish or cabbage with aged brine (6–8 w/w). This study aimed to reveal the effects of paocai raw materials on fermentation microorganisms by metagenomics sequencing technology, and on volatile organic compounds (VOCs) by gas chromatography–mass spectroscopy, using red radish or cabbage fermented for six rounds with aged brine. The results showed that in the same fermentation period, the microbial diversity in cabbage was higher than that in red radish. *Secundilactobacillus paracollinoides* and *Furfurilactobacillus siliginis* were the characteristic bacteria in red radish paocai, whereas 15 species of characteristic microbes were found in cabbage. Thirteen kinds of VOCs were different between the two raw materials and the correlation between the microorganisms and VOCs showed that cabbage paocai had stronger correlations than radish paocai for the most significant relationship between 4-isopropylbenzyl alcohol, α-cadinol, terpinolene and isobutyl phenylacetate. The results of this study provide a theoretical basis for understanding the microbiota and their relation to the characteristic flavors of the fermented paocai.

Keywords: red radish; cabbage; fermented foods; microbial ecology; flavor components

Citation: Jiang, L.; Xian, S.; Liu, X.; Shen, G.; Zhang, Z.; Hou, X.; Chen, A. Metagenomic Study on Chinese Homemade Paocai: The Effects of Raw Materials and Fermentation Periods on the Microbial Ecology and Volatile Components. *Foods* **2022**, *11*, 62. https://doi.org/10.3390/foods11010062

Academic Editors: Valentina Bernini and Juliano De Dea Lindner

Received: 5 November 2021
Accepted: 22 December 2021
Published: 28 December 2021

1. Introduction

Fermentation is a widely used technology to preserve foods, improve nutritional value and extend shelf-life, and fermented vegetables are popular and traditional in Asian countries [1]. Among the fermented vegetables, "paocai" is a beloved food with a history of more than 3000 years, which are typically consumed as side dishes or appetizers and are characterized by their tender, crisp texture and rich health benefits owing to the lactic-acid bacteria (LAB) [2–4]. Vegetables for making paocai are usually immersed in 6–8% (w/w) sodium chloride solution and allowed to undergo spontaneous anaerobic or microaerobic fermentation by the epiphytic microbes (mostly LAB) present on the raw materials for 6–10 days [5].

The microbial composition and flavor in paocai fermentation systems are related to various factors, especially the raw materials. Previous studies have found that the microbial niches in raw ingredients determine the eventual microbial community [6], this conclusion has been confirmed in some studies on microorganisms in Sichuan paocai; for instance, *Rosenbergiella*, *Staphylococcus*, *Hyphopichia* and *Kodamaea* dominated the fermentation of chili pepper [7], while *Lactobacillus*, *Leuconostoc*, *Achromobacter* and *Pediococcus* were the main microbial organisms during the fermentation of cabbage pickle [8]. This is due to the fact that the spontaneous fermentation of vegetable products is highly dependent on the naturally occurring LAB present on the raw materials. A large amount of different LAB can be accumulated in the aged brine in the process of continuous periodic fermentation, and the total viable count can be up to 8.79 Log CFU/mL [9]. It is worth noting that the microorganisms were shown to rapidly build a stable fermentation system at the initial stages and finally obtain paocai products with different flavors. Due to the complex microbial ecology in the fermentation system, two or three strains of artificially isolated LAB are often used in industry to accelerate the fermentation process and to achieve a uniform quality of the products [10]. However, based on consumer surveys, the quality of the final products is not as good as homemade paocai. Homemade paocai usually uses aged brine as the starter, which is brine fermented for many years and even decades [11,12]. The microorganisms in the fermentation system metabolize the nutrients in the raw materials to produce acids, that makes the fermented vegetables have a unique flavor [13], which is another source of paocai flavor besides that of the raw materials. Therefore, understanding the effects of raw materials and fermentation periods on the microbial ecology and volatile components is of far-reaching significance for understanding how microorganisms construct stable fermentation systems.

Metagenomics sequencing technology can extract all microbial DNA from the fermentation system, construct a metagenomic PE library and use genomics research strategy to study the genetic composition and gene function of all microorganisms contained in environmental samples. This technique can avoid amplification and sequencing bias, which commonly occurred in other sequencing techniques [14]. Metagenomics, although still underexploited, can reveal the succession and function of microorganisms and establish the relationship between microorganisms and substance metabolism [15–17]. However, to the extent of authors' knowledge, little insights on relationship between "raw material–microorganism-VOCs" in the process of paocai fermentation are available, especially based on metagenomics sequencing technology.

Therefore, in this study, red radish and cabbage, the most common paocai raw materials in Sichuan, were used as raw materials for multiround fermentation, and the composition of microbial and volatile components in paocai fermentation systems were analyzed by metagenomics and gas chromatography–mass spectrometry (GC–MS), in order to reveal the effects of vegetable raw materials on the microbial community structure during Chinese homemade paocai multiple fermentation rounds and their relationship with volatile compounds.

2. Materials and Methods

2.1. Paocai Preparations

Red radish (*Red raphanus* L.) and cabbage (*Brassica oleracea* L.) were purchased from the local market in Ya'an, Sichuan Province, China. The fresh vegetables were cleaned under the same conditions to ensure the consistency of the initial flora on the surface of the vegetables. Paocai was made according to the local traditional method, specifically: one kilogram of each vegetable was cut into 2–3 cm pieces and set in 4 L pottery jars with garlic (3%), hot red peppers (3.5%), ginger (3%), and Chinese prickly ash (3%) (all percentages were calculated according to the volume of brine in the pottery jars, in w/v). Two liters of brine was prepared with cooled, boiled water containing 8% salt, 1% Chinese baijiu and 20% aged Sichuan paocai brine (in terms of w/v, collected from the homemade paocai of a local family used for more than 5 years). The pottery jars were sealed by adding water to the jar edge and stored at 24 ± 2 °C for six rounds of fermentation for about 9 d, 7 d, 5 d, 4 d and 3 d, respectively.

2.2. DNA Extraction, Library Construction and Metagenomic Sequencing

A total of 50 mL of initial aged brine and paocai brine at the end of each fermentation round were enriched on a 0.22 μm sterile microfiltration membrane. According to the manufacturer's recommendation, the microbial DNA from brine samples was extracted with the E.Z.N.A.® DNA Kit (Omega Bio-tek, Norcross, GA, USA). The concentration and purity of the DNA were determined with Turner Biosystems (Sunnyvale, CA, USA) and a NanoDrop2000 spectrophotometer (Thermo Fisher Scientific, TBS-380, Shanghai, China), and the integrity of the DNA was detected by 1% w/w agarose gel electrophoresis.

DNA was fragmented to an average size of about 400 bp using Covaris M220 (Gene Company Limited, Hong Kong, China) for paired-end library construction, and the paired-end library was constructed using NEXTflex^{TM} Rapid DNA-Seq (Bioo Scientific, Austin, TX, USA).

Adapters containing the full complement of sequencing primer hybridization sites were ligated to the blunt end of fragments. Paired-end sequencing was performed on Illumina NovaSeq (Illumina Inc., San Diego, CA, USA) at Majorbio Bio-Pharm Technology Co., Ltd. (Shanghai, China) using NovaSeq 6000 according to the manufacturer's instructions. Sequence data associated with this project have been deposited in the NCBI Short Read Archive database (accession number: PRJNA731323).

2.3. Metagenome Data Integration

The raw reads from the metagenome sequencing were used to generate clean reads by removing adaptor sequences, trimming and removing low-quality reads (length < 50 bp, or with a quality value < 20 or having N-bases) using the fastp [18] (https://github.com/OpenGene/fastp, version 0.20.0, accessed on 1 December 2020) on the free online platform of Majorbio's Cloud Platform (https://cloud.majorbio.com/, accessed on 1 December 2020). Megahit (parameters: kmer_min = 47, kmer_max = 97, step = 10) (http://www.l3-bioinfo.com/products/megahit.html, version 1.1.2, accessed on 6 December 2020) based on the principle of succinct de Bruijn graphs was used to assemble the optimized sequences [19]. Among the splicing results, contigs $\geq$ 300 bp were selected as the final assembly result. MetaGene was used to predict the open reading frames (ORFs) of contigs in the stitching results (http://metagene.cb.k.u-tokyo.ac.jp/, accessed on 11 December 2020) [20].

The predicted ORFs with lengths over 100 bp were retrieved and translated to amino acid sequences using the NCBI translation table (https://www.ncbi.nlm.nih.gov/Taxonomy/taxonomyhome.html/index.cgi?chapter=tgencodes#SG1, accessed on 15 December 2020). CD-HIT (http://www.bioinformatics.org/cd-hit/, version 4.6.1, accessed on 15 December 2020) was used to classify all sequences with 90% sequence identity and 90% coverage as nonredundant gene catalog [21]. Reads after quality control were mapped to the representative genes with 95% identity using SOAPaligner (https://github.com/ShujiaHuang/SOAPaligner, version 2.21, accessed on 20 December 2020) [22], and gene abundance in each sample was evaluated.

2.4. Species Annotation

Diamond (https://github.com/bbuchfink/diamond, version 0.8.35, accessed on 21 December 2020) was employed for taxonomic annotations by aligning nonredundant gene catalogs against the NCBI NR database using blastp as implemented in DIAMOND v0.9.19 with an e-value cutoff of 1×10^{-5} [23].

2.5. Determination of VOCs Using GC–MS

The analysis of the VOCs was performed by HS–SPME combined with GC–MS (7890A GC, Agilent, USA, 5975C MS; Agilent Technologies, Santa Clara, CA, USA). The paocai brine (5 mL) were accurately weighed into 15 mL headspace glass vials containing 1 g of salt, the vials were immediately placed in a heating block to equilibrate for 30 min at 50 °C and the VOCs were extracted using a 65 μm 50/30 um DVB/CAR/PDMS optical fiber for 10 min at 50 °C. VOCs were desorbed from the SPME fiber at 250 °C for 5 min in the injector in splitless mode. The carrier gas was helium, and it was supplied at a flow rate of 1.2 mL/min (constant linear velocity). The column temperature was programmed as follows: maintained at 36 °C for 3 min, increased at 5 °C/min to 65 °C, increased at 3 °C/min to 155 °C, increased at 10 °C/min to 200 °C, and maintained at 200 °C for 3 min. The MS condition was set as: 280 °C for the transfer line, 230 °C for the ionization, 70 eV for the ionization energy [10]. The quantification analysis was done by using cyclohexanone (27.424 g/L, *w*/*v*) as an internal standard. The NIST11.L standard mass spectral database was used to identify VOCs based on retention time and mass spectral similarity match. Each sample was analyzed in triplicates at each sampling time.

2.6. Statistical Analysis

The nonmetric multidimensional scaling analysis (NMDS), the analysis of similarities (ANOSIM), the heatmap and the bar plot were performed in the Vegan package of R (version 3.3.1). The differences between classes were analyzed using Linear discriminant analysis Effect Size (LEfSe) (http://huttenhower.sph.harvard.edu/galaxy/root?tool_id=lefse_upload, Version 1.0, accessed on 1 December 2020). SIMCA-P (Version14.0 (Umetrics, Umeå, Sweden) was used to analyze VOCs by partial least squares discriminant analysis (PLS-DA).

3. Results and Discussion

3.1. Metagenomic Sequencing Statistics and Quality Control

A total of 39 samples were sequenced by the Illumina NovaSeq 6000 sequencing systems, and 1708.39×10^6 raw reads, about 257.97 GB, with an average of 6.61 GB of each sample were generated. After the quality processing and eliminating the sample host gene, the sequence utilization of each sample was more than 98.48, including 787.85×10^6 clean reads in the red radish samples and 776.19×10^6 clean reads in the cabbage samples. A total of 966,355 contigs were assembled with Megahit, and the single sample's contigs were between 9431 to 47,483. N50 and N90 were two indices for the distribution of contig lengths within a draft assembly, and the longer the assembly of samples the better. In all samples, the minimum value of N50 reached 1677 bp, the average value was 6680 bp, the N90 minimum was 421 bp, and the average value was 528 bp. Furthermore, a total of 1,778,731 genes were obtained by ORF prediction of the contigs, and the sequence length was 1,150,522,526 bp (Table S1). All predicted genes were clustered using CD-HIT (parameters: identity, 95; coverage, 90), resulting in a total of 122,012 nonredundant gene set. Figure 1A shows the sequence length distribution of nonredundant gene sets, mostly concentrated between 201~600 bp, with an average sequence length of 557.88 bp, and the information was used for subsequent species and functional annotations to reveal microbial community of samples.

3.2. Microbial Structure

The sequences comprising the total reads corresponded to five domains, eight kingdoms, 47 phyla, 99 classes, 207 orders, 353 families, 706 genera, and 2068 species, of which four domains were known. These four were identified to be bacterial, fungal, archaeal and virus domains, of which the bacterial domain had the highest proportion at 99.31% and the lowest was for the fungal domain. In addition, eight kingdoms in all samples were detected, of which the major phyla were *Firmicutes* and *Ascomycota* with *Firmicutes* the most abundant, contributing to over 98% in all samples, consistent with previous reports of other fermentation vegetables [24,25]. Further, it was interesting to note that with the progress of fermentation, the content of *Ascomycota* decreased gradually in the red radish samples, but there was no obvious change in the cabbage samples. The major classes, order and family were *Bacilli* and *Saccharomycetes*, *Lactobacillales* and *Saccharomycetales* and *Lactobacillaceae* and *Saccharomycetaceae*, consistent with the phyla identified.

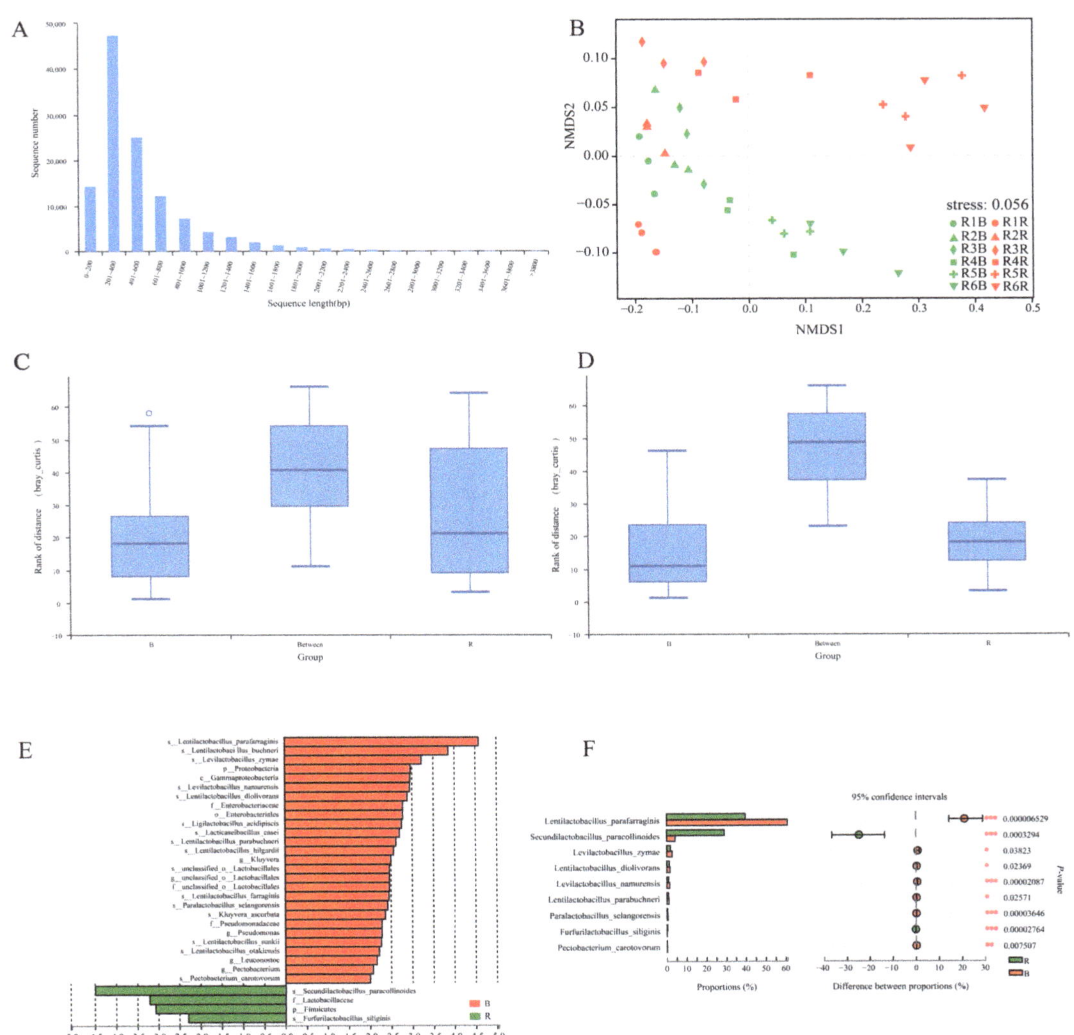

Figure 1. (**A**) Sequence length distribution of nonredundant gene catalog. (**B**) NMDS analysis at species level. Analysis of similarities between different groups by ANOSIM (**C**,**D**) of red radish paocai and cabbage paocai in the 3rd and 4th round, separately. (**E**) LDA score of the two kinds of red radish paocai and cabbage paocai. (**F**) Wilcoxon rank-sum test of different species in red radish paocai and cabbage paocai. * $0.01 < p \leq 0.05$, ** $0.001 < p \leq 0.01$, *** $p \leq 0.001$.

As for the genus level, a total of 706 microbe genera were detected, of which 319, 521 and 664 genera were detected in the initial red radish and cabbage brines, respectively. Among them, 303 genera occurred in all paocai samples, wherein 38 genera only appeared in the red radish samples, 171 genera only appeared in the cabbage samples and 303 genera appeared in both paocai samples (Figure 2A). The microbial diversity of paocai was found to be relatively simple, and there were only three genera with abundance more than 1% of the diversity, and they belonged to the *Lactobacillus*, *Pediococcus* and an unclassified *Lactobacillaceae* genus. The average abundance of *Lactobacillus* in the initial brine, red radish paocai and cabbage paocai reached 92.35%, 85.10% and 81.84%, respectively. Furthermore,

Lactobacillus was the predominant bacteria in the multiround fermentation, and was found to be in agreement with other paocai studies [8]. However, with the progress of fermentation, *Lactobacillus* gradually decreased, accounting for only 67.4% of red radish paocai at the end of the sixth round of fermentation and 69.5% of cabbage paocai. Although previous studies have different conclusions about whether *pediococcus* is the dominant bacterium in fermented vegetables [12,26,27], the proportion of *Pediococcus* genus was found in general to be increasing along with the fermentation. Specifically, its abundance was found to increase from 2.82% to 28.37% in the red radish and 2.58% to 26.28% in the cabbage paocai brines (Figure 2B), respectively. These data indicate that although they belong to the same phylum (Firmicutes), *Pediococcus* may be more stress-resistant or competitive than *Lactobacillus* in the later stage of fermentation.

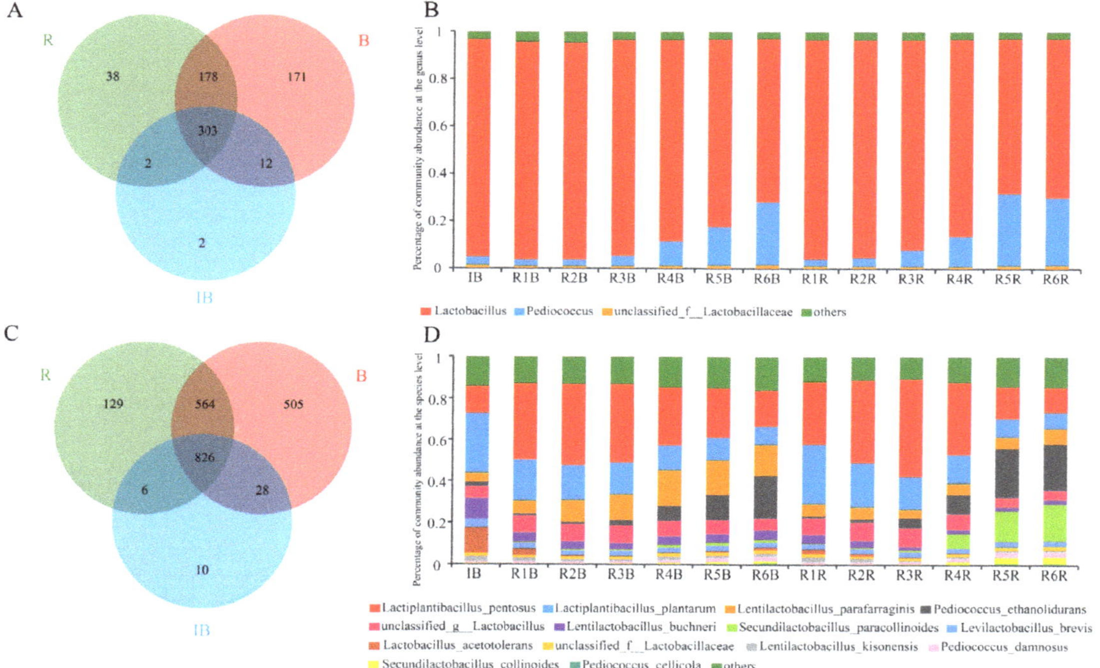

Figure 2. Venn plot (**A**) and bar plot of relative abundance (**B**) of microbial communities at the species level in the initial brine and two kinds of paocai and (**C**,**D**) at the genus level in the initial brine and two kinds of paocai. R indicates red radish paocai, B indicates cabbage paocai, IB indicates initial brine. Note: others in B and D means other microbes.

At the species level, a total of 2068 microbial species were detected by sequencing, of which 870, 1525 and 1923 species were detected in the initial, red radish and cabbage brines. Among the detected species 826 of them occurred in all paocai samples (Figure 2C), whereas 129 species appeared only in the red radish samples, and 505 species only in the cabbage samples. A further 564 species commonly appeared in all the red radish and cabbage paocai samples, but only 11 of them were recorded with relative abundances of more than 1%. In addition, a higher number of microbial species were present in the cabbage paocai than in the red radish paocai with a brine of the same age; moreover, with the increase of fermentation time, the number of microorganisms in the red radish paocai increased first and then decreased, while the number of microorganisms in the cabbage paocai did not obviously change, but after six rounds of fermentation, the number of microorganisms in the cabbage paocai was significantly higher than that in the red radish paocai.

As for the individual detected species, the *Lentilactobacillus buchneri* (9.9%) and *Lactobacillus acetotolerans* (12.5%) in the initial brine were found to be the dominant species. *Lactiplantibacillus paraplantarum* species were found to be gradually decreasing with the progression of fermentation rounds from 29.1% to 8.6% in the cabbage paocai and from 29.1% to 7.9% in the red radish paocai samples. With regard to the *Pediococcus ethanlidurans*, *Pediococcus damnosus* and *Secundilactobacillus paracollinoides* species, their proportions were found to be gradually increasing with fermentation rounds. Specifically, *P. ethanlidurans* species registered a significant increase from the first to the last round of fermentation from 0.7% to 20.67% in the red radish and 0.94% to 22.39% in the cabbage paocai samples, respectively. This occurrence indicates that *P. ethanlidurans* might have a better growth competitiveness compared to *L. paraplantarum* under the final round of fermentation conditions characterized by a high salinity and acidity.

Further, *Lactiplantibacillus pentosus*, which is used as the starter in the production of homemade paocai to improve the quality [5], was observed to be the dominant strain, and increased gradually until the third round of fermentation, then decreased continuously until the last round of paocai fermentation (from 38.8% to 18.0% and from 47.4% to 12.2% in red radish samples and cabbage samples, respectively). This could be due to the preparation process and some limitations as reported in previous studies [28,29]. The microbial community in the later stages of fermentation for both the red radish paocai and cabbage paocai was found to be similar. The proportion of *Lentilactobacillus parafarraginis* was found to be relatively higher in the cabbage paocai (15.2%) than that in the red radish paocai (7.3%), whereas the relative abundance of *S. paracollinoides* was higher in the red radish paocai (17.7%) than that in the cabbage paocai (1.7%) (Figure 2D).

3.3. Microbial Composition Comparison

The heatmap analysis was utilized to reveal the relative relationship of the top 40 microbial genera detected with the colors representing the species abundance (Figure 3A). It can be clearly observed from the heatmaps that in both evaluated samples, the abundance of *Lactobacillus* and *Pediococcus* in bacteria and *Kazachstania* in eukaryotes was relatively stable during all six rounds of fermentation. Moreover, *Weissella* was found to be relatively rich during the whole fermentation process, but its relative abundance decreased as the fermentation progressed. This decrease in *Weissella*'s abundance can be attributed to the increase in the types of species and the subsequent competition for nutrients, as was also observed in other studies [30,31]. In addition, the abundance of *Geotrichum* and *Kazachstania* decreased gradually during the fermentation in the two fermented vegetables, whereas the abundance of *Enterococcus*, *Streptococcus*, *Clostridium* and *Bacillus* showed a trend toward a gradual increase in the fermentation process, with no change in their relative abundance in both fermentation systems even though they exhibited significantly opposite microorganisms.

In fact, metagenomics sequencing allows the obtained data to be compared with a database to extract the information of microorganisms at the species level [32,33]. Within the top 40 abundant species detected (Figure 3B), 30 species belonged to the genus of LAB. Within the LAB species, a varying behavior of abundance was observed along the fermentation process with species such as *L. paraplantarum*, *Levilactobacillus brevis*, *L. acetotolerans* decreasing in abundance during the fermentation in both samples.

A UPGMA cluster analysis of the samples was performed based on the identified genus and species, demonstrating the higher similarity between the red radish samples.

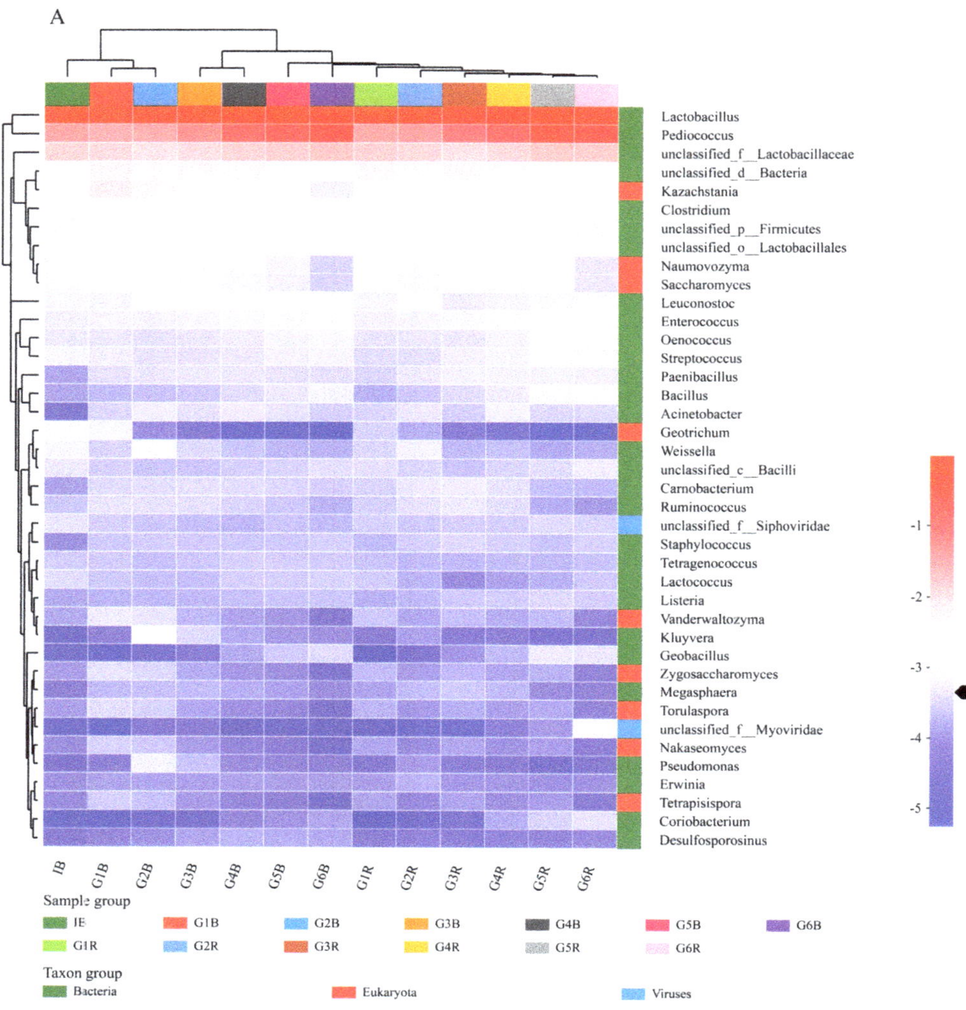

Figure 3. *Cont.*

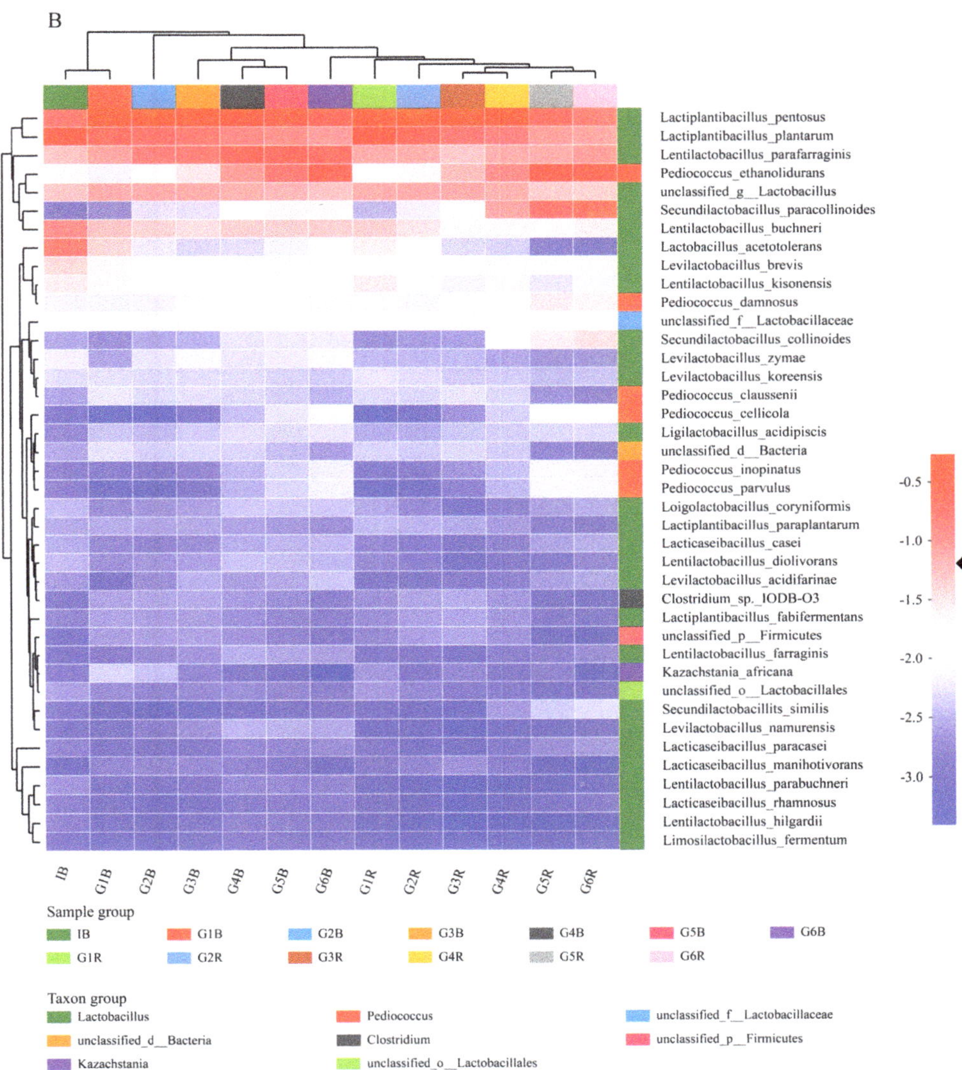

Figure 3. (**A**) The heatmap analysis of the top 40 microorganisms at the genus level. (**B**) The heatmap analysis of the top 40 microorganisms at the species level.

To further understand the microbiota structure during paocai fermentation, NMDS and LDA Effect Size (LEfSe) was used to analyze the microbial differences. The NMDS analysis (Figure 1B) showed that the microbial community of the red radish and cabbage paocais displayed an obvious spatial structure during the fermentation process, where the composition of microorganisms in different vegetable fermentation systems exhibited greater differences (stress = 0.016 < 0.05). With the progress of the fermentation, the distance between spatial positioning of different fermented vegetable samples of the same round gradually increased, indicating that the influence of raw materials on the microbial composition increased with the increase of fermentation time. These observed results were consistent with the ANOSIM analysis, with the R value closer to 1 confirming the greater differences between groups with a p-value less than 0.05 indicative of the high reliability of the test (Figure 1C,D). Towards the end of the third and fourth rounds of paocai fermentation, the different samples were significantly separated (R = 0.46, $p = 0.009$) with a further increase in distance at the end of the fifth and the sixth fermentation rounds (R = 0.89, $p = 0.003$). The aforementioned data further confirm that the microbial composition of different vegetable paocais presented continuous changes in the multiround fermentation with the differences significantly increasing along the fermentation time.

LEfSe (Figure 1E) was used as an algorithm for high-dimensional biomarker discovery and identifying genomic features characterizing the differences between two or more biological conditions [34]. In this study, LEfSe was performed to evaluate the differences between the bacterial community structures of the different paocais. In the paocai, a total of 31 taxa were found to represent a remarkable difference in their relative abundance, with an LDA (log10) > 2. At the species level, *S. paracollinoides* and *Furfurilactobacillus siliginis* were found to be the characteristic bacteria of the red radish paocai, whereas a total of 15 species of microbes were found characteristic of the cabbage paocai, of which 13 species belonged to the Firmicutes (namely, *L. parafarraginis*, *L. buchneri*, *Levilactobacillus zymae*, *Levilactobacillus namurensis*, *Lentilactobacillus diolivorans*, *Ligilactobacillus acidipiscis*, *Lacticaseibacillus casei*, *Lentilactobacillus parabuchneri*, *Lentilactobacillus hilgardii*, *Lentilactobacillus farraginis*, *Paralactobacillus selangorensis*, *Lentilactobacillus sunkii*, *Lentilactobacillus otakiensis*) and two species to that of Proteobacteria (namely, *Kluyvera ascorbata*, *Pectobacterium carotovorum*). Further, the differential microorganisms were screened by a Wilcoxon ranksum test (Figure 1F), which revealed that only nine species were significantly different in the red radish paocai and cabbage paocai, with *L. parafarraginis*, *S. paracollinoides*, *L. namurensis*, *P. selangorensis* and *F. siliginis* being the significantly different species in the paocais tested. In fact, all the nine species detected were also screened by LEfSe and regarded as biomarkers for the subsequent analysis.

3.4. *Profiles of VOCs during Different Paocai Fermentations*

The compositions of VOCs in the paocai samples are shown in Table S2. A total of 62 and 59 volatile substances were identified in the red radish and cabbage paocais. As shown in Figure 4A, the main VOCs detected were alcohols, hydrocarbons, esters, sulfide and ketones during the whole fermentation process. Alcohols were the most abundant compounds among those detected and ranged from 74.1% to 79.3%, with a total of 17 different components. The second largest group was heterocycles which ranged from 9.8% to 13.1% and included 23 component types.

Figure 4B shows 22 major VOCs that accounted for more than one of the two paocai volatile profiles. In general, the types of VOCs identified were similar, both types included compounds like linalool, α-terpineol, eucalyptol, D-carvone and other substances. Among these, linalool registered the highest proportions reaching up to 35% and 41.6% in the red radish and cabbage paocais, respectively, similar to previous studies [8]. Linalool contributes to the unique flavor of paocai and is described as having a strong scent, similar to bergamot oil or French lavender, with a mixture of woody, floral aromas with a touch of spiciness [35]. Other alcohols were also evaluated to have a positive effect on paocai's flavor with α-terpineol contributing a floral typically lilac odor; eucalyptol with a eucalyptus, herbal and camphor odor; and nerol with notes of roses [36,37].

Red radish and cabbage belong to the Cruciferae family of plants, and typically have a strong aroma, bitter taste and pungent taste similar to that of mustard. The compounds responsible for these special flavors were found to be mainly the secondary defense metabolite, glucosinolate and its hydrolysate isothiocyanate [38]. The composition and content of isothiocyanates is variable based on the type of vegetables subsequently imparting characteristic flavor components and compositions in the fermented paocais. Both tested paocais were found to contain butyl isothiocyanate and amyl isothiocyanate, with significantly higher values in the red radish paocai. Among the sulfides, dimethyl disulfide and dimethyl trisulfide were uniquely found in the red radish paocai, whereas 3-butenylisothiocyanate was found to be the unique sulfide in the cabbage paocai.

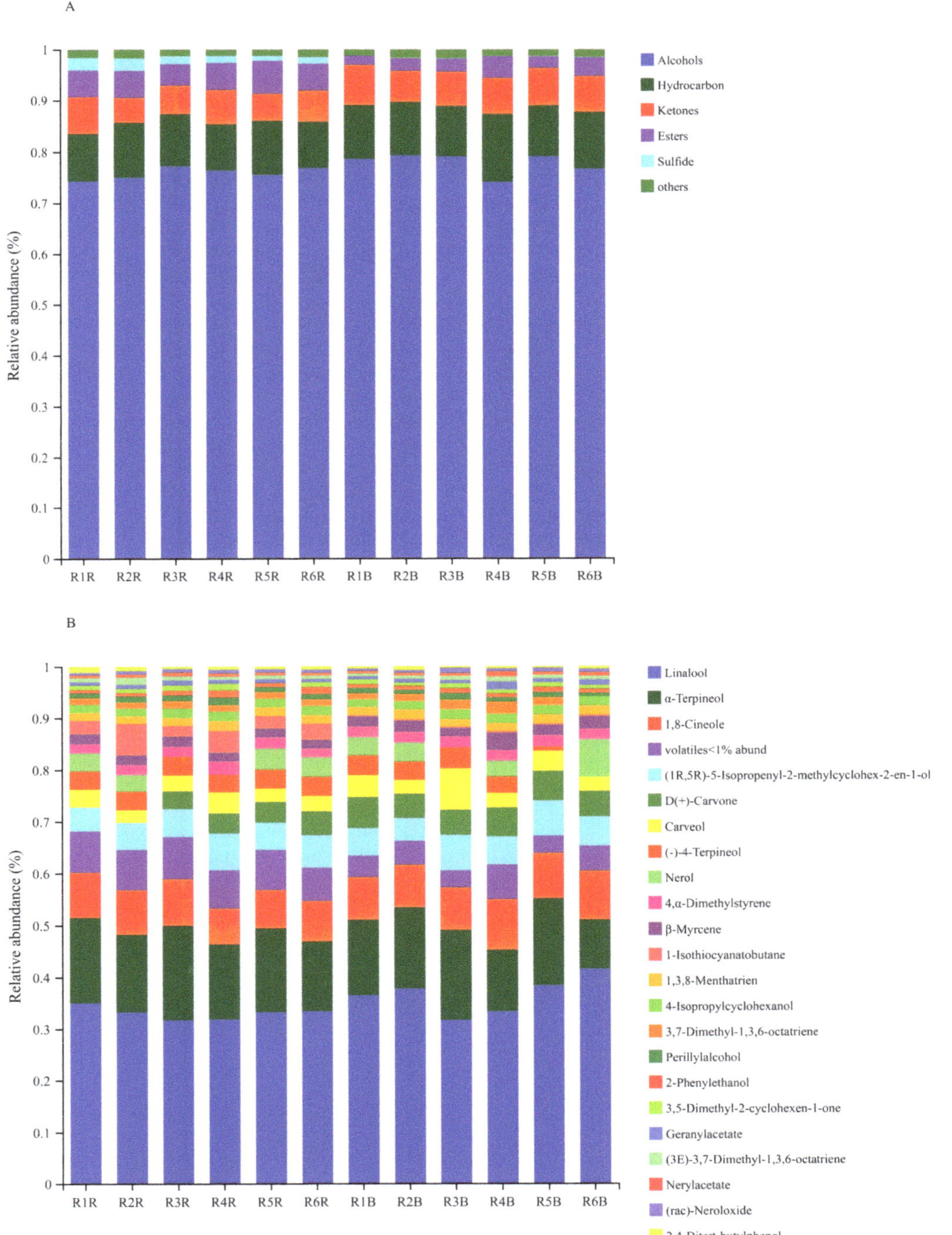

Figure 4. (**A**) Changes of VOCs in red radish paocai and cabbage paocai during fermentation. (**B**) Major VOCs with relative abundance of more than 1% of the red radish paocai and cabbage paocai.

Among the various potential detected substances, those with a low content (lower than the threshold values) were excluded in the statistical analysis to narrow down the range of key volatiles. In detail, 36 potential volatile substances were screened out, which included 14 alcohols, 3 ketones, 5 esters, 10 hydrocarbons, 3 sulfides, and 1 phenol compounds.

To further screen the different VOCs in two kinds of paocai, a PLS-DA model was established, as shown in Figure 5A,B. Variable projection importance (VIP) was used to evaluate the contribution of variables. Substances with VIP value $\geq$1.0 were selected and combined with the conditions with a significant difference between the two groups ($p < 0.05$). The point distance between the VOCs of paocai fermented with two different raw materials was obvious, which indicated that the VOCs of paocai fermented with different raw materials were different. The total contribution rate of the first two principal components on variable X was 52.7% ($R_2X = 0.527$), and the contribution rate on variable Y was 95.7% ($R_2Y = 0.957$). At the same time, the prediction ability based on cross-validation was 94.0% ($Q_2 = 0.94$), indicating that the model had a good explanatory degree. The stability and reliability of this model were verified by the cross-validation model with 200 permutation validations. As shown in Figure 5B, all Q_2 values on the left were lower than the original point on the right, and the regression line of point Q_2 intersected with the vertical axis below zero, and the intercept of -0.306 was less than 0.05, indicating that this PLS-DA model did not exhibit over-fitting and had good representativeness. A total of 13 compounds were screened out as the difference substances and included compounds like (−)-4-terpineol, α-terpineol, nerol, 4-isopropylbenzyl alcohol, carveol, α-cadinol, terpinolene, isobutyl phenylacetate, 1-isothiocyanatobutane, dimethyl disulfide, 1,4-dithiane, dimethyl trisulfide and 2,4-ditert-butylphenol.

Further to explore the correlation between microorganisms and VOCs, a Spearman correlation analysis was carried out with a relative abundance exceeding 5% and the nine biomarkers previously identified. It can be seen from Figure 5C that the microorganisms have different effects on the VOCs of paocai fermented by two kinds of vegetables. Nerol was also significantly correlated with six kinds of microorganisms in the red radish paocai to different degrees. In the cabbage paocai, the correlations between nerol, α-cadinol, isobutyl phenylacetate and the microorganisms were consistent, that is, they have a significant positive correlation with seven species and a significant negative correlation with three species at the same time, indicating that these 13 microorganisms have the same regulatory effect on the changes of these three substances. In addition, 1-isothiocyanatobutane, 1,4-dithiane and dimethyl trisulfide, which account for a high proportion of compounds in the red radish paocai, were related to six kinds of microorganisms, especially *L. paraplantarum* species. However, 2,4-ditert-butylphenol in the cabbage paocai and α-cadinol in the red radish paocai were not related to microorganisms.

From the perspective of the relationship between microorganisms and VOCs, *F. siliginis*, *L. parafarraginis*, *L. namurensis*, *L. zymae* and *S. paracollinoides* were found to be the key microorganisms for distinguishing the two kinds of paocai, and 4-isopropylbenzyl alcohol, α-cadinol, terpinolene and isobutyl phenylacetate were the key VOCs for distinguishing the two kinds of paocai made from red radish and cabbage.

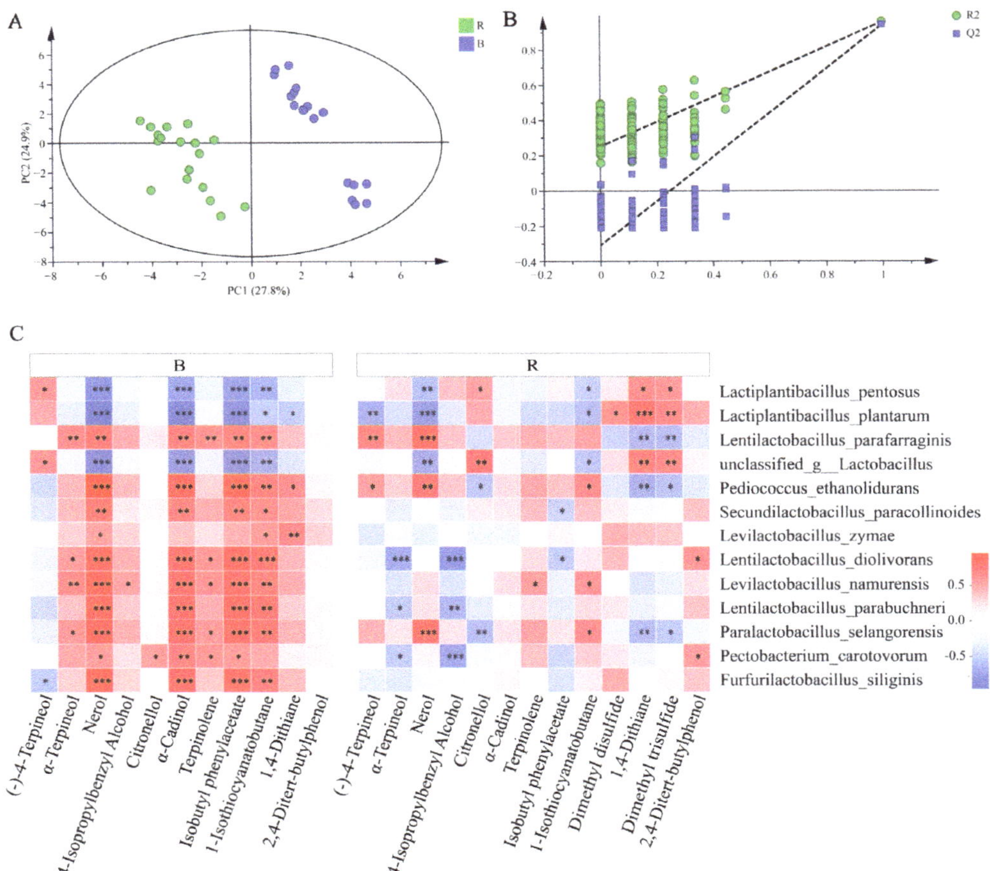

Figure 5. (**A**) PLS-DA score scatter plot and (**B**) permutation tests of VOCs. (**C**) Heatmap of correlation between microbial species (relative abundance exceeding 5% and the nine biomarkers previously identified) and different volatile flavor substances in the two kinds of paocai. R indicates red radish paocai, B indicates cabbage paocai. * $0.01 < p \leq 0.05$, ** $0.001 < p \leq 0.01$, *** $p \leq 0.001$.

4. Conclusions

In this paper, metagenomic sequencing and a HS–SPME method based on a GC–MS method were used to systematically study the changes of microbial flora and VOCs in paocai fermented with different vegetables in six rounds of fermentation, combined with a multivariate statistical method to explore the intrinsic associations of "raw materials—microorganism—VOCs". The results revealed that after multiple rounds of fermentation, a higher number of microbial species were present in the cabbage paocai than in the red radish paocai with a brine of the same age. At the species level, *S. paracollinoides* and *F. siliginis* were found to be the characteristic microbial species of the red radish paocai, as well as 13 species belonging to the Firmicutes phylum (namely, *L. parafarraginis*, *L. buchneri*, *L. zymae*, *L. namurensis*, *L. diolivorans*, *L. acidipiscis*, *L. casei*, *L. parabuchneri*, *L. hilgardii*, *L. farraginis*, *P. selangorensis*, *L. sunkii* and *L. otakiensis*) and two species belonging to Proteobacteria (namely, *K. ascorbata* and *P. carotovorum*) for the cabbage paocai. By constructing a PLS-DA model, it was found that there were 13 kinds of different VOCs in the paocai from two raw materials. In addition, the correlation between the microorganisms and

VOCs in the cabbage paocai was stronger than that in the red radish paocai, and the most significant differences were found to be in 4-isopropylbenzyl alcohol, α-cadinol, terpinolene and isobutyl phenylacetate. The metagenomic sequencing in this study allowed the detection, identification and screening of the microbial ecology as well as the characteristic microorganisms of two different fermentation systems of the vegetable paocais. Further, the study also effectively allowed the identification of the VOCs and the establishment of their relationship to the microorganisms, which provided insights into the microbial flavor profiles of the studied paocais.

Supplementary Materials: The following are available online at https://www.mdpi.com/article/10.3390/foods11010062/s1, Table S1: Statistics profiles of the metagenome, Table S2: Content of volatile organic compounds in the two kinds of paocai.

Author Contributions: Conceptualization, L.J. and S.X.; methodology, L.J.; data curation, S.X.; resources, X.L.; writing—original draft preparation, L.J., S.X., G.S. and Z.Z.; visualization, X.H.; supervision, A.C.; project administration, A.C.; funding acquisition, A.C. All authors have read and agreed to the published version of the manuscript.

Funding: This work was supported by the Research on Key Technologies of Food Processing of Characteristic Biological Resources in Arid Valley of Southwest China (No. 2017YFC0505106-3).

Institutional Review Board Statement: Not applicable.

Informed Consent Statement: Not applicable.

Data Availability Statement: The data presented in this study are openly available in NCBI database at accession PRJNA731323. deposited in https://www.ncbi.nlm.nih.gov/sra/PRJNA731323 (accessed on 21 December 2020).

Acknowledgments: We thank Fei Xu and Sichuan Liji Lebao Food Co., Ltd. for their technical support.

Conflicts of Interest: The authors declare no competing financial interest.

References

1. Liu, D.Q.; Tong, C. Bacterial community diversity of traditional fermented vegetables in China. *LWT—Food Sci. Technol.* **2017**, *86*, 40–48. [CrossRef]
2. Chen, G.; Yu, W.H.; Zhang, Q.S.; Song, P.; Zhang, B.B.; Liu, Z.; You, J.G.; Li, H. Research of Sichuan Paocai and Lactic Acid Bacteria. *Adv. J. Food Sci. Technol.* **2014**, *6*, 1–5. [CrossRef]
3. Liang, H.P.; Chen, H.Y.; Ji, C.F.; Lin, X.P.; Zhang, W.X.; Li, L. Dynamic and Functional Characteristics of Predominant Species in Industrial Paocai as Revealed by Combined DGGE and Metagenomic Sequencing. *Front. Microbiol.* **2018**, *9*, 2416. [CrossRef]
4. Zhang, S.S. *Chinese Pickles*; Higher Education Press: Beijing, China, 1994.
5. Yan, P.-M.; Xue, W.-T.; Tan, S.-S.; Zhang, H.; Chang, X.-H. Effect of inoculating lactic acid bacteria starter cultures on the nitrite concentration of fermenting Chinese paocai. *Food Control* **2008**, *19*, 50–55. [CrossRef]
6. Song, H.S.; Whon, T.W.; Kim, J.; Lee, S.H.; Kim, J.Y.; Kim, Y.B.; Choi, H.-J.; Rhee, J.-K.; Roh, S.W. Microbial niches in raw ingredients determine microbial community assembly during kimchi fermentation. *Food Chem.* **2020**, *318*, 126481. [CrossRef] [PubMed]
7. Xu, X.; Wu, B.; Zhao, W.; Lao, F.; Chen, F.; Liao, X.; Wu, J. Shifts in autochthonous microbial diversity and volatile metabolites during the fermentation of chili pepper (*Capsicum frutescens* L.). *Food Chem.* **2021**, *335*, 127512. [CrossRef] [PubMed]
8. Xiao, Y.; Xiong, T.; Peng, Z.; Liu, C.; Huang, T.; Yu, H.; Xie, M. Correlation between microbiota and flavours in fermentation of Chinese Sichuan Paocai. *Food Res. Int.* **2018**, *114*, 123–132. [CrossRef] [PubMed]
9. Zhao, N.; Yang, B.; Lu, W.; Liu, X.; Zhao, J.; Ge, L.; Zhu, Y.; Lai, H.; Paul Ross, R.; Chen, W.; et al. Divergent role of abiotic factors in shaping microbial community assembly of paocai brine during aging process. *Food Res. Int.* **2020**, *137*, 109559. [CrossRef]
10. Chen, A.J.; Luo, W.; Peng, Y.T.; Niu, K.L.; Liu, X.Y.; Shen, G.H.; Zhang, Z.Q.; Wan, H.; Luo, Q.Y.; Li, S.S. Quality and microbial flora changes of radish paocai during multiple fermentation rounds. *Food Control* **2019**, *106*, 106733. [CrossRef]
11. Yang, J.; Cao, J.; Xu, H.; Hou, Q.; Yu, Z.; Zhang, H.; Sun, Z. Bacterial diversity and community structure in Chongqing radish paocai brines revealed using PacBio single-molecule real-time sequencing technology. *J. Sci. Food Agric.* **2018**, *98*, 3234–3245. [CrossRef] [PubMed]
12. Cao, J.; Yang, J.; Hou, Q.; Xu, H.; Zheng, Y.; Zhang, H.; Zhang, L. Assessment of bacterial profiles in aged, home-made Sichuan paocai brine with varying titratable acidity by PacBio SMRT sequencing technology. *Food Control* **2017**, *78*, 14–23. [CrossRef]
13. Wang, D.D.; Chen, G.; Tang, Y.; Li, H.; Shen, W.; Wang, M.; Liu, S.; Qin, W.; Zhang, Q. Effects of temperature on paocai bacterial succession revealed by culture-dependent and culture-independent methods. *Int. J. Food Microbiol.* **2020**, *317*, 108463. [CrossRef]

14. Bokulich, N.A.; Mills, D.A. Improved Selection of Internal Transcribed Spacer-Specific Primers Enables Quantitative, Ultra-High-Throughput Profiling of Fungal Communities. *Appl. Environ. Microbiol.* **2013**, *79*, 2519–2526. [CrossRef]
15. Agyirifo, D.S.; Wamalwa, M.; Otwe, E.P.; Galyuon, I.; Runo, S.; Takrama, J.; Ngeranwa, J. Metagenomics analysis of cocoa bean fermentation microbiome identifying species diversity and putative functional capabilities. *Heliyon* **2019**, *5*, e02170. [CrossRef]
16. Verce, M.; De Vuyst, L.; Weckx, S. Shotgun Metagenomics of a Water Kefir Fermentation Ecosystem Reveals a Novel Oenococcus Species. *Front. Microbiol.* **2019**, *10*, 479. [CrossRef] [PubMed]
17. De Filippis, F.; Parente, E.; Ercolini, D. Metagenomics insights into food fermentations. *Microb. Biotechnol.* **2017**, *10*, 91–102. [CrossRef]
18. Chen, S.; Zhou, Y.; Chen, Y.; Gu, J. fastp: An ultra-fast all-in-one FASTQ preprocessor. *Bioinformatics* **2018**, *34*, i884–i890. [CrossRef]
19. Li, D.; Liu, C.-M.; Luo, R.; Sadakane, K.; Lam, T.-W. MEGAHIT: An ultra-fast single-node solution for large and complex metagenomics assembly via succinct de Bruijn graph. *Bioinformatics* **2015**, *31*, 1674–1676. [CrossRef]
20. Noguchi, H.; Park, J.; Takagi, T. MetaGene: Prokaryotic gene finding from environmental genome shotgun sequences. *Nucleic Acids Res.* **2006**, *34*, 5623–5630. [CrossRef] [PubMed]
21. Fu, L.; Niu, B.; Zhu, Z.; Wu, S.; Li, W. CD-HIT: Accelerated for clustering the next-generation sequencing data. *Bioinformatics* **2012**, *28*, 3150–3152. [CrossRef]
22. Li, R.; Li, Y.; Kristiansen, K. SOAP: Short Oligonucleotide Alignment Program. *Bioinformatics* **2008**, *24*, 713–714. [CrossRef] [PubMed]
23. Buchfink, B.; Xie, C.; Huson, D.H. Fast and sensitive protein alignment using DIAMOND. *Nat. Methods* **2015**, *12*, 59–60. [CrossRef] [PubMed]
24. Xiao, Y.S.; Huang, T.; Huang, C.L.; Hardie, J.; Peng, Z.; Xie, M.Y.; Xiong, T. The microbial communities and flavour compounds of Jiangxi yancai, Sichuan paocai and Dongbei suancai: Three major types of traditional Chinese fermented vegetables. *LWT—Food Sci. Technol.* **2020**, *121*, 108865. [CrossRef]
25. Liu, Z.; Peng, Z.; Huang, T.; Xiao, Y.; Li, J.; Xie, M.; Xiong, T. Comparison of bacterial diversity in traditionally homemade paocai and Chinese spicy cabbage. *Food Microbiol.* **2019**, *83*, 141–149. [CrossRef]
26. Xiong, T.; Guan, Q.; Song, S.; Hao, M.; Xie, M. Dynamic changes of lactic acid bacteria flora during Chinese sauerkraut fermentation. *Food Control* **2012**, *26*, 178–181. [CrossRef]
27. Liu, A.; Liu, G.; Huang, C.; Shen, L.; Li, C.; Liu, Y.; Liu, S.; Hu, B.; Chen, H. The bacterial diversity of ripened Guang'yuan Suancai and In Vitro evaluation of potential probiotic lactic acid bacteria isolated from Suancai. *LWT—Food Sci. Technol.* **2017**, *85*, 175–180. [CrossRef]
28. Kim, M.; Chun, J. Bacterial community structure in kimchi, a Korean fermented vegetable food, as revealed by 16S rRNA gene analysis. *Int. J. Food Microbiol.* **2005**, *103*, 91–96. [CrossRef]
29. An, F.; Sun, H.; Wu, J.; Zhao, C.; Li, T.; Huang, H.; Fang, Q.; Mu, E.; Wu, R. Investigating the core microbiota and its influencing factors in traditional Chinese pickles. *Food Res. Int.* **2021**, *147*, 110543. [CrossRef]
30. Lee, K.; Lee, Y. Effect of Lactobacillus plantarum as a starter on the food quality and microbiota of kimchi. *Food Sci. Biotechnol.* **2010**, *19*, 641–646. [CrossRef]
31. Wang, Z.; Shao, Y. Effects of microbial diversity on nitrite concentration in pao cai, a naturally fermented cabbage product from China. *Food Microbiol.* **2018**, *72*, 185–192. [CrossRef]
32. Segata, N.; Waldron, L.; Ballarini, A.; Narasimhan, V.; Jousson, O.; Huttenhower, C. Metagenomic microbial community profiling using unique clade-specific marker genes. *Nat. Methods* **2012**, *9*, 811–814. [CrossRef] [PubMed]
33. Conlan, S.; Kong, H.H.; Segre, J.A. Species-Level Analysis of DNA Sequence Data from the NIH Human Microbiome Project. *PLoS ONE* **2012**, *7*, e47075. [CrossRef] [PubMed]
34. Segata, N.; Izard, J.; Waldron, L.; Gevers, D.; Miropolsky, L.; Garrett, W.S.; Huttenhower, C. Metagenomic biomarker discovery and explanation. *Genome Biol.* **2011**, *12*, R60. [CrossRef]
35. Pereira, I.; Severino, P.; Santos, A.C.; Silva, A.M.; Souto, E.B. Linalool bioactive properties and potential applicability in drug delivery systems. *Colloids Surf. B* **2018**, *171*, 566–578. [CrossRef] [PubMed]
36. Kupska, M.; Chmiel, T.; Jedrkiewicz, R.; Wardencki, W.; Namiesnik, J. Comprehensive two-dimensional gas chromatography for determination of the terpenes profile of blue honeysuckle berries. *Food Chem.* **2014**, *152*, 88–93. [CrossRef] [PubMed]
37. Verzera, A.; Dima, G.; Tripodi, G.; Ziino, M.; Lanza, C.M.; Mazzaglia, A. Fast Quantitative Determination of Aroma Volatile Constituents in Melon Fruits by Headspace-Solid-Phase Microextraction and Gas Chromatography-Mass Spectrometry. *Food Anal. Meth.* **2011**, *4*, 141–149. [CrossRef]
38. Hanschen, F.S.; Rohn, S.; Mewis, I.; Schreiner, M.; Kroh, L.W. Influence of the chemical structure on the thermal degradation of the glucosinolates in broccoli sprouts. *Food Chem.* **2012**, *130*, 1–8. [CrossRef]

Article

Bacterial Community of Grana Padano PDO Cheese and Generical Hard Cheeses: DNA Metabarcoding and DNA Metafingerprinting Analysis to Assess Similarities and Differences

Miriam Zago [1], Lia Rossetti [1], Tommaso Bardelli [2], Domenico Carminati [1], Nelson Nazzicari [1] and Giorgio Giraffa [1,*]

1 Research Centre for Animal Production and Aquaculture (CREA-ZA), Council for Agricultural Research and Economics, 26900 Lodi, Italy; miriam.zago@crea.gov.it (M.Z.); lia.rossetti@crea.gov.it (L.R.); domenico.carminati@crea.gov.it (D.C.); nelson.nazzicari@crea.gov.it (N.N.)

2 Research Centre for Plant Protection and Certification (CREA-DC), Council for Agricultural Research and Economics, 26900 Lodi, Italy; tommaso.bardelli@crea.gov.it

* Correspondence: giorgio.giraffa@crea.gov.it; Tel.: +39-0371-45011

Citation: Zago, M.; Rossetti, L.; Bardelli, T.; Carminati, D.; Nazzicari, N.; Giraffa, G. Bacterial Community of Grana Padano PDO Cheese and Generical Hard Cheeses: DNA Metabarcoding and DNA Metafingerprinting Analysis to Assess Similarities and Differences. *Foods* **2021**, *10*, 1826. https://doi.org/10.3390/foods10081826

Academic Editors: Valentina Bernini and Juliano De Dea Lindner

Received: 30 June 2021
Accepted: 6 August 2021
Published: 7 August 2021

Abstract: The microbiota of Protected Designation of Origin (PDO) cheeses plays an essential role in defining their quality and typicity and could be applied to protect these products from counterfeiting. To study the possible role of cheese microbiota in distinguishing Grana Padano (GP) cheese from generical hard cheeses (HC), the microbial structure of 119 GP cheese samples was studied by DNA metabarcoding and DNA metafingerprinting and compared with 49 samples of generical hard cheeses taken from retail. DNA metabarcoding highlighted the presence, as dominant taxa, of *Lacticaseibacillus rhamnosus*, *Lactobacillus helveticus*, *Streptococcus thermophilus*, *Limosilactobacillus fermentum*, *Lactobacillus delbrueckii*, *Lactobacillus* spp., and *Lactococcus* spp. in both GP cheese and HC. Differential multivariate statistical analysis of metataxonomic and metafingerprinting data highlighted significant differences in the Shannon index, bacterial composition, and species abundance within both dominant and subdominant taxa between the two cheese groups. A supervised Neural Network (NN) classification tool, trained by metagenotypic data, was implemented, allowing to correctly classify GP cheese and HC samples. Further implementation and validation to increase the robustness and improve the predictive capacity of the NN classifier will be needed. Nonetheless, the proposed tool opens interesting perspectives in helping protection and valorization of GP and other PDO cheeses.

Keywords: Grana Padano cheese; generical hard cheeses; bacterial diversity; DNA metabarcoding; DNA (meta)fingerprinting; predictive models; neural network

1. Introduction

Protected Designation of Origin (PDO) products represent the excellence of European agricultural food production and are the result of the interplay between environmental (e.g., climate) and human factors (e.g., production techniques handed down over time), which are typical of a given territory. To this regard, many cheeses benefit from the PDO quality label. PDO cheeses are subject to specific production conditions, which producers spontaneously adhere to by joining the Consortia, which establish the transformation criteria through specific shared rules and guidelines. In this way, the consumer is guaranteed in terms of transparency and traceability, while benefiting from high-quality products [1]. The overall quality of cheeses, including PDO cheeses, is the result of many concomitant factors, such as the quality of the raw material, the farming methods, and the processing technology, which, in some cheeses (such as Grana Padano, Parmigiano Reggiano, Silter, Pélardon, Poro), involves the use of undefined microbial cultures. The interaction between these elements contributes to shape the qualitative and quantitative microbiological content of the ripened product, giving it a specific composition [2–8]. On the other hand, the increasing frequency of imitations and frauds involving PDO foods, including cheeses [9], addressed the search for finer and more sensitive methods of discrimination.

In 2020, 210,000 tons (or approximately 5,200,000 cheeses) of Grana Padano (GP) were produced. With a growing export trend (40% of total production, +3.3% compared to 2019), GP is increasingly consumed both in EU and non-EU countries (https://www.granapadano.it, accessed on 5 July 2021) [10]. Grana Padano is a cooked, long ripened PDO cheese made from raw and partially skimmed milk, which is fermented by lactic bacteria present in natural starters ('sieroinnesto'). The sieroinnesto, which is one of the typical elements of Grana Padano, is prepared by backslopping, i.e., using part of the drained whey from the previous day's cheese making, which is left for 18–24 h at 42–45 °C until a final acidity of approx. 60 °SH/100 (pH 3.3–3.6) is reached [2,11,12]. On the other hand, GP is also one of the most imitated and counterfeited cheeses. This has stimulated the search for increasingly sensitive analytical approaches to GP mapping and characterization. A recent investigation carried out on a limited number of cheese samples using untargeted metabolomics revealed differences in chemical fingerprints between PDO and non-PDO grana-like cheeses [13]. In other studies, geographic or technological differences were observed in the microbiota of cheese, suggesting it as a possible tool to establish cheese authenticity and diversity [14–17]. In a previous study, the bacterial taxa present in GP were highlighted by 16S rRNA gene sequencing (DNA metabarcoding) [18]. In the present study, the usefulness of the cheese microbiota to distinguish GP cheese from generical hard cheeses, i.e., cheeses whose appearance could be confused with Grana Padano PDO cheese, was investigated. The structure and the genotypic fingerprinting of the bacterial taxa of 119 GP samples were evaluated by 16S rRNA gene sequencing (DNA metabarcoding) and RAPD-PCR from total cheese DNA (RAPD-PCR metafingerprinting), respectively, and compared with 49 samples of generical hard cheeses retrieved from retail.

2. Materials and Methods

2.1. Sampling of Cheeses

One hundred nineteen samples (2-kg slice each), representative of all producers of GP and the entire geographical area of production, were collected. Cheeses had been produced in August 2018 and had a ripening time of 6–7 months. Forty-nine samples of generical hard cheeses (from now on 'HC'), labelled as EU and extra-EU products, were retrieved from retail. All cheese samples were stored at −20 °C. After thawing at 4 °C for 18 h, slices (10 g each) including three different sections of the cheese (i.e., outer, central, and inner) were sampled and put into sterile containers.

2.2. Total DNA Extraction

The 10-g cheese slices were treated as described by Zago et al. [18]. Briefly, the samples were homogenized twice in a Stomacher 400 Circulator (Seward Laboratory, London, UK) with sterile sodium citrate (NaCt 2% w/v, pH 7.5). The cheese homogenate was centrifuged (14,400× g for 7 min at 4 °C) for fat removing. The pellet was then resuspended in Triton X-100 (2.5% v/v), washed twice with phosphate-buffered saline (pH 7.5), and centrifuged (8700× g, 7 min, 4 °C). Finally, after the pellet was resuspended in Tris-EDTA (0.1 M, pH 8), total dsDNA was extracted using a QIAcube HT automated station (Qiagen, Milan, Italy) using QIAamp 96 QIAcube HT kit (Qiagen, Milano, Italy). Total dsDNA was quantified fluorometrically (Qubit™, Life Technologies, Monza, Italy).

2.3. DNA Metabarcoding

Total DNA extracted from the 119 GP and the 49 HC samples was subjected to DNA metabarcoding analysis (IGATech, Udine, Italy) by sequencing of the variable V3–V4 regions of the 16S rRNA gene using an Illumina MiSeq platform, as described previously [18].

2.4. RAPD-PCR Metafingerprinting

RAPD-PCR from total DNA of each sample was carried out according to Zago et al. [18]. PCR products were separated by QIAxcel electrophoresis using dedicated DNA Screening Gel Cartridges (Qiagen, Milan, Italy). Two QX Alignment Markers (15 bp–5 kb and 15 bp–600 bp, in a 1:1 ratio) and the QX DNA Size Marker (100 bp–2.5 kb) were included on each run. The repeatability and reproducibility of this method was evaluated by repeated amplification and analysis of four different DNA cheese samples, with both primers for every analysis performed.

2.5. Data Analysis and Bioinformatics Processing

Reads were de-multiplexed based on the Illumina indexing system. Following the QIIME pipelines, the USEARCH algorithm (version 8.1.1756, 32-bit) allowed the following steps: chimera filtering; grouping of replicate sequences; sorting sequences per decreasing abundance; and OTU identification, with a species-level taxonomic resolution. When the taxonomy assignment did not reach the species level, the genus or family name were reported. After removing OTUs <5 reads [19], alpha (α) diversity (richness and Shannon indexes), beta (β) diversity (principal component analysis-PCA and principal coordinate analysis-PCoA), and the rarefaction curves were estimated, on the resulting OTU table, by means of R (http://www.r-project.org/index.html; accessed on 7 April 2021), using "vegan" [20] and "agricolae" [21]. Statistical differences (p value ≤ 0.05) and evaluation of the influence of the two cheese samplings on the microbial indexes were evaluated by ANOVA followed by the Tukey HSD test. Relative abundance for each OTU across all cheeses was calculated, and "subdominant" and "dominant" OTUs were discriminated according to Zago et al. [18]. Taxonomic analysis was carried out through "reshape2" and "ggplot2" packages [22,23]. RAPD-PCR profiles were imported and analyzed by BioNumerics^{TM} (version 7.6, Applied Maths, Sint-Martens-Latem, Belgium), as previously described [18].

A Neural Network (NN), trained to discriminate GP cheese and HC samples, was implemented with the aim to assess the feasibility of automatic classification for cheese samples and to compare the discriminatory power of metabarcoding and metafingerprinting analyses. The NN was implemented from scratch using Keras python library [24]. The input layer received PCoA data and contained 64 fully connected ReLU units; the second layer had ReLU 32 nodes; the third layer had 16. The last layer had only one sigmoid unit and performed the final binary classification. NN were trained on 80% of the available data, picked as a random stratification (thus maintaining the GP cheese/HC ratio).

Adam optimizer was used to minimize "binary cross-entropy" loss function. The following other performance metrics were measured: Area Under Curve (AUC), true and false positive counts (TP, FP), and true and false negatives count (TN, FN). We also derived Precision, True Positive and True Negative Rates (TPR, TNR), and Accuracy and Binary accuracy. All reported performance metrics were averaged over 10 repetitions of the training process, so as to avoid possible biases due to the random selection of validation set samples. Given the unbalance of tally classes, class weights were computed via the 'compute_class_weight' function from the sklearn [25] package and passed to the optimizer.

As a post-hoc investigation, we decided to repeat the NN training tuning class weights so as to explore the limits of the detection power of a hypothetical automated NN-based screening system. Specifically, it measured the possibility to raise True Positive Rate (TPR), i.e., the fraction of HC that were correctly identified, without significantly lowering the True Negative Rate (TNR), i.e., the fraction of GP cheese that were correctly classified. Given that, by definition, there is a tradeoff between TPR and TNR, we fixed the target thresholds to have both metrics higher than 0.9.

3. Results

3.1. DNA Metabarcoding Analysis

One hundred and nineteen GP cheese samples and 49 samples of generical hard cheeses (i.e., cheeses whose appearance could be confused with Grana Padano PDO cheese, or HC) were analyzed. Overall, 35,842,968 reads were sequenced, with 218,554 reads per sample on average (range 80,636–805,340). A total of 477 OTUs, of which 130 were further split into 48 dominant (≥1% total reads) and 82 subdominant (0.1–1% total reads) taxa, were identified (data not shown).

3.1.1. Species Abundance

According to the aim of this study, the samples were divided into GP cheese and HC. Considering the relative abundance of the dominant bacterial species (48 taxa; ≥1% total reads) found both in GP and in HC, *Lacticaseibacillus rhamnosus* was the prevalent species, followed by *Lactobacillus helveticus, Streptococcus thermophilus, Limosilactobacillus fermentum, Lactobacillus delbrueckii, Lactobacillus* spp., and *Lactococcus* spp., with average values between 2% and 46% in GP cheese samples. Notably, the dominant species were less abundant in HC samples compared to GP cheese (Figure 1). In the cheese microbiota, many contaminating bacteria deriving from raw milk or the process environment, such as potentially pathogenic lactic acid bacteria (LAB) (i.e., *Streptococcus uberis* and *Lactococcus garviae*) and other non-LAB taxa (i.e., *Micrococcus, Staphylococcus, Acinetobacter, Pseudomonas,* and many enterobacteria), were also detected (Figure 1).

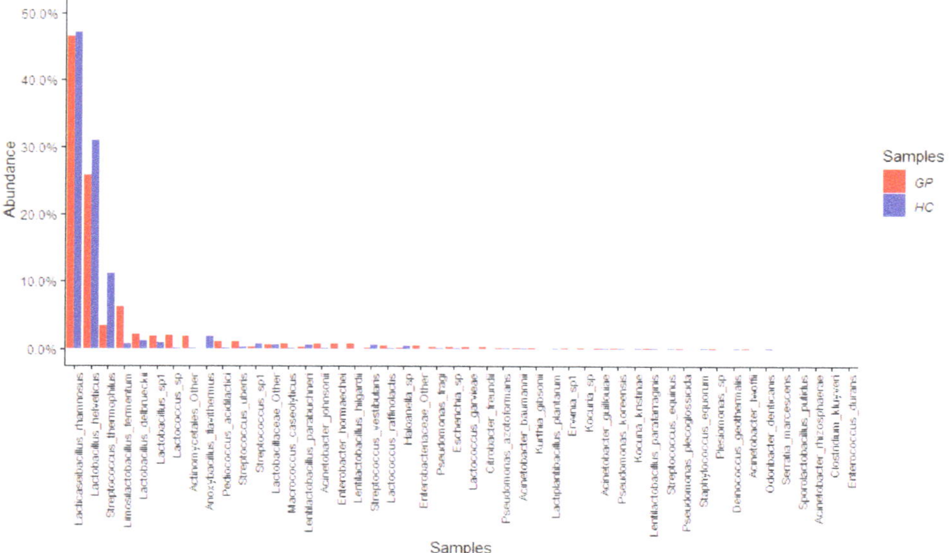

Figure 1. Average values of relative abundance of the 48 dominant taxa retrieved in Grana Padano (GP) and similar hard cheese (HC) samples.

A different distribution between GP cheese and HC samples of the 82 subdominant species (0.1–1% total reads) was also observed (Figure S1).

3.1.2. Alpha and Beta Diversity

Richness (i.e., the number of different species in each sample) and Shannon indexes are reported (Figure 2). A greater uniformity of species was retrieved in GP cheese, compared to HC, as stated by the statistical significance of the Shannon diversity index (d.f. = 1, F = 16.89, $p < 0.001$), whereas there was not a significant difference in the number of species as indicated by Richness (d.f. = 1, F = 2.02, $p = 0.157$). The PCA showed no differences between GP cheese and HC, which were clustered together (Figure 3a). Conversely, the PCoA reported some differences between GP cheese and HC, as the latter were isolated on both sides of the panel (d.f. = 1, F = 4.70, $p < 0.001$; Figure 3b).

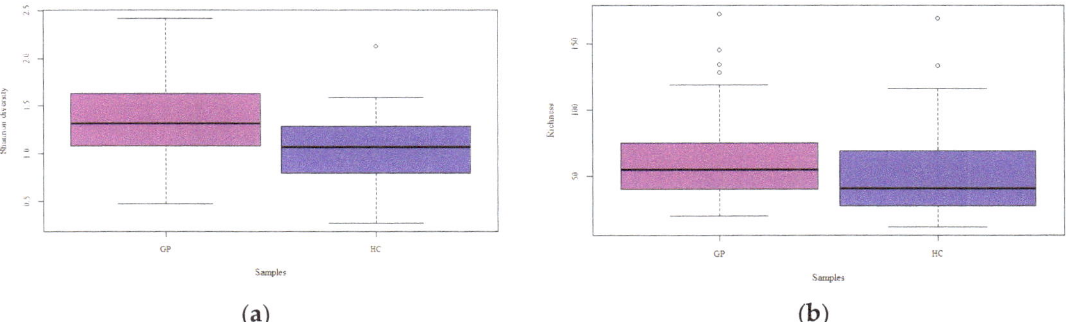

Figure 2. (**a**) Shannon diversity and (**b**) Richness of Grana Padano cheese (GP) and generical hard cheese (HC) samples. Different letters indicate significant ($p < 0.05$) differences between samplings.

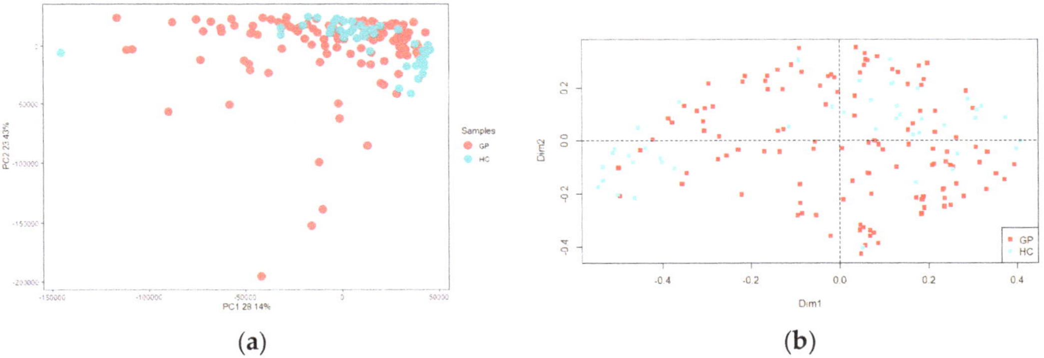

Figure 3. (**a**) Principal component analysis (PCA) and (**b**) Principal coordinate analysis (PCoA) based on operational taxonomic unit (OTU) relative abundance of Grana Padano cheese (GP, red) and generical hard cheeses (HC, green). The first component (horizontal) of PCA accounts for 28.14% of variance, and the second component (vertical) of PCA accounts for 23.43% of variance.

3.2. RAPD-PCR Metafingerprinting Analysis

After RAPD-PCR amplification of the total microbial DNA extracted from all the cheeses, the resulting patterns were analyzed to detect sample-associated profiles and/or specific bands, with a repeatability and reproducibility of 70%. The presence of a panel of shared bands (labelled I in the left side of Figure S2) was highlighted. It included five series of common bands, present in more than 65% of the whole set of samples (GP cheese and HC). Moreover, a panel of unshared bands (or single bands), marked II in the right side of Figure S2, was observed. The matrix obtained from the data was analyzed by PCA, and statistically significant differences were observed between samples (d.f. = 1, F = 3.49, $p < 0.001$), which allowed to clearly separate the GP cheese and HC at the top and bottom of the panel, respectively (Figure 4a). The same data set was subjected to PCoA, and this trend was confirmed (d.f. = 1, F = 18.43, $p < 0.001$; Figure 4b).

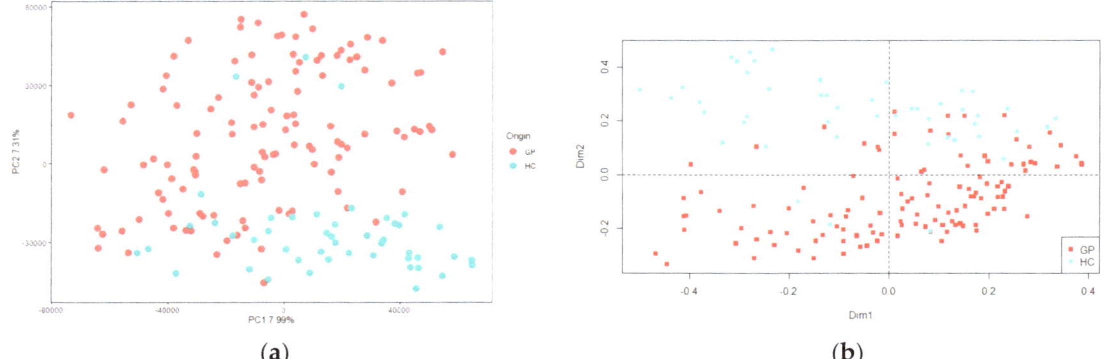

(a) (b)

Figure 4. (**a**) Principal component analysis (PCA) and (**b**) Principal coordinate analysis (PCoA) based on band matching and Pearson correlation analysis of the metafingerprinting data of Grana Padano cheese (GP, red) and generical hard cheeses (HC, green). The first component (horizontal) of PCA accounts for 7.99% of variance and the second component (vertical) of PCA accounts for 7.31% of variance.

3.3. Implementation of a Classifier

DNA metabarcoding and RAPD-PCR metafingerprinting PCoA data were used to implement a classifier that could provide a fast, reliable, and automatic categorization of cheese samples into GP cheese or HC. Results for several performance metrics, measured on a randomly picked 20% of the data acting as validation set, are reported in Table 1. The same metrics measured on the training set are reported in Table S1. In general, the metafingerprinting dataset performed better than the metabarcoding dataset, e.g., for Area Under Curve (0.934 vs. 0.657, respectively) and Accuracy (0.928 vs. 0.637, respectively). Metafingerprinting showed a very low average level of False Positives (0.262), indicating that, on average, on about four out of five runs, no False Positives were detected, and in the fifth run, only one GP cheese was erroneously classified as HC.

Table 1. Performance statistics on the validation set for a Neural Network classifier trained on either DNA metabarcoding or RAPD-PCR metafingerprinting PCoA data. The table reports statistics averaged over 10 rounds of training on an 80/20 stratified split of training and validation set. The rightmost column reports metrics when the default class weight for class GP cheese is adjusted by a multiplicative coefficient equal to 0.25, chosen so that both TPR and TNF resulted above a threshold of 0.9.

Statistics	Metabarcoding	Fingerprinting	Fingerprinting Weight Adjusted
FN	3.952	2.188	0.840
FP	8.022	0.262	2.320
TN	15.978	23.738	21.680
TP	5.048	7.812	9.160
AUC	0.657	0.934	0.972
Loss	1.147	0.338	0.312
Precision	0.386	0.968	0.798
TPR	0.561	0.781	0.916
TNR	0.666	0.989	0.903
Accuracy	0.637	0.928	0.907
Balanced Accuracy	0.613	0.885	0.910

FN: number of False Negatives. FP: number of False Positives. TN: True Negatives. TP: True Positives. AUC: Area Under Curve. Loss: final value of loss function, optimized during the training phase. Precision: TP/(TP+FP). TPR, True Positive Rate: TP/(TP+FN). TNR, True Negative Rate: TN/(TN+FP). Accuracy: (TP+TN)/(TP+TN+FP+FN). Balanced Accuracy: (TPR+TNR)/2.

As a post-hoc analysis, we decided to repeat the NN training on the metafingerprinting PCoA dataset using class weights more skewed toward the detection of HC samples, with the declared target of having both TPR and TNR metrics above the threshold of 0.9. We tested several configurations of weights (data not reported) and found that using an adjustment coefficient equal to 0.25 on the GP cheese class weight, the desired performance was achieved. On the DNA metabarcoding PCoA dataset, no weight adjustment was found to be able to bring both TPR and TNR above the 0.9 target threshold.

4. Discussion

Differences in microbial profiling could be useful to establish which microbial components may be responsible for the authentication of PDO cheeses [16,17,26]. On this basis, the microbial diversity of a large sampling that included the entire production area of the GP cheese was investigated and compared with that of 49 generical hard cheeses (HC) taken from retail. Two metataxonomic methods were applied, i.e., 16S rRNA gene sequencing and metagenotyping using RAPD-PCR from total cheese DNA, to identify and implement a robust classifier to differentiate GP cheese from HC. Metataxonomic data revealed a total of about 477 OTUs, including GP cheese and HC, but an overall greater relative abundance in the former ones. This trend occurred among both dominant (among which *Lcb. rhamnosus*, *L. helveticus*, *S. thermophilus*, *Limosilactobacillus fermentum*, *L. delbrueckii*, *Lactobacillus* spp., and *Lactococcus* spp. prevailed) and subdominant species. While detected LAB taxa within both dominant and subdominant taxa belong to the typical microbiota generally recovered from hard cheeses, the finding of residual psychrotrophic bacteria is not uncommon. Indeed, raw milk for Grana Padano is normally kept under refrigerated conditions before collection and subsequent transport to processing sites [18]. The presence of bacterial DNA of potentially pathogenic LAB and non-LAB taxa is not uncommon, but it has no safety significance, as these bacteria are inactivated by the combined effects of cooking temperature, the curd acidification during the early phases of cheesemaking, and the long ripening times under the harsh conditions (low moisture, low activity water) that characterize the production of hard cheeses [27–29]. The Shannon index, which accounts for both microbial richness and evenness, was significantly influenced by the two samplings, highlighting a greater uniformity of species in GP cheese but a similar richness. Differential analysis based on Bray–Curtis dissimilarity between all samples and calculated using species relative abundancies delineated a significant, although incomplete, separation between the two groups.

Cheese production conditions within a PDO area were selected for specific microbial populations. The microbiota of cheeses, especially those (such as Grana Padano) obtained with artisanal processes and from raw milk, plays a fundamental role in defining the qualitative characteristics and safety parameters of the final products. Its composition, structure, and modulation are the result of different selective pressures, e.g., microbiological quality of milk, technology, type of starter used, and processing environment [30]. Therefore, differences in one or more of these factors can be decisive in shaping specific microbiological profiles in cheese. Grana Padano cheese is obtained from raw milk produced in a large production area, which includes most of the provinces of northern Italy included in the Po Valley. Although animal feed is substantially similar, the microbiological composition of milk aimed at GP cheese production can be influenced by management practices at the farm level and by seasonal, climatic, and environmental variations. For example, the presence of some LAB species, such as *Lactiplantibacillus plantarum*, *Lentilactobacillus parabuchneri*, *Lentilactobacillus parafarraginis*, *Lentilactobacillus hilgardii*, used also as silage starters, can be related to corn silage fed to cows producing milk for GP [31]. The microbial content of raw milk is subsequently modulated by the selective action of the technology and the practice of using undefined whey starter cultures [32,33]. The greater microbial heterogeneity of raw milk for GP, coming from a large production area, and the 'balancing' action exerted by the application of a very similar technology and the addition to raw milk of whey starter cultures, which are usually prepared with comparable methods among the different dairies, are critical to explain the higher species abundance and uniformity observed in GP cheese. On the other hand, HC not being subject to the constraints of a PDO can be produced from milk outside the production area (with its specific microbial content) and with technologies that, although substantially similar, often rely on the use of selected starter cultures and/or the application of thermization, pasteurization, bactofugation, or microfiltration of milk [34]. These selective pressures could be decisive in explaining the lower OTU abundance and species uniformity in HC. The above trends were corroborated, and the separation between the two groups of samples, which was highlighted by the Bray–Curtis dissimilarity test, was noted.

DNA fingerprinting methods, such as RAPD-PCR, are often applied to evaluate the endemicity, or the prevalence, of a given strain. Furthermore, this technique was found useful in discriminating soil microbial communities and estimating their relatedness [35,36]. In a previous study, RAPD-PCR proved useful to highlight a pattern of bands present in all the samples, as well as more specific bands, which aggregated groups of samples, or distinguished single samples, within the microbial community of GP [11]. In the present work, RAPD-PCR was applied to fingerprint the overall bacterial community of GP cheese and HC samples. The obtained metagenotypes were evaluated as possible tools to differentiate the two sampling groups, assuming, unlike metataxonomic analysis, that this technique was able to identify strain- or group-specific differences within complex microbial communities. Data processing of RAPD-PCR profiles using PCA and Bray–Curtis dissimilarity analysis allowed to obtain a clear and statistically very significant separation between GP cheese and HC samples. Metataxonomic and meta-fingerprinting data were then used as inputs to train and validate a two-class (GP cheese vs. HC) classifier based on a neural network as computational model. While metataxonomic data did not allow for reliable classification, the discriminatory power of metafingerprinting enabled to build an extremely robust model (very high binary accuracy). When trained by metagenotyping data, the model correctly classified GP and HC samples. The origin of this differentiation is not currently known, although it is likely that strain-specific peculiarities in the microbial community, determined by previously outlined ecological, geographic, or technological selective pressures, might explain it. The molecular (meta)fingerprint of the entire microbial community could be promising to assist to authenticate GP cheese and to distinguish it from imitation products as part of an attempt to hinder any counterfeits. Further validation will be needed to increase robustness of the classifier and confirm, with unknown samples, its discriminating ability.

Supplementary Materials: The following are available online at https://www.mdpi.com/article/10.3390/foods10081826/s1, Figure S1: Average values of relative abundance of the 82 subdominant taxa retrieved in Grana Padano (GP) samples and similar hard cheeses (HC) samples; Figure S2: Band matching cluster analysis showing the relationship between Grana Padano and similar hard cheese samples and M13-/BOXA1R-RAPD-PCR bands; Table S1: Performance statistics on the training set for a Neural Network classifier trained on either metabarcoding or fingerprinting PCoA data.

Author Contributions: Conceptualization, M.Z. and G.G.; methodology, M.Z. and L.R.; software, M.Z. and N.N.; validation, N.N.; formal analysis, L.R. and T.B.; investigation, L.R. and T.B.; writing—original draft preparation, M.Z.; N.N., and G.G.; writing—review and editing, M.Z.; D.C., and G.G.; supervision, M.Z. and G.G.; funding acquisition, G.G. and D.C. All authors have read and agreed to the published version of the manuscript.

Funding: This research was funded by the Ministry of Agricultural, Food, and Forestry Policies (MIPAAF) as part of the Project entitled "New Technologies for Cheese Production (NEWTECH)" (D.M. 0021568 of 28 July 2017).

Conflicts of Interest: The authors declare no conflict of interest.

References

1. Regulation, E.C. Council Regulation EC no 1151 of November 21, 2012. *Off. J. Eur. Union* **2015**, *343*, 1–29.
2. Gatti, M.; Bottari, B.; Lazzi, C.; Neviani, E.; Mucchetti, G. Invited review: Microbial evolution in raw-milk, long-ripened cheeses produced using undefined natural whey starters. *J. Dairy Sci.* **2014**, *97*, 573–591. [CrossRef]
3. Santarelli, M.; Bottari, B.; Lazzi, C.; Neviani, E.; Gatti, M. Survey on the community and dynamics of lactic acid bacteria in Grana Padano cheese. *Syst. Appl. Microbiol.* **2013**, *36*, 593–600. [CrossRef] [PubMed]
4. Aldrete-Tapia, A.; Escobar-Ramírez, M.C.; Tamplin, M.L.; Hernandez-Iturriaga, M. High-throughput sequencing of microbial communities in Poro cheese, an artisanal Mexican cheese. *Food Microbiol.* **2014**, *44*, 136–141. [CrossRef] [PubMed]
5. Delcenserie, V.; Taminiau, B.; Delhalle, L.; Nezer, C.; Doyen, S.; Crevecoeur, S.; Roussey, D.; Korsak, N.; Daube, G. Microbiota characterization of a Belgian protected designation of origin cheese, Herve cheese, using metagenomic analysis. *J. Dairy Sci.* **2014**, *97*, 6046–6056. [CrossRef] [PubMed]
6. Silvetti, T.; Capra, E.; Morandi, S.; Cremonesi, P.; Decimo, M.; Gavazzi, F.; Giannico, R.; De Noni, I.; Brasca, M. Microbial population profile during ripening of Protected Designation of Origin (PDO) Silter cheese, produced with and without autochthonous starter culture. *LWT Food Sci. Technol.* **2017**, *84*, 821–831. [CrossRef]
7. Bertani, G.; Levante, A.; Lazzi, C.; Bottari, B.; Gatti, M.; Neviani, E. Dynamics of a natural bacterial community under technological and environmental pressures: The case of natural whey starter for Parmigiano Reggiano cheese. *Food Res. Int.* **2020**, *129*, 108860. [CrossRef]
8. Penland, M.; Falentin, H.; Parayre, S.; Pawtowski, A.; Maillard, M.-B.; Thierry, A.; Mounier, J.; Coton, M.; Deutsch, S.-M. Linking Pelardon artisanal goat cheese microbial communities to aroma compounds during cheese-making and ripening. *Int. J. Food Microbiol.* **2021**, *345*, 109130. [CrossRef]
9. ICQRF Report 2021. In *Department of Central Inspectorate for Fraud Repression and Quality Protection of the Agri-Food Products and Foodstuffs*; Ministry of Agricultural, Food and Forestry Policies: Rome, Italy, 2021.
10. Bava, L.; Bacenetti, J.; Gislon, G.; Pellegrino, L.; D'Incecco, P.; Sandrucci, A.; Tamburini, A.; Fiala, M.; Zucali, M. Impact assessment of traditional food manufacturing: The case of Grana Padano cheese. *Sci Total Environ.* **2018**, *626*, 1200–1209. [CrossRef]
11. Summer, A.; Formaggioni, P.; Franceschi, P.; Di Frangia, F.; Righi, F.; Malacarne, M. Cheese as functional food: The example of Parmigiano Reggiano and Grana Padano. *Food Technol Biotechnol.* **2017**, *55*, 277–289. [CrossRef]
12. Gobbetti, M.; Fox, P.; Neviani, E. The most traditional and popular Italian cheeses. In *The Cheeses of Italy: Science and Technology*; Gobbetti, M., Fox, P., Neviani, E., Eds.; Springer Int. Publishing AG: Cham, Switzerland, 2018; pp. 99–267. [CrossRef]
13. Rocchetti, G.; Lucini, L.; Gallo, A.; Masoero, F.; Trevisan, M.; Giuberti, G. Untargeted metabolomics reveals differences in chemical fingerprints between PDO and non-PDO Grana Padano cheeses. *Food Res. Int.* **2018**, *113*, 407–413. [CrossRef] [PubMed]
14. Smit, G.; Smit, B.A.; Engels, W.J.M. Flavour formation by lactic acid bacteria and biochemical flavour profiling of cheese products. *FEMS Microbiol. Rev.* **2005**, *29*, 591–610. [CrossRef] [PubMed]
15. Barron, L.J.R.; Aldai, N.; Virto, M.; de Renobales, M. Cheeses with Protected land and tradition-related labels: Traceability and authentication. In *Global Cheesemaking Technology: Cheese Quality and Characteristics*, 1st ed.; Papademas, P., Bintsis, T., Eds.; John Wiley's and Sons Ltd.: Hoboken, NJ, USA, 2018; pp. 100–119. [CrossRef]
16. Yeluri Jonnala, B.R.; McSweeney, P.L.H.; Sheehan, J.J.; Cotter, P.D. Sequencing of the cheese microbiome and its relevance to industry. A review. *Front Microbiol.* **2018**, *9*, 1020. [CrossRef] [PubMed]
17. Kamimura, B.A.; De Filippis, F.; Sant'Ana, A.S.; Ercolini, D. Large-scale mapping of microbial diversity in artisanal Brazilian cheeses. *Food Microbiol.* **2019**, *80*, 40–49. [CrossRef]
18. Zago, M.; Bardelli, T.; Rossetti, L.; Nazzicari, N.; Carminati, D.; Galli, A.; Giraffa, G. Evaluation of bacterial communities of Grana Padano cheese by DNA metabarcoding and DNA metafingerprinting analysis. *Food Microbiol.* **2021**, *93*, 103613. [CrossRef] [PubMed]
19. Probst, M.; Gómez-Brandón, M.; Bardelli, T.; Egli, M.; Insam, H.; Ascher-Jenull, J. Bacterial communities of decaying Norway spruce follow distinct slope exposure and time-dependent trajectories. *Environ Microbiol.* **2018**, *20*, 3657–3670. [CrossRef]
20. Oksanen, J.; Blanchet, G.F.; Friendly, M.; Kindt, R.; Legendre, P.; McGlinn, D.; Minchin, P.R.; O'Hara, R.B.; Simpson, G.L.; Solymos, P.; et al. *Vegan: Community Ecology Package*; R Package, Version 2.5-5; R Foundation: Vienna, Austria, 2019.
21. De Mendiburu, F. Package 'Agricolae.' R Package Version 1.2-8. Available online: http://CRAN.R-project.org/package=agricolae (accessed on 4 July 2019).
22. Caporaso, J.; Kuczynski, J.; Stombaugh, J.; Bittinger, K.; Bushman, F.D.; Costello, E.K.; Fierer, N.; Peña, A.G.; Goodrich, J.K.; Gordon, J.I.; et al. QIIME allows analysis of high-throughput community sequencing data. *Nat. Meth.* **2010**, *7*, 335–336. [CrossRef]
23. Wickham, H. *Reshape2: Flexibly Reshape Data: A Reboot of the Reshape Package*; R Package Version; R Foundation: Vienna, Austria, 2012.
24. Chollet, F. Keras. Available online: https://github.com/fchollet/keras. (accessed on 10 April 2015).
25. Pedregosa, F.; Varoquaux, G.; Gramfort, A.; Michel, V.; Thirion, B.; Grisel, O.; Blondel, M.; Prettenhofer, P.; Weiss, R.; Dubourg, V.; et al. Scikit-learn: Machine learning in Python. *J. Mach. Learn Res.* **2011**, *12*, 2825–2830.
26. Marino, M.; Dubsky de Wittenau, G.; Saccà, E.; Cattonaro, F.; Spadotto, A.; Innocente, N.; Radovic, S.; Piasentier, E.; Marroni, F. Metagenomic profiles of different types of Italian high-moisture Mozzarella cheese. *Food Microbiol.* **2019**, *79*, 123–131. [CrossRef]
27. Pellegrino, L.; Resmini, P. Cheesemaking conditions and compositive characteristics supporting the safety of the raw milk cheese Italian grana. *Sci. Tecn. Latt. Cas.* **2001**, *52*, 105–114.

28. Panari, G.; Perini, S.; Guidetti, R.; Pecorari, M.; Merialdi, G.; Albertini, A. Indagine sul comportamento di germi potenzialmente patogeni nella tecnologia del formaggio Parmigiano-Reggiano. *Sci. Tecn. Latt. Cas.* **2001**, *52*, 13–22.
29. Bachmann, H.P.; Spahr, U. The fate of potentially pathogenic bacteria in Swiss hard and semihard cheeses made from raw milk. *J. Dairy Sci.* **1995**, *78*, 476–483. [CrossRef]
30. Ercolini, D. Secrets of the cheese microbiome. *Nat. Foods* **2020**, *1*, 466–467. [CrossRef]
31. Avila, C.L.S.; Carvalho, B.F. Silage fermentation-updates focusing on the performance of micro-organisms. *J. Appl. Microbiol.* **2019**, *128*, 966–984. [CrossRef]
32. Pogačić, T.; Mancini, A.; Santarelli, M.; Bottari, B.; Lazzi, C.; Neviani, E.; Gatti, M. Diversity and dynamic of lactic acid bacteria strains during aging of a long ripened hard cheese produced from raw milk and undefined natural starter. *Food Microbiol.* **2013**, *36*, 207–215. [CrossRef]
33. Bava, L.; Zucali, M.; Tamburini, A.; Morandi, S.; Brasca, M. Effect of different farming practices on lactic acid bacteria content in cow milk. *Animals* **2021**, *11*, 522. [CrossRef]
34. Cattaneo, S.; Hogenboom, J.A.; Masotti, F.; Rosi, V.; Pellegrino, L.; Resmini, P. Grated Grana Padano cheese: New hints on how to control quality and recognize imitations. *Dairy Sci. Technol.* **2008**, *88*, 595–605. [CrossRef]
35. Franklin, R.B.; Taylor, D.R.; Mills, A.L. Characterization of microbial communities using randomly amplified polymorphic DNA (RAPD). *J. Microbiol. Meth.* **1999**, *35*, 225–235. [CrossRef]
36. Khodakova, A.S.; Smith, R.J.; Burgoyne, L.; Abarno, D.; Linacre, A. Random whole metagenomic sequencing for forensic discrimination of soils. *PLoS ONE* **2014**, *9*, e104996. [CrossRef] [PubMed]

 *foods* 

Article

Detection of Enteric Viruses and Core Microbiome Analysis in Artisanal Colonial Salami-Type Dry-Fermented Sausages from Santa Catarina, Brazil

Roberto Degenhardt [1,2], Doris Sobral Marques Souza [1,3], Leidiane A. Acordi Menezes [1,4], Gilberto Vinícius de Melo Pereira [5], David Rodríguez-Lázaro [6,7], Gislaine Fongaro [3] and Juliano De Dea Lindner [1,*]

1 Food Technology and Bioprocess Research Group, Department of Food Science and Technology, Federal University of Santa Catarina (UFSC), Florianópolis 88034-000, SC, Brazil; roberto.degenhardt@gmail.com (R.D.); doris.sobral@gmail.com (D.S.M.S.); leidianeacordi@gmail.com (L.A.A.M.)
2 Biological and Health Sciences Department, West of Santa Catarina State University (UNOESC), Joaçaba 89600-000, SC, Brazil
3 Laboratory of Applied Virology, Department of Microbiology, Immunology and Parasitology, Federal University of Santa Catarina (UFSC), Florianópolis 88034-000, SC, Brazil; gislainefongaro@gmail.com
4 Neoprospecta Microbiome Technologies, Sapiens Park, Florianópolis 88056-000, SC, Brazil
5 Department of Bioprocess Engineering and Biotechnology, Federal University of Paraná (UFPR), Curitiba 80060-000, PR, Brazil; gilbertovinicius@gmail.com
6 Microbiology Division, Faculty of Sciences, University of Burgos, 9070 Burgos, Spain; drlazaro@ubu.es
7 Center for Emerging Pathogens and Global Health, University of Burgos, 9070 Burgos, Spain
* Correspondence: juliano.lindner@ufsc.br

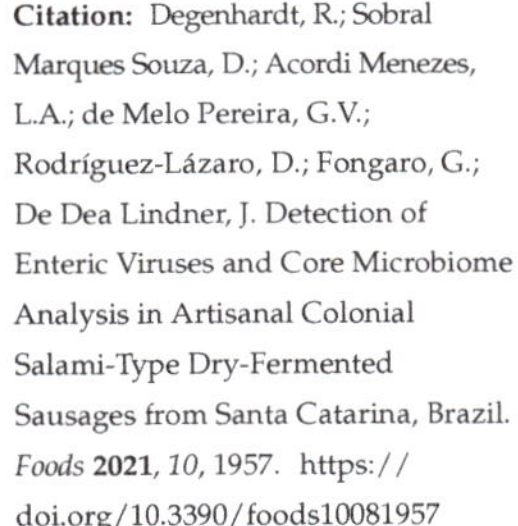

Citation: Degenhardt, R.; Sobral Marques Souza, D.; Acordi Menezes, L.A.; de Melo Pereira, G.V.; Rodríguez-Lázaro, D.; Fongaro, G.; De Dea Lindner, J. Detection of Enteric Viruses and Core Microbiome Analysis in Artisanal Colonial Salami-Type Dry-Fermented Sausages from Santa Catarina, Brazil. *Foods* **2021**, *10*, 1957. https://doi.org/10.3390/foods10081957

Academic Editor: Arun K. Bhunia

Received: 16 July 2021
Accepted: 18 August 2021
Published: 22 August 2021

Abstract: Microbial fermentation plays an important role in the manufacturing of artisanal sausages and can have major effects on product quality and safety. We used metagenomics and culture-dependent methods to study the presence of Hepatitis E virus (HEV) and Rotavirus-A (RV-A), and fungal and bacterial communities, in artisanal Colonial salami-type dry-fermented sausages in Santa Catarina state, Brazil. Lactic acid bacteria (LAB) and yeast dominated the microbiome. *Latilactobacillus sakei* and *Debaryomyces hansenii* were ubiquitous and the most abundant species. The DNA of some foodborne pathogens was found in very low concentrations although viable cells of most of these species were undetectable by cultivation methods. The characteristics of the raw material and hygiene of the artisanal sausage manufacturing process resulted in high loads of beneficial microorganisms and the absence of HEV and RV-A viruses as determined by RT-qPCR assays. In conclusion, high LAB load in sausages was more relevant to preventing pathogen growth than the ripening time and/or physicochemical characteristics. However, the presence of *Clostridium* spp. and other pathogens in some samples must be taken into account for the development of future preservation methods; appropriate LAB starter cultures and health surveillance are required in the production process to prevent foodborne outbreaks.

Keywords: swine and pork production chain; Hepatitis E virus; Rotavirus-A; metagenomic analysis; food safety

1. Introduction

Fermented sausages (dry and semi-dry) make up a substantial proportion of pork produced and consumed in western Santa Catarina State. This is in part a consequence of the European origin and cultural characteristics of the population in this region in the south of Brazil. Several small and medium-scale production companies are dedicated to the transformation of pork meat, employing artisanal production techniques including natural fermentation. However, these production practices can allow the survival and multiplication of spoilage and pathogenic microorganisms if hygiene and sanitary conditions and process parameters are not strictly controlled.

There is a great diversity of dry-fermented sausages produced in the world. Different raw meat types, formulations, additives, processes, casing, and drying periods are used [1]. Colonial sausages produced in Santa Catarina State are characterized by low acidity and high moisture content when they are offered for sale, but they continue the ripening process outdoors in the market, gradually losing moisture [2]. These uncontrolled conditions can lead to a risk for the multiplication of undesirable microorganisms.

The production of artisanal sausages involve fermentation by the indigenous microbiota of meat (composed mainly by lactic acid bacteria—LAB) producing organic acids, and consequently a decrease in the pH that allows preservation of the product. *Lactobacillus*, *Pediococcus*, *Enterococcus*, and *Leuconostoc* are believed to be the principal genera involved in sausage fermentation [3,4], and diverse LAB are found in food products according to the manufacturing practices used. The preservation ability of LAB is based on competition for nutrients and the production of antimicrobial metabolites, hydrogen peroxide and bacteriocins. The selective influence of diverse factors (e.g., pH, redox potential, water activity (a_w), temperature, relative humidity, oxygen availability) in fermented sausage will contribute to determining the microbial consortium present.

The microbial communities in fermented sausage play significant roles in the flavor, texture, quality, and safety of sausages [5,6]. Understanding the ecology of fermented sausages is clearly key to understanding the physical and chemical changes that occur during fermentation and ripening [7]. High-throughput sequencing of the microbiota from salami-type dry-fermented sausages of different origins indicates that LAB and coagulase-negative cocci, including both micrococci and coagulase-negative staphylococci (CNS) are the predominant groups in spontaneously fermented artisanal sausages [8]. *Debaryomyces*, *Penicillium*, and *Aspergillus* are the yeast and mold genera most commonly found [6,9–11]. Studies involving metagenomics for this type of fermented product are scarce in Brazil, and therefore, the ecology of Brazilian artisanal dry-fermented sausages needs further investigation.

Hepatitis E virus (HEV) and Rotavirus—A (RV-A) belong to the human enteric viruses group. They are non-enveloped and icosahedral viruses, and are very persistent in the environment [12]. HEV has a positive-sense single-stranded RNA and RV-A a double-stranded and segmented RNA (11 segments) that facilitates reassortment events among species [12]. Both are zoonotic viruses that infect many animal species as well as humans. They are mainly transmitted by the fecal-oral route, although HEV is also transmitted parenterally (rare), or by the ingestion of contaminated meat from viremic animals [12,13].

HEV and RV-A infect domestic pigs, which serve as reservoirs, and are emerging risks for human infection [13,14]. These viruses are excreted at high concentrations in animal stools (mainly from swine) so animal waste can contaminate water supplies, and when used in crops may contaminate food. Indeed, they persist as infectious particles in the environment, e.g., water and sewage, for several weeks at lower than body temperatures [13,15–17]. HEV circulation has already been reported in humans and domestic swine in Brazil [17–19].

The present study aimed to detect and quantify HEV and RV-A in artisanal Colonial salami-type dry-fermented sausages. We also analyzed the presence of fermenting, spoiling, and pathogenic microorganisms in the sausages. Both culture-dependent and -independent (metagenomic) approaches were used.

2. Materials and Methods

2.1. Colonial Salami-Type Dry-Fermented Sausage Production

Colonial salami-type dry-fermented sausages were collected from 13 production sites (Table 1) under the state inspection system (SIE) in the Vale do Rio do Peixe zone, Santa Catarina state, Brazil (ca. 27°10′ S; 51°30′ W). The sausages were prepared using the same manufacturing process in all sites. Pigs were slaughtered and the carcasses underwent temperature equalization for up to 24 h in cooling chambers to establish rigor mortis. The carcasses were exposed and shanks and palettes were used to prepare the sausages. The meat was comminuted in a meat grinder and seasoned with NaCl, curing agent (NO_2), and spices.

Table 1. Origin of Colonial salami-type dry-fermented sausages in Santa Catarina State, ripening time and physical-chemical characteristics.

Sample	City	Ripening (Days)	Casing	pH	Moisture (%)	NaCl (%)	WPS (%)	a_w
L01	Piratuba	40	N	5.50	30.02	4.30	12.50	0.891
L02	Herval Velho	16	S	6.30	50.53	3.70	6.80	0.950
L03	Jaborá	1	S	6.10	51.14	3.17	5.80	0.957
L04	Concórdia	7	S	5.10	30.20	6.15	16.90	0.829
L05	Luzerna	2	N	5.80	50.22	4.20	7.70	0.943
L06	Videira	9	N	5.50	41.03	4.32	9.50	0.927
L07	Salto Veloso	12	N	6.40	40.85	4.75	10.40	0.917
L08	Salto Veloso	6	N	5.60	38.10	3.01	7.30	0.946
L09	Tangará	6	N	6.00	56.69	4.18	6.90	0.949
L10	Caçador	16	N	6.00	28.48	5.04	15.00	0.857
L11	Videira	2	N	6.00	43.22	4.49	9.40	0.928
L12	Iomerê	10	N	5.70	39.02	4.92	11.20	0.908
L13	Lacerdópolis	7	S	5.40	39.06	4.41	10.10	0.920

WPS: water phase salt; a_w: water activity; N: natural; S: synthetic.

The production of these sausages does not involve addition of commercial starter cultures for pan-curing. The fermentation is due to the natural microbiota of the meat and environment. After homogenization in a mixer, the meat is extruded into a natural bovine or a synthetic cellulose casing. The sausages are placed in a smoker where the drying is started, and the curing is completed under smoking. After smoking, the sausages are ready for sale. Thereafter, drying took place in the market, subject to the environmental conditions of the market stall site. The 13 sausage samples used had drying times of between one and 40 days, with half of the samples having been dried for more than seven days.

2.2. Physical-Chemical Evaluation

The physical-chemical characteristics of the sausages (Table 1) were determined using the official methods of the Ministry of Agriculture, Livestock and Supply (MAPA) [20]. The pH was determined using a potentiometer (Quimis, model Q400A, Diadema, SP, Brazil). The moisture content was determined by a gravimetric method, placing the samples in a drying oven (SP Labor, model SP-100/150, SP, Brazil) at 105 °C for 24 h [21]. The NaCl content was determined by the mercury metric method [20]. The a_w was determined by grinding the samples and placing them into a digital hygrometer at 25 °C (Aqualab Model-Series 3 TE, Decagon Devices Inc., Pullman, USA). Water phase salt (WPS) was calculated from the values of NaCl and moisture content, according to Koral and Köse [22], using the following equation:

$$WPS\% = (NaCl\%/(NaCl\% + moisture\%)) \times 100$$

2.3. Microbiological Analyses

The samples were tested for several microbial groups: total coliforms (TC), thermotolerant coliforms (TTC), coagulase-positive staphylococci (CPS), sulfite-reducing clostridia (SRC), LAB, Enterococci, *Salmonella* and *Listeria monocytogenes*. TC were counted by spread plating on Violet Red Bile Lactose agar (VRBL, Oxoid Limited, Hampshire, United Kingdom), incubated aerobically at 37 °C for 24 h [23]; TTC by pour plating on double-layered VRBL agar, incubated aerobically at 44 °C for 24 h; CPS by spread plating on Baird-Parker agar (BPA, Neogen, Lansing, MI, USA), incubated aerobically at 35 °C for 24–48 h [24]; SRC by pour plating on Iron Sulfite agar (ISA, Neogen), incubated anaerobically at 37 °C for 24–48 h [25]; LAB by pour plating on De Man, Rogosa and Sharpe agar (MRS, Neogen), incubated anaerobically at 30 °C for 72 h [26]; Enterococci by spread plating on m-Enterococcus agar (Neogen), incubated anaerobically at 35 ± 2 °C for 24–48 h; *Salmonella* by incubating a 25 g pre-enrichment sample in buffered peptone water (BPW, Neogen) at 34–38 °C for 18 h, selective enrichment by adding 0.1 mL of the BPW to modified semi-solid Rappaport-Vassiliadis medium (MSRV, Neogen) at 41.5 ± 1 °C for 24 h and 1 mL of BPW to Muller-Kauffmann Tetrathionate-Novobiocin broth (MKTTn, Neogen) at 37 ± 1 °C for 24 h, plating-out in xylose lysine deoxycholate agar (XLD, Neogen) at 37 °C for 24 h and briliant green agar (BGA, Neogen) at 37 °C for 24 h, with confirmation by biochemical and serological testing [27]; *Listeria* spp. by incubating a 25 g pre-enrichment sample in demi Fraser broth (DF, Neogen) at 30 °C for 24 h, selective enrichment by incubating 0.1 mL of the DF in Fraser broth (Neogen) at 35 ± 1 °C for 24–48 h, then plating-out on Listeria agar (ALOA, Laborclin, Pinhais, Paraná, Brazil) at 37 °C for 24 h, with confirmation by biochemical testing [28].

Yeasts and molds were counted by spread plating on dichloran glycerol agar (DG18, Neogen) containing chloramphenicol (0.1 g/L) and incubating at 25 °C for 5–7 days [29]. Macroscopic observation was used to distinguish between yeasts and molds among the colonies growing on the plates.

2.4. Viral Detection by RT-qPCR

Virus was detected by mechanical disruption of tissues followed by silica-membrane RNA extraction, as described by García et al. [14] and Rodríguez-Lázaro et al. [30]. Briefly, 25 g samples of sausage (Table 1) were collected, chopped using a sterile scalpel blade and homogenized; aliquots of 0.25 g were taken from the samples, and they were inoculated with ~10^5 PFU (Plaque Forming Units) of MNV-1 (Murine Norovirus-1) as sample process control virus (SPCV). Each sample (including SPCV) was homogenized in 1 mL of lysis buffer RLT (Qiagen, Hilden, Germany) containing 0.14M β-mercapto-ethanol in a tissue homogenizer, and incubated for 5 min at room temperature. The samples were then centrifuged at 10,000× *g* for 20 min at 4 °C and the supernatants (800 μL) were transferred to new tubes for nucleic acid extraction with the RNeasy® Mini kit (Qiagen, Germany), following the manufacturer's protocols. The resulting nucleic acid samples were subjected to RT-qPCR, both undiluted and diluted tenfold to reduce the inhibitory effects of any contaminants in the samples.

The TaqMan technique was used for RT-qPCR, as previously described by Baert et al. [31] for MNV-1 as SPCV, Jothikumar et al. [32] for HEV and [33] for RV-A. All amplifications were performed in a StepOne Plus® Real-Time PCR System (Applied Biosystems, Waltham, MA, USA), and each sample was analyzed in triplicate. Ultrapure water was used as the non-template control for each assay.

2.5. Metagenomic Analyses

DNA was extracted from samples with the DNeasy Power Soil Pro kit (Qiagen, Germany), following the manufacturer's instructions. Total DNA extracted was used as the template for Next Generation Sequencing (NGS), on the Illumina MiSeq platform (Illumina Inc., San Diego, CA, USA).

The microbiome of the samples was studied by high throughput sequencing for bacterial taxonomic identification based on conserved and variable regions V3/V4 of the 16S rRNA gene with primers 341F (CCTACGGGRSGCAGCAG) and 806R (GGACTACHVGGGTWTCTAAT) [34,35]. Extracted DNA was subjected to PCR using a previously described protocol [36] developed by Neo-prospecta Microbiome Technologies (Florianopolis, Brazil). The first PCR used sequences based on Illumina's TruSeq adapters, which allows a second PCR with primers with indexed sequences. PCRs were carried out in triplicate using Taq Platinum (Invitrogen, Waltham, MA, EUA) with the following conditions: PCR1—95 °C/5 min, 25 cycles (95 °C/45 s, 55 °C/30 s, 72 °C/45 s) and final extension

72 °C/2 min; PCR2—95 °C/5 min, 10 cycles (95 °C/45 s, 66 °C/30 s, 72 °C/45 s) and final extension 72 °C/2 min.

The community of filamentous fungi and yeasts was studied by amplifying the ITS1 region using ITS1F (GAACCWGCGGARGGATCA) and ITS2R (GCTGCGTTCTTCATCGATGC) primers [37], and the same conditions for PCR1. For PCR2, the conditions were: 95 °C/5 min, 15 cycles (95 °C/45 s, 66 °C/30 s, 72 °C/45 s) and final extension 72 °C/2 min. The final reaction products were purified using a protocol developed by Neoprospecta involving magnetic beads. The samples were grouped in libraries quantified by Real-Time PCR using KAPA Library Quantification (KAPA Biosystems, Woburn, MA, USA). Library pools were sequenced in a MiSeq sequencer system (Illumina Inc., USA) using V2- 300 cycle kit, in a paired-end run, without normalization of libraries. The raw data (DNA sequences in fastq files) were analyzed through bioinformatics workflow considering accumulated error in the sequencing to be at most 1%. To identify the species of microorganisms present in the samples, the DNA sequences obtained were compared with a proprietary database containing well-characterized DNA sequences.

Sequencing data for each sample were processed with the Quantitative Insights into Microbial Ecology (*Qiime*) software package [38]. The sequencing output was analyzed by a read quality filter; reads with an average Phred score < 20 were removed and then 100% identical reads were clustered. Clusters with fewer than five reads were excluded from further analysis, to remove putative chimeric sequences. The remaining good-quality sequences were further clustered at 97% similarity to define operational taxonomic units (OTU). OTUs were classified by comparison with a custom 16S rRNA/ITS1 database (NEORefDB, Neoprospecta). Sequences with at least 99% identity to the reference database were taxonomically assigned. Each OTU was given an arbitrary unit corresponding to the relative abundance (r.a.) calculated as the number of reads pertaining to each species as a percentage of the total number of sequences read for each sample. Species with a r.a. below 1% were grouped in the category "others". The r.a. values were compared between OTUs using heatmaps prepared on *Qiime*.

2.6. Statistical Analyses

Differences in virus parameters between samples collected from different sites and seasons were analyzed by nonparametric analysis of variance (Kruskal–Wallis) performed with Statistic 7.0. Pearson correlation and linear regression tests, ANOVA, and Student's *t*-tests were performed using GraphPad Prism 5.0 (USA). Differences were considered statistically significant at a p-value ≤ 0.05.

3. Results and Discussion

3.1. Physical-Chemical Characteristics

Selected physical-chemical characteristics (pH, moisture, NaCl, WPS and a_w) of the sausage samples are summarized in Table 1. The sausages can be classified into two subgroups according to the criteria of Brazilian legislation [39]: (i) dry sausages, with moisture $\leq$ 40% and $a_w \leq 0.92$, and (ii) semi-dry sausages, for which these parameters are not stipulated. A particular feature of these locally produced sausages is that they are offered for sale as semi-dry sausages and, due to the environmental conditions, they start to present dry sausage characteristics as the ripening continues. The dry sausages (L01, L04, L07, L10, L12) had undergone seven or more days of ripening, whereas the semi-dry sausages (L02, L03, L05, L06, L08, L09, and L11) had undergone up to six, except for samples L02 and L06; these two sausages had been maintained under refrigeration at the site of sale, reducing dehydration.

All the samples presented pH between 5.1 and 6.4 (Table 1), characteristic of the sausages produced in the region. This weak acidity is also characteristic of Italian sausages, produced in the Mediterranean region only with pork meat. Low pH and a_w are the two main obstacles to spoilage and the development of pathogenic microorganisms in fermented sausages [40]. If these factors are weak, as is the case in our samples due to the relatively high pH of the samples (≥5.1), the association between a_w and the competitive microbiota (mainly LAB) becomes very important to the safety of the product. The WPS is not frequently used to assess the potential for multiplication of spoilage and/or pathogenic microorganisms in meat sausages in Brazil, but it is a potentially valuable indicator for small producers, due to its relationship with water activity.

The minimum salt concentration in the samples was 3.01 (sample L08) and the maximum was 6.15 (sample L04). The upper limit of sensorial consumer acceptability for salt in cured meats is 3.5% to 5.0% with 1.5% to 2.0% being optimal. The salt concentration was therefore acceptable in the samples, except for sample L04. The higher concentration of salt in this case appears to be a consequence of excessive drying: the moisture content was only 30.2%. In artisanal cured products, salt is still the main resource used for conservation, and this is the main reason for the high salt content in these sausages, mostly between 3.7% and 5.04% (Table 1). These values are compatible with those required for the preservation of sausages and related products [41].

3.2. *Viral Detection*

The samples of artisanal Colonial salami-type dry-fermented sausages from different production sites (Table 1) were tested by RT-qPCR for HEV and RV-A. Neither virus was detected in any of the sausage samples, and the mean viral extraction efficiency (measured as SPCV -MNV-1 recovery) was 30%.

HEV is more frequently detected in feces than serum and liver samples from infected animals [42], but there is nevertheless a risk of foodborne hepatitis E infection from consumption of pork products. Boxman et al. [43] found HEV RNA in 14.6% of pork sausages collected from a Dutch market from 2017 to 2019, and HEV RNA was detected in 18.5% of salami (n = 92) samples; sequence-based typing was successful for 33 samples, and all were genotype 3c. Berto et al. [44] found HEV RNA in four samples of liver sausage in France, and three of them had infectious particles of HEV as assessed by cell culture assay. Colson et al. [45] found HEV RNA of genotype 3 in seven samples of raw figatelli purchased in French supermarkets. However, HEV RNA was not detected in any of the 22 sausages from different production sites analyzed in this work. This contrasts with Souza et al. [17], who detected HEV in effluent after swine manure digestion by psychrophilic anaerobic biodigesters (PABs) in Concórdia city, in Santa Catarina state. This demonstrates the circulation of HEV in the pig-finishing farms in the zone where the Colonial sausage we studied was produced.

Lowering pH during meat fermentation is believed to inhibit the growth of certain pathogens. Wolff et al. [46] tested the pH resistance of a genotype 3 strain of HEV. Only minimal inactivating effects were found at pH conditions common in sausages during curing (4.5 to 6.5), simulated using D/L-lactic acid. High salt concentrations are also used to preserve meat products and to inactivate foodborne pathogens. Wolff et al. [47] tested the effects of NaCl, sodium nitrite, and sodium nitrate concentrations on the infectivity of HEV. Conditions consistent with those in fermented sausages were simulated. Treatments with up to 20% NaCl for 24 h at 23 °C, with and without the addition of 0.015% nitrite or 0.03% nitrate, did not inactivate the virus, demonstrating that HEV is highly stable at salt concentrations used to preserve raw meat products. Therefore, the acid pH and high salinity of the products we tested were unlikely to be responsible for the absence of HEV.

Organs from naturally HEV-infected pigs were tested by García et al. [14] and the HEV virus was not detected in loin samples: meat even from infected animals may therefore be only a low risk for HEV transmission. The main ingredient used in the preparation of Colonial salami is the loin, and this may explain the absence of HEV RNA from the samples tested, despite evidence of the circulation of HEV in pig farms in the region [17]. Other organs would need to be analyzed to assess the HEV contamination of the pigs in the region.

RV-A RNA was not detected in the Colonial sausage samples analyzed, evidencing the good sanitary conditions during slaughter and sausages preparation. RV-A zoonotic transmission to humans usually occurs via direct contact with infected animals or fomites [15]. RV-A causes diarrheic diseases in swine and meat contamination could occur in the slaughterhouse or during sausage preparation by handlers and fomites. RV-A is the most prevalent rotaviral cause of diarrhea in swine [48]. Other RV species that can infect pigs were not investigated in this work.

3.3. Microbiological and Metagenomic Analyses

The combined culture-dependent and -independent approaches are best for analyses of the microbiota of fermented foods [49]. Culture-dependent analysis showed high levels of LAB and yeasts and the presence of enterococci and molds in all samples. There were low or undetectable counts of undesirable microbiota, in accordance with Brazilian standards (Table 2). In general, the samples showed similar cell count values that were independent of the ripening time. Likewise, there was no correlation between ripening time and microbiota in the metagenomic analyses. The pH in all samples was higher than 5.0, indicating that LAB community was mostly bacteria with low acid production capacity, resulting in low acidity fermented sausages. Although low acidity does not favor the inhibition of undesirable microbiota, it enhances the water-retention capacity of meat proteins, contributing to the maintenance of moisture during ripening [50]. The yeast community is important for the development of flavor and inhibition of toxigenic molds.

Table 2. Microbial analyses of Colonial salami-type dry-fermented sausages.

Sample	TC *	TTC *	CPS *	SRC *	LAB *	Enterococci *	Yeasts *	Molds *	*Salmonella* **	*Listeria* spp. **
L01	2.04	2.00	2.00	1.00	7.60	5.89	8.34	<2.00	A	P
L02	<1.00	<1.00	<2.00	<1.00	6.91	4.29	6.64	7.00	A	P
L03	1.70	1.70	<2.00	<1.00	8.32	3.11	3.11	7.30	A	A
L04	3.89	3.81	<2.00	1.48	6.60	5.68	6.81	4.30	A	A
L05	4.73	4.73	2.48	<1.00	7.84	5.78	8.30	7.48	A	P
L06	2.26	<1.00	<2.00	<1.00	7.84	3.63	7.28	7.00	A	A
L07	1.00	1.00	4.73	<1.00	7.86	6.20	6.81	7.30	A	A
L08	<1.00	<1.00	<2.00	<1.00	7.91	3.67	6.78	7.00	A	A
L09	2.08	1.60	2.90	<1.00	8.38	6.25	7.20	4.48	A	A
L10	<1.00	<1.00	<2.00	<1.00	6.79	3.15	6.55	7.30	A	A
L11	2.86	2.28	2.00	<1.00	6.75	4.70	7.47	7.30	A	A
L12	2.00	2.00	2.30	<1.00	8.48	3.87	7.95	4.00	A	P
L13	<1.00	<1.00	<2.00	1.00	7.96	2.30	6.83	<2.00	A	P

* cfu·log^{-1}; ** Presence (P) or absence (A) in 25 g of sausage; TC: total coliforms; TTC: thermotolerant coliforms; CPS: coagulase-positive staphylococci; SRC: sulfite-reducing clostridia; LAB: lactic acid bacteria.

The 16S amplicon target sequencing identified more than 60 species each with a r.a. above 0.1% (Figure 1 shows the 31 species with over 1% r.a., in one or more samples). Species with a very small number of reads (r.a. < 1%) were classified into the category "others".

LAB and CNS were the most abundant groups in our samples (Figure 1), consistent with previous reports for dry sausages from Spain and Italy [6,8]. LAB was predominant in 11 of the 13 sausage samples: *Latilactobacillus sakei* was the most abundant taxon in samples L01, L03, L05, L06, L08, L10, L12, and L13 (r.a. of 43.56%, 65.85%, 49.57%, 38.38%, 42.70%, 62.75%, 40.30%, and 92.14%, respectively); *Latilactobacillus curvatus* was predominant in sample L02 (r.a. 60.31%); *Lactococcus lactis* in sample L04 (r.a. 33.33%), and *Pediococcus pentosaceus* in sample L11 (r.a. 36.39%).

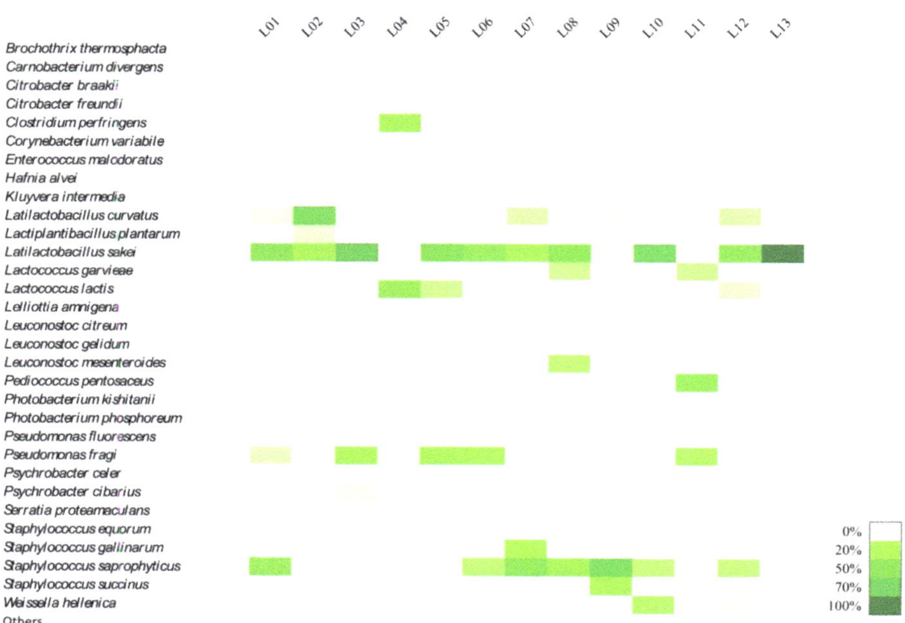

Figure 1. Relative abundance (%) of bacterial species in artisanal Colonial salami-type dry-fermented sausages. Species with a very small number of reads (relative abundance < 1%) were classified into the category "others".

Most samples (Figure 1) contained three or four species with r.a. above 5%, although in sample L13 *Lb. sakei* was massively predominant and the microbial community was not very diverse. *Lb. curvatus* was found at r.a. of between 5% and 12% in samples L01, L07, L09, and L12, *Lactococcus lactis* at between 9% and 14% in samples L05 and L12, *Lactococcus garvieae* at around 13% in samples L08 and L011, *Leuconostoc mesenteroides* at 17% in L08, *Lactiplantibacillus plantarum* at 7% in L02 and *Weissella hellenica* at 18% in L10. *Lb. sakei*, *Lb. curvatus* and *Lp. plantarum* are the species of LAB commonly found in sausages with no added starters. Other species of *Weissella*, *Leuconostoc*, *Lactococcus*, and *Pediococcus* were also found, but as part of the subdominant microbiota in our samples. The fermentation in these sausages is spontaneous, so the microorganisms detected come from the meat and the surrounding environment [6].

The manufacturing process of artisanal Colonial salami-type dry-fermented sausages from Santa Catarina includes spontaneous fermentation, cure and drying as conservation. Thus, the obstacles to the development of spoilage and pathogenic microbiota are the addition of preservatives (NaCl and sodium nitrite), pH, a_w, and the competitive microbiota, mostly LAB. The protective effects of LAB are due in part to their production of bacteriocins with antimicrobial activity against pathogens [51]. Hebert et al. [52] showed that *Lb. curvatus* CRL705 isolated from Argentinean artisanal fermented sausage produces bacteriocin with anti-listeria activity. Palavecino et al. [6] describe coagulase-negative staphylococci (CNS) with antimicrobial activity. The presence of CNS (*Staphylococcus saprophyticus*, *Staphylococcus galinarum*, and *Staphylococcus equorum*) in our samples (Figure 1) may be one of the reasons for the absence of pathogens common in meat products, such as *Clostridium botulinum*, *Staphylococcus aureus*, *Listeria monocytogenes*, and *Escherichia coli* [53].

In general, the sausage samples (Table 2) that showed the best hygiene indicator results (TC, TTC, CPS, SRC, and *Salmonella*) were drier (e.g., L10) and had longer ripening time (e.g., L02). This is in contrast to the observation for the high LAB semi-dry sausage subgroup, which corroborates the metagenomic data (Figure 1). *Listeria* spp. were detected in 38.4%

of the samples, but *L. monocytogenes* was not detected at a significant r.a. in any sample (Figure 1). *Listeria* spp. are detected in many fermented sausages with low acidity and high a_w [54–56]. *Listeria* spp. counts determined by culture-dependent methods are generally low in fermented sausages [57,58], and culture-dependent methods employ enrichment and cell concentration steps. Therefore, the possibility of isolating a microorganism is greater than for the metagenomic analysis when it is present at low r.a.

In addition to LAB, species of the *Micrococcaceae* family, mainly belonging to the genus *Staphylococcus*, are commonly found in artisanal sausages. *S. saprophyticus, Staphylococcus xylosus, S. equorum*, and *Staphylococcus carnosus* are often predominant. *Staphylococcus succinus* is also frequently observed in sub-dominant populations [7,59]. For samples L07 and L09 (Figure 1), the predominant microorganism was *S. saprophyticus* (r.a. 42.60% and 61.65%, respectively). This species was also found in samples L01, L06, L08, L10, and L12 (r.a. from 15% to 26%). *S. succinus* was found in sample L09 at an r.a. off 20%.

Brochothrix thermosphacta, Carnobacterium spp., clostridia, and *Pseudomonas* spp. are among the bacteria most frequently causing spoilage of refrigerated pork meat [60]. *Pseudomonas fragi* (Figure 1) was found in samples L01, L03, L05, L06, and L011 with r.a. between 10% and 23%. *Carnobacterium divergens* (r.a. 1.87%) and *Clostridium perfringens* (r.a. 27.76%) were found in sample L04, in which the abundance of lactobacilli was low. *B. thermosphacta* was detected in samples (Figure 1) with a low abundance of lactobacilli (L03, L05, and L11). However, for sample L09, *S. saprophyticus* was predominant, and although the r.a of lactobacilli were low, no sequences belonging to *B. thermosphacta, C. divergens* (r.a. 1.87%), *Pseudomonas* spp. or clostridia were detected.

Taxonomical assignment of the yeasts (Figure 2) showed the dominance of *Debaryomyces hansenii* in L02, L04, L06, L07, L09, L10, L11, and L12 (r.a. 63.49%, 44.25%, 82.44%, 80.01%, 65.71%, 35.20%, 28.92% and 10.85%, respectively) followed by *Candida zeylanoides*, predominant in samples L01, L03, and L12 (r.a. 73.91%, 19.13%, and 58.23%, respectively). *Yarrowia galli* predominant in L05 (r.a. 42.15%), and *Kodamaea ohmeri* in L08 (r.a. 53.31%). *D. hansenii, C. zeylanoides* and *Yarrowia lipolytica* were found in all the samples, although they were less abundant than the other yeast groups. *Y. galli, Pichia kudriavzevii, K. ohmeri, Candida parapsilosis* and *Candida tropicalis* were also found with relevant r.a. in the sub-dominant populations of the samples.

D. hansenii is the most common yeast in fermented sausages and contributes to two important characteristics: production of volatile compounds during ripening, especially esters, branched alcohols, and aldehydes [61,62], and bioprotective antifungal activity against toxigenic penicillia [63]. Species belonging to the genera *Candida, Yarrowia*, and *Pichia* are also frequently found [6,51,64].

Development of mold on the surface of the casings during ripening of fermented sausages is common, and molds are present in the environment at the production and ripening sites. Molds are important in the development of flavor, but some fungi of the genera *Aspergillus* and *Penicillium* can produce undesirable mycotoxins [65]. For example, a combination of high moisture, low temperature and the predominance of penicillia has been found to be conducive to the production of penicillic acid, a carcinogen [66]. The filamentous fungus found at the highest abundance was *Aspergillus cibarius* in L02, L11 and L12 (3.38%, 42.51%, and 35.30%, respectively) followed by *Penicillium thymicola* (r.a. 8.78% in L12), and *Geotrichum candidum* (r.a. 4.90% in L04). *Aspergillus* and *Penicillium* are the main genera found in fermented sausages, while *Geotrichum* is found with lower frequency [6,51,67–69].

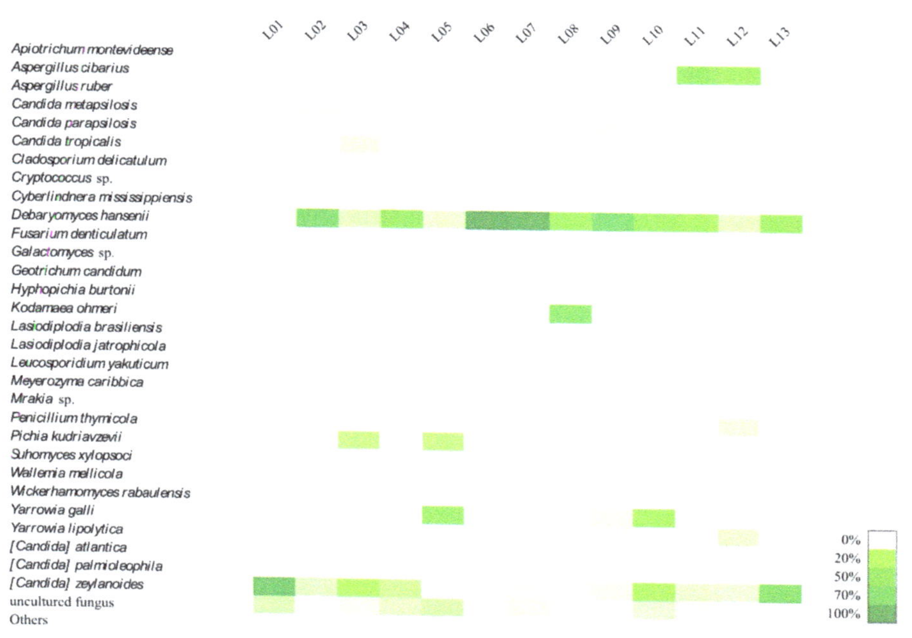

Figure 2. Relative abundance (%) of fungal species in artisanal Colonial salami-type dry-fermented sausages. Species with a very small number of reads (relative abundance < 1%) were classified into in the category "others".

4. Conclusions

The present study suggests that LAB may provide a larger contribution to prevent pathogen growth than ripening time and/or physicochemical characteristics (pH, moisture, NaCl, and a_w) in dry-fermented, low-acid artisanal Colonial sausages. Although we provide some evidence for a protective role of LAB, and enteric zoonotic viruses (HEV and RV-A) were not detected in the sausage samples, the presence of *Cl. perfringens* is an indication of the need for sanitary improvements in the production process to prevent foodborne outbreaks.

Author Contributions: Conceptualization, R.D., D.S.M.S., L.A.A.M., G.F. and J.D.D.L.; Data curation, R.D., D.S.M.S. and L.A.A.M.; Formal analysis, R.D., D.S.M.S. and L.A.A.M.; Funding acquisition, G.F. and J.D.D.L.; Methodology, D.S.M.S., L.A.A.M. and G.F.; Project administration, J.D.D.L.; Supervision, J.D.D.L.; Writing—original draft, R.D., D.S.M.S. and L.A.A.M.; Writing—review & editing, G.V.d.M.P., D.R.-L., G.F. and J.D.D.L. All authors have read and agreed to the published version of the manuscript.

Funding: This research was funded by CNPq project number 556276/2009-4 (CT-Hidro 22/2009) and by CAPES/PNPD 88887.473179/2020-00 (Postdoc fellowship).

Institutional Review Board Statement: Not applicable.

Informed Consent Statement: Not applicable.

Data Availability Statement: The Data associated with this article can be requested by e-mail.

Conflicts of Interest: The role of the Neoprospecta Microbiome Technologies company in the manuscript was to provide the metagenomic analyses. The authors declare no conflict of interest.

References

1. Toldrá, F.; Nip, W.K.; Hiu, Y.H. Dry-Fermented Sausages: An Overview. In *Handbook of Fermented Meat and Poultry*, 1st ed.; Toldrá, F., Ed.; Blackwell Publishing: Hoboken, NJ, USA, 2007; pp. 321–325, ISBN 0–8247–4780–1.

2. Degenhardt, R.; Sant' Anna, E.S. Pesquisa de *Listeria* sp, em embutidos cárneos fermentados produzidos na região Meio-Oeste de Santa Catarina, Brasil. *Boletim do Centro de Pesquisa de Processamento de Alimentos* **2007**, *25*, 133–140. [CrossRef]
3. Fontana, C.; Fadda, S.; Cocconcelli, P.S.; Vignolo, G. Lactic Acid Bacteria in Meat Fermentations. In *Lahtinen*, 4th ed.; Ouwehand, A.C., Salminen, S., Von Wright, A., Eds.; CRC Press. Taylor & Francis Group: London, UK, 2012; pp. 247–264.
4. Fraqueza, M.J. Antibiotic resistance of lactic acid bacteria isolated from dry-fermented sausages. *Int. J. Food Microbiol.* **2015**, *212*, 76–88. [CrossRef] [PubMed]
5. Hu, Y.; Zhang, L.; Zhang, H.; Wang, Y.; Chen, Q.; Kong, B. Physicochemical properties and flavour profile of fermented dry sausages with a reduction of sodium chloride. *LWT* **2020**, *124*, 109061. [CrossRef]
6. Palavecino Prpich, N.Z.; Camprubí, G.E.; Cayré, M.E.; Castro, M.P. Indigenous Microbiota to Leverage Traditional Dry Sausage Production. *Int. J. Food Sci.* **2021**, *30*, 6696856. [CrossRef]
7. Dalla Santa, O.R.; Chacón Alvarez, D.; Dalla Santa, H.S.; Zanette, C.M.; Freitas, R.J.S.; Macedo, R.E.F.; Terra, N.N. Microbiota of sausages obtained by spontaneous fermentation produced in the South of Brazil. *Food Sci. Technol.* **2012**, *32*, 653–660. [CrossRef]
8. Aquilanti, L.; Garofalo, C.; Osimani, A.; Clementi, F. Ecology of lactic acid bacteria and coagulase negative cocci in fermented dry sausages manufactured in Italy and other Mediterranean countries: An overview. *Int. Food Res. J.* **2016**, *23*, 429–445.
9. Cocolin, L.; Urso, R.; Rantsiou, K.; Cantoni, C.; Comi, G. Dynamics and characterization of yeasts during natural fermentation of Italian sausages. *FEMS Yeast Res.* **2006**, *6*, 692–701. [CrossRef] [PubMed]
10. Mendonça, R.C.; Gouvêa, D.M.; Hungaro, H.M.; Sodré, A.D.F.; Querol-Simon, A. Dynamics of the yeast flora in artisanal country style and industrial dry cured sausage (yeast in fermented sausage). *Food Control.* **2013**, *29*, 143–148. [CrossRef]
11. Sunesen, L.O.; Stahnke, L.H. Mould starter cultures for dry sausages—Selection, application and effects. *Meat Sci.* **2003**, *65*, 935–948. [CrossRef]
12. Knipe, D.M.; Howley, P.M. (Eds.) *Fields Virology*, 6th ed.; Lippincott Williams & Wilkins a Wolters Kluwer Business: Philadelphia, PA, USA, 2013.
13. Doceul, V.; Bagdassarian, E.; Demange, A.; Pavio, N. Zoonotic hepatitis E virus: Classification, animal reservoirs and transmission routes. *Viruses* **2016**, *8*, 270. [CrossRef]
14. García, N.; Hernández, M.; Gutierrez-Boada, M.; Valero, A.; Navarro, A.; Muñoz-Chimeno, M.; Fernández-Manzano, A.; Escobar, F.M.; Martínez, I.; Bárcena, C.; et al. Occurrence of Hepatitis E Virus in Pigs and Pork Cuts and Organs at the Time of Slaughter, Spain. *Front. Microbiol.* **2020**, *10*, 1–10. [CrossRef]
15. Doro, R.; Farkas, S.L.; Martella, V.; Banyai, K. Zoonotic transmission of rotavirus: Surveillance and control. *Expert Rev. Anti-Infect. Ther.* **2015**, *13*, 1337–1350. [CrossRef]
16. Krog, J.S.; Hjulsager, C.K.; Larsen, M.A.; Larsen, L.E. Triple-reassortant influenza A virus with H3 of human seasonal origin, NA of swine origin, and internal A (H1N1) pandemic 2009 genes is established in Danish pigs. *Influenza Other Respir. Viruses* **2017**, *11*, 298–303. [CrossRef]
17. Souza, D.S.M.; Tápparo, D.C.; Rogovski, P.; Cadamuro, R.D.; de Souza, E.B.; da Silva, R.; Degenhardt, R.; De Dea Lindner, J.; Viancelli, A.; Michelon, W.; et al. Hepatitis E Virus in Manure and Its Removal by Psychrophilic anaerobic Biodigestion in Intensive Production Farms, Santa Catarina, Brazil, 2018. *Microorganisms* **2020**, *8*, 2045. [CrossRef]
18. Amorim, A.R.; Mendes, G.S.; Pena, G.P.A.; Santos, N. Hepatitis E virus infection of slaughtered healthy pigs in Brazil. *Zoonoses Public Health* **2018**, *65*, 501–504. [CrossRef]
19. Tengan, F.M.; Figueiredo, G.M.; Nunes, A.K.; Manchiero, C.; Dantas, B.P.; Magri, M.C.; Prata, T.V.G.; Nascimento, M.; Mazza, C.C.; Abdala, E.; et al. Seroprevalence of hepatitis E in adults in Brazil: A systematic review and meta-analysis. *Infect. Dis. Poverty* **2019**, *8*, 1–10. [CrossRef] [PubMed]
20. BRASIL Ministério da Agricultura e do Abastecimento. *Instrução Normativa n° 20, de 31 de julho de 2000, que oficializa os Métodos Analíticos Físico-químicos, para controle dos produtos cárneos e seus ingredientes–Sal e Salmoura. Diário Oficial da União, 27 jul. Seção 1, p. 10;* Ministério da Agricultura e do Abastecimento: Brasília, Brazil, 2020.
21. AOAC. *Official Methods of Analysis*; Association of Official Analytical Chemists: Gaithersburg, MD, USA, 2002.
22. Koral, S.; Köse, S. The Effect of Using Frozen Raw Material and Different Salt Ratios on the Quality Changes of Dry Salted Atlantic Bonito (Lakerda) at Two Storage Conditions. *Food Health* **2018**, *4*, 213–230. [CrossRef]
23. International Standards Organization. *Microbiology of Food and Animal Feeding Stuffs–Horizontal Method for the Enumeration of Coliforms—Colony-Count Technique*; ISO 4832:2012; International Standards Organization: Geneva, Switzerland, 2012.
24. International Standards Organization. *Microbiology of Food and Animal Feeding Stuffs–Horizontal Method for the Enumeration of Coagulase-Positive Staphylococci (Staphylococcus Aureus and Other Species) Part 1: Technique Using Baird-Parker Agar Medium*; ISO 6888–1:2019; International Standards Organization: Geneva, Switzerland, 2019.
25. International Standards Organization. *Microbiology of Food and Animal Feeding Stuffs–Horizontal Method for the Enumeration of Sulfite-Reducing Bacteria Growing under Anaerobic Conditions*; ISO 15213:2003; International Standards Organization: Geneva, Switzerland, 2003.
26. International Standards Organization. *Microbiology of Food and Animal Feeding Stuffs–Horizontal Method for the Enumeration of Mesophilic Lactic Acid Bacteria—Colony-Count Technique at 30 Degrees °C*; ISO 15214:1998; International Standards Organization: Geneva, Switzerland, 1998.

27. International Standards Organization. *Microbiology of Food and Animal Feeding Stuffs–Horizontal Method for the Detection, Enumeration and Serotyping of Salmonella—Part 1: Detection of Salmonella spp.*; ISO 6579–1:2017; International Standards Organization: Geneva, Switzerland, 2017.
28. International Standards Organization. *Microbiology of Food and Animal Feeding Stuffs–Horizontal Method for the Detection and Enumeration of Listeria Monocytogenes and of Listeria spp. Part 1: Detection Method*; ISO 11290–1:2017; International Standards Organization: Geneva, Switzerland, 2017.
29. International Standards Organization. *Microbiology of Food and Animal Feeding Stuffs–Horizontal Method for the Enumeration of Yeasts and Moulds—Part 2: Colony Count Technique in Products with Water Activity Less Than or Equal to 0.95, ISO 21527–2:2008 standard.*; International Standards Organization: Geneva, Switzerland, 2008.
30. Rodríguez-Lázaro, D.; Cook, N.; Ruggeri, F.M.; Sellwood, J.; Nasser, A.; Nascimento, M.S.J.; van der Poel, W.H.M. Virus hazards from food, water and other contaminated environments. *FEMS Microb. Rev.* **2012**, *36*, 786–814. [CrossRef] [PubMed]
31. Baert, L.; Wobus, C.E.; Van Coillie, E.; Thackray, L.B.; Debevere, J.; Uyttendaele, M. Detection of murine norovirus 1 by using plaque assay, transfection assay, and real-time reverse transcription-PCR before and after heat exposure. *Appl. Environ. Microbiol.* **2008**, *74*, 543–546. [CrossRef]
32. Jothikumar, N.; Cromeans, T.L.; Robertson, B.H.; Meng, X.J.; Hill, V.R. A broadly reactive one-step real-time RT-PCR assay for rapid and sensitive detection of hepatitis E virus. *J. Virol. Methods* **2006**, *131*, 65–71. [CrossRef] [PubMed]
33. Zeng, S.Q.; Halkosalo, A.; Salminen, M.; Szakal, E.D.; Puustinen, L.; Vesikari, T. One-step quantitative RT-PCR for the detection of rotavirus in acute gastroenteritis. *J. Virol. Methods* **2008**, *153*, 238–240. [CrossRef]
34. Caporaso, J.G.; Lauber, C.L.; Walters, W.A.; Berg-Lyons, D.; Huntley, J.; Fierer, N.; Knight, R. Ultra-high-throughput microbial community analysis on the Illumina HiSeq and MiSeq platforms. *ISME J.* **2012**, *6*, 1621–1624. [CrossRef]
35. Wang, Y.; Qian, P.Y. Conservative fragments in bacterial 16S rRNA genes and primer design for 16S ribosomal DNA amplicons in metagenomic studies. *PLoS ONE* **2009**, *4*, e7401. [CrossRef] [PubMed]
36. Christoff, A.P.; Cruz, G.N.F.; Sereia, A.F.R.; Yamanaka, L.E.; Silveira, P.P.; De Oliveira, L.F.V. End-to-end assessment of fecal bacteriome analysis: From sample processing to DNA sequencing and bioinformatics results. *BioRxiv* **2019**, 646349. [CrossRef]
37. Schmidt, P.A.; Bálint, M.; Greshake, B.; Bandow, C.; Römbke, J.; Schmitt, I. Illumina metabarcoding of a soil fungal community. *Soil Biol. Biochem.* **2013**, *65*, 128–132. [CrossRef]
38. Caporaso, J.G.; Kuczynski, J.; Stombaugh, J.; Bittinger, K.; Bushman, F.D.; Costello, E.K.; Knight, R. QIIME allows analysis of high-throughput community sequencing data. *Nat. Methods* **2010**, *7*, 335–336. [CrossRef]
39. BRASIL Ministério da Agricultura e do Abastecimento. *Instrução Normativa n°22, de 31 de julho de 2000, aprova os Regulamento Técnicos de Identidade e Qualidade de Copa, de Jerked Beef, de Presunto tipo Parma, de Presunto Cru, de Salame, de Salaminho, de Salaminho tipo Alemão, de Salame tipo Calabrês, de Salame tipo Friolano, de Salame tipo Napolitano, de Salame tipo Hamburguês, de Salame tipo Italiano, de Salame tipo Milano, de Lingüiça Colonial e Pepperoni. Diário Oficial da União, 3 ago. Seção 1, p. 15*; Ministério da Agricultura e do Abastecimento: Brasília, Brazil, 2020.
40. Ingham, S.C.; Engel, R.A.; Fanslau, M.A.; Schoeller, E.L.; Searls, G.; Buege, D.R.; Zhu, J. Fate of *Staphylococcus aureus* on Vacuum-Packaged Ready-to-Eat Meat Products Stored at 21 °C. *J. Food Protec.* **2005**, *68*, 1911–1915. [CrossRef]
41. Tim, H. Sodium–Technological functions of salt in the manufacturing of food and drink products. *Br. Food J.* **2002**, *104*, 126–152. [CrossRef]
42. De Souza, A.J.S.; Gomes-Gouvêa, M.S.; Soares, M.d.C.P.; Pinho, J.R.R.; Malheiros, A.P.; Carneiro, L.A.; Dos Santos, D.R.L.; Pereira, W.L.A. HEV infection in swine from Eastern Brazilian Amazon: Evidence of co-infection by different subtypes. *Comp. Immunol. Microbiol. Infect. Dis.* **2012**, *35*, 477–485. [CrossRef]
43. Boxman, I.L.; Jansen, C.C.; Zwartkruis-Nahuis, A.J.; Hägele, G.; Sosef, N.P.; Dirks, R.A. Detection and quantification of hepatitis E virus RNA in ready to eat raw pork sausages in the Netherlands. *Int. J. Food Microb.* **2020**, *333*, 108791. [CrossRef] [PubMed]
44. Berto, A.; Grierson, S.; Hakze-van der Honing, R.; Martelli, F.; Johne, R.; Reetz, J.; Banks, M. HEV in Pork Liver sausages, France. *Emerg. Infec. Dis.* **2013**, *19*, 3–5. [CrossRef]
45. Colson, P.; Borentain, P.; Queyriaux, B.; Kaba, M.; Moal, V.; Gallian, P.; Gerolami, R. Pig Liver Sausage as a Source of Hepatitis E Virus Transmission to Humans. *J. Infect. Dis.* **2010**, *202*, 825–834. [CrossRef]
46. Wolff, A.; Günther, T.; Albert, T.; Schilling-Loeffler, K.; Gadicherla, A.K.; Johne, R. Stability of hepatitis E virus at different pH values. *Int. J. Food Microbiol.* **2020**, *325*, 108625. [CrossRef]
47. Wolff, A.; Günther, T.; Albert, T.; Johne, R. Effect of Sodium Chloride, Sodium Nitrite and Sodium Nitrate on the Infectivity of Hepatitis E Virus. *Food Environ. Virol.* **2020**, *12*, 350–354. [CrossRef] [PubMed]
48. Dhama, K.; Chauhan, R.S.; Mahendran, M.; Malik, S.V.S. Rotavirus diarrhea in bovines and other domestic animals. *Vet. Res. Commun.* **2009**, *33*, 1–23. [CrossRef]
49. Terzić-Vidojević, A.; Veljović, K.; Tolinački, M.; Živković, M.; Lukić, J.; Lozo, J.; Golić, N. Diversity of non-starter lactic acid bacteria in autochthonous dairy products from Western Balkan Countries-technological and probiotic properties. *Food Res. Int.* **2020**, *136*, 109494. [CrossRef]
50. Lund, P.A.; De Biase, D.; Liran, O.; Scheler, O.; Mira, N.P.; Cetecioglu, Z.; O'Byrne, C. Understanding how microorganisms respond to acid pH is central to their control and successful exploitation. *Front. Microbiol.* **2020**, *11*, 2233. [CrossRef]
51. Franciosa, I.; Alessandria, V.; Dolci, P.; Rantsiou, K.; Cocolin, L. Sausage fermentation and starter cultures in the era of molecular biology methods. *Int. J. Food Microbiol.* **2018**, *279*, 26–32. [CrossRef] [PubMed]

52. Hebert, E.M.; Saavedra, L.; Taranto, M.P.; Mozzi, F.; Christian Magni, C.; Nader, M.E.F.; de Valdez, G.F.; Sesma, F.; Vignolo, G.; Raya, R.R. Genome Sequence of the Bacteriocin-Producing *Lactobacillus curvatus* Strain CRL. *J. Bacteriol.* **2012**, *194*, 538–539. [CrossRef]
53. Mainar, M.S.; Xhaferi, R.; Samapundo, S.; Devlieghere, F.; Leroy, F. Opportunities and limitations for the production of safe fermented meats without nitrate and nitrite using an antibacterial *Staphylococcus sciuri* starter culture. *Food Control.* **2016**, *69*, 267–274. [CrossRef]
54. Colak, H.; Hampikyan, H.; Ulusoy, B.; Bingol, E.B. Presence of *Listeria monocytogenes* in Turkish style fermented sausage (sucuk). *Food Control.* **2007**, *18*, 30–32. [CrossRef]
55. De Cesare, A.; Mioni, R.; Manfreda, G. Prevalence of *Listeria monocytogenes* in fresh and fermented Italian sausages and ribotyping of contaminating strains. *Int. J. Food Microbiol.* **2007**, *120*, 124–130. [CrossRef]
56. Meloni, D. Presence of *Listeria monocytogenes* in Mediterranean-style dry fermented sausages. *Foods* **2015**, *4*, 34–50. [CrossRef]
57. Auvolat, A.; Besse, N.G. The challenge of enumerating *Listeria monocytogenes* in food. *Food Microb.* **2016**, *53*, 135–149. [CrossRef]
58. Thevenot, D.; Delignette-Muller, M.L.; Christieans, S.; Vernozy-Rozand, C. Prevalence of *Listeria monocytogenes* in 13 dried sausage processing plants and their products. *Int. J. Food Microb.* **2005**, *102*, 85–94. [CrossRef]
59. Mainar, M.S.; Stavropoulou, D.A.; Leroy, F. Exploring the metabolic heterogeneity of coagulase-negative staphylococci to improve the quality and safety of fermented meats: A review. *Int. J. Food Microb.* **2017**, *247*, 24–37. [CrossRef] [PubMed]
60. Stellato, G.; La Storia, A.; De Filippis, F.; Borriello, G.; Villani, F.; Ercolini, D. Overlap of spoilage-associated microbiota between meat and the meat processing environment in small-scale and large-scale retail distributions. *Appl. Environ. Microb.* **2016**, *82*, 4045–4054. [CrossRef]
61. Andrade, M.J.; Córdoba, J.J.; Casado, E.M.; Córdoba, M.G.; Rodríguez, M. Effect of selected strains of *Debaryomyces hansenii* on the volatile compound production of dry fermented sausage "salchichón". *Meat Sci.* **2010**, *85*, 256–264. [CrossRef]
62. Cano-García, L.; Rivera-Jiménez, S.; Belloch, C.; Flores, M. Generation of aroma compounds in a fermented sausage meat model system by *Debaryomyces hansenii* strains. *Food Chem.* **2014**, *151*, 364–373. [CrossRef]
63. Núñez, F.; Lara, M.S.; Peromingo, B.; Delgado, J.; Sánchez-Montero, L.; Andrade, M.J. Selection and evaluation of *Debaryomyces hansenii* isolates as potential bioprotective agents against toxigenic penicillia in dry-fermented sausages. *Food Microb.* **2015**, *46*, 114–120. [CrossRef] [PubMed]
64. Encinas, J.P.; López-Díaz, T.M.; García-López, M.L.; Otero, A.; Moreno, B. Yeast populations on Spanish fermented sausages. *Meat Sci.* **2000**, *54*, 203–208. [CrossRef]
65. Mižáková, A.; Pipová, M.; Turek, P. The occurrence of moulds in fermented raw meat products. *Czech. J. Food Sci.* **2002**, *20*, 89–94. [CrossRef]
66. Ciegler, A.; Mintzlaff, H.J.; Weisleder, D.; Leistner, L. Potential production and detoxification of penicillic acid in mold-fermented sausage (salami). *Appl. Microb.* **1972**, *24*, 114–119. [CrossRef] [PubMed]
67. Castellari, C.; Quadrelli, A.M.; Laich, F. Surface mycobiota on Argentinean dry fermented sausages. *Int. J. Food Microbiol.* **2010**, *142*, 149–155. [CrossRef] [PubMed]
68. Sonjak, S.; Ličen, M.; Frisvad, J.C.; Gunde-Cimerman, N. The mycobiota of three dry-cured meat products from Slovenia. *Food Microbiol.* **2011**, *28*, 373–376. [CrossRef]
69. Zadravec, M.; Vahčić, N.; Brnić, D.; Markov, K.; Frece, J.; Beck, R.; Pleadin, J. A study of surface moulds and mycotoxins in Croatian traditional dry-cured meat products. *Int. J. Food Microbiol.* **2020**, *317*, 108459. [CrossRef]

Article

Mediterranean Spontaneously Fermented Sausages: Spotlight on Microbiological and Quality Features to Exploit Their Bacterial Biodiversity

Federica Barbieri [1], Giulia Tabanelli [2,3,*], Chiara Montanari [3], Nicolò Dall'Osso [1], Vida Šimat [4], Sonja Smole Možina [5], Alberto Baños [6], Fatih Özogul [7], Daniela Bassi [8], Cecilia Fontana [8] and Fausto Gardini [1,3]

1 Department of Agricultural and Food Sciences, University of Bologna, 47521 Cesena, Italy; federica.barbieri16@unibo.it (F.B.); nicolo.dallosso2@unibo.it (N.D.); fausto.gardini@unibo.it (F.G.)
2 Department of Agricultural and Food Sciences, University of Bologna, 40127 Bologna, Italy
3 Interdepartmental Center for Industrial Agri-Food Research, University of Bologna, 47521 Cesena, Italy; chiara.montanari8@unibo.it
4 University Department of Marine Studies, University of Split, 21000 Split, Croatia; vida@unist.hr
5 Department of Food Science and Technology, Biotechnical Faculty, University of Ljubljana, 1000 Ljubljana, Slovenia; sonja.smole-mozina@bf.uni-lj.si
6 Department of Microbiology, DOMCA S.A.U., 18620 Alhendín, Spain; abarjona@domca.com
7 Department of Seafood Processing Technology, Faculty of Fisheries, Cukurova University, Adana 01330, Turkey; fozogul@cu.edu.tr
8 Department for Sustainable Food Process (DISTAS), Università Cattolica del Sacro Cuore, 26100 Cremona, Italy; daniela.bassi@unicatt.it (D.B.); cecilia.fontana@unicatt.it (C.F.)
* Correspondence: giulia.tabanelli2@unibo.it; Tel.: +39-347-032-8294

Citation: Barbieri, F.; Tabanelli, G.; Montanari, C.; Dall'Osso, N.; Šimat, V.; Smole Možina, S.; Baños, A.; Özogul, F.; Bassi, D.; Fontana, C.; et al. Mediterranean Spontaneously Fermented Sausages: Spotlight on Microbiological and Quality Features to Exploit Their Bacterial Biodiversity. *Foods* **2021**, *10*, 2691. https://doi.org/10.3390/foods10112691

Academic Editors: Valentina Bernini and Juliano De Dea Lindner

Received: 20 October 2021
Accepted: 27 October 2021
Published: 3 November 2021

Abstract: The wide array of spontaneously fermented sausages of the Mediterranean area can represent a reservoir of microbial biodiversity and can be an important source of new technological and functional strains able to preserve product properties, counteracting the impoverishment of their organoleptic typical features due to the introduction of commercial starter cultures. We analysed 15 artisanal salamis from Italy, Spain, Croatia and Slovenia to evaluate the microbiota composition, through culture-dependent and culture-independent techniques (i.e., metagenomic analysis), chemical–physical features, biogenic amines and aroma profile. The final pH varied according to origin and procedures (e.g., higher pH in Italian samples due to long ripening and mold growth). Lactic acid bacteria (LAB) and coagulase-negative cocci (CNC) were the dominant population, with highest LAB counts in Croatian and Italian samples. Metagenomic analysis showed high variability in qualitative and quantitative microbial composition: among LAB, *Latilactobacillus sakei* was the dominant species, but *Companilactobacillus* spp. was present in high amounts (45–55% of the total ASVs) in some Spanish sausages. Among staphylococci, *S. epidermidis*, *S. equorum*, *S. saprophyticus*, *S. succinus* and *S. xylosus* were detected. As far as biogenic amines, tyramine was always present, while histamine was found only in two Spanish samples. These results can valorize the bacterial genetic heritage present in Mediterranean products, to find new candidates of autochthonous starter cultures or bioprotective agents.

Keywords: natural fermentation; dry fermented sausages; microbial biodiversity; lactic acid bacteria; CNC; 16S metagenomics

1. Introduction

Fermented meat products represent an important industrial sector in Europe, particularly in the Mediterranean countries (MC). In addition to the economic importance of this supply chain, the cured products constitute a valued cultural heritage strongly linked to the identity of a population or specific production areas. Particularly, a wide variety of sausages are produced using typical regional recipes and ancient processes [1,2]. Many authors reported that the microbiological features and technological attributes of fermented foods, including dry fermented sausages, are affected by the geographical origin, due to the specific manufacturing process conditions and different raw materials and formulations [3–7].

For years, sausages have undergone massive technological, economic, social, and even nutritional transformations. Innovations in the meat industry, meeting production and safety standards, helped to reduce waste and save production time, energy, and costs [1].

Particularly, in the last decades, the use of starter cultures has been introduced in the meat industry to guide fermentation, enhancing product safety but losing biodiversity [8]. However, in the different areas of the MC, the presence of numerous local products still obtained through spontaneous fermentation is recognized as a formidable treasure chest of unexplored microbial biodiversity.

Spontaneous fermentation relies on the presence of indigenous microorganisms conferring peculiar characteristics to the products in terms of both technological and organoleptic traits [9]. The meat environment and the process applied to facilitate the growth of some microbial groups, among which lactic acid bacteria (LAB) and coagulase-negative cocci (CNC) play a major role. Besides, filamentous fungi and yeasts can exert important effects, preventing excessive dehydration and the oxidation of the lipid fraction due to oxygen [9]. The development of the peculiar sausage flavor, aroma and texture relies on the interactions among microorganisms, raw materials and processing technology, due to biochemical and physicochemical reactions in which several bacteria, yeast and fungi cooperate within the meat matrix and its surface [10,11].

The study of traditional spontaneously fermented product characteristics and their microbiota can be a strategy to explore new technological and functional strains and to add value to local productions while preserving authenticity and traditional features. In this context, producers in the rural areas of MCs could take advantage of competitive opportunities and development, safeguarding the traceability and diversification of fermented sausages.

This work aimed to generate new knowledge about the microbial ecology in spontaneously fermented sausages produced in four MCs (Italy, Spain, Croatia, and Slovenia), using a combination of cultivation-dependent and metagenomic techniques. In addition, the safety (i.e., biogenic amine content) and chemical–physical features of the products, as well as their aroma profile, were evaluated.

2. Materials and Methods

2.1. Sample Collection

A total of fifteen samples of artisanal fermented sausages, produced without starter addition, were collected at the end of production from four different MCs: two samples of Italian salami produced by two different small companies in the Marche region (IM1 and IM2); one salame Alfianello coming from the Lombardia region (IAL); two traditional Slovenian smoked salami (SN and SWO, that differed only for the presence of nitrate/nitrite); seven traditional salchichón and chorizo produced in different locations from Andalusia (Spain): a salchichón Alhendín (ESA), a salchichón Bérchules (ESB), a salchichón Écija (ESE), a salchichón Olvera (ESO), a chorizo Bérchules (ECB), a chorizo Écija (ECE), and a chorizo Olvera (ECO); and three samples from Croatia, a traditional unsmoked salami (HNS), a traditional smoked salami (HS) and a Salami Zminjska Klobasica (HZK).

2.2. pH and Composition Analysis

The pH of the 15 ripened samples was detected by using a pH-meter Basic 20 (Crison Instruments, Barcelona, Spain). Aw detection was performed with an Aqualab CX3-TE (Labo- Scientifica, Parma, Italy). All measures were performed in triplicate. A FoodScan instrument (Foss, Hilleroed, Denmark) was used to carry out the sausage centesimal composition. This technique uses near-infrared transmission (NIT), with a wavelength between 850–1050 nm, for inhomogeneous samples. The absorbance data obtained are processed with a mathematical function and a calibration model to calculate the expected value to provide the % of fat, moisture, protein, collagen and salt.

2.3. Microbial Counts

For bacterial enumeration, the samples were prepared by removing aseptically the casing, and a slice of approx. 10 g of the sausage was transferred into a stomacher bag, mixed with 90 mL of 0.9% (*w*/*v*) NaCl sterile solution and homogenized in a Lab Blender Stomacher (Seward Medical, London, UK) for 2 min. Appropriate decimal dilutions were prepared and plated onto selective media: (i) Mannitol Salt Agar (MSA) for enumeration of CNC (30 °C for 72 h); (ii) de Man-Rogosa-Sharpe (MRS) agar for LAB (30 °C for 48 h in anaerobic conditions); (iii) Slanetz and Bartley medium for enterococci (44 °C for 24 h); (iv) Sabouraud Dextrose Agar, added with 200 mg/l of chloramphenicol for yeasts and molds (28 °C for 72 h); (v) pour plating on Violet Red Bile Glucose Agar for *Enterobacteriaceae* (37 °C for 24 h); (vi) Violet Red Bile Agar with MUG (4-methylumbelliferyl-ß-D-glucuronide) for the enumeration of *Escherichia coli* (42 °C for 24 h). All media were provided by Oxoid (Basingstoke, UK). The results reported are the means of 3 different sausages. Each sausage was sampled in triplicate.

The presence of *Listeria monocytogenes* and *Salmonella* was evaluated according to the methods EN ISO 11290–1 and EN ISO 6579, respectively [12,13].

2.4. DNA Extraction and Sequencing

Total genomic DNA was directly extracted from 200 mg of frozen sausage samples, which were treated with lysozyme at 37 °C for 1 h, followed by mechanic lysis through TissueLyser II (Qiagen) with a frequency of 30 Hz for 5 min. The DNA was then purified using a Wizard genomic DNA purification kit (Promega, Mannheim, Germany) according to the manufacturer's recommendations. The purified DNA resuspended in water was quantified using a Qubit 4 Fluorimeter (ThermoFisher Scientific, Waltham, MA, USA). After the normalization of DNA concentration, the sequencing was carried out through the Illumina MiSeq platform, generating 300 bp pair-end sequencing reads. The library for Illumina sequencing was generated from V3–V4 variable regions of ribosomal 16S rRNA to characterize the bacterial population of the samples.

2.5. Bioinformatic Analysis

FASTQ sequence files from Illumina reads were analysed using DADA2 version 1.8 [14] with the R 3.5.1 environment, which implements a new quality-aware model of Illumina amplicon errors without constructing OTUs [14].

The parameters applied, as described in https://benjjneb.github.io/dada2/tutorial.html (access date on 26 July 2021), were the following: trimLeft equal to 30 and truncLen option set to 270 and 200 for the forward and reverse fastq files, respectively. The comparison between the amplicon sequence variant (ASV) predicted from DADA2 against SILVA database (version 138 updated according to the reclassification of the genus *Bacillus* and *Lactobacillus*) was used for the taxonomic assignment. ASVs belonging to taxa classified as external sample [15] contamination were not included in the composition analysis for microbial population or for ASVs with low abundance setting, specifically a threshold of relative abundance equal to 0.5%. The assignment at the species level for the remaining ASVs was confirmed.

2.6. Biogenic Amine Determination

The samples were extracted with trichloroacetic acid, according to Pasini et al. [16]. The extracts were subjected to a dansyl chloride derivatization (Sigma Aldrich, Gallarate, Italy), according to Martuscelli et al. [17]. An HPLC Agilent Technologies 1260 Infinity with the automatic injector (G1329B ALS 1260, loop of 20 µL), equipped with a UV detector (G1314F VWD 1260) set at 254 nm, was used. For the chromatographic separation a C18 Waters

Spherisorb ODS-2 (150 × 4.6 mm, 3 μm) column was used with the following gradient elution: 0–1 min acetonitrile/water 35:65, 1–6 min acetonitrile/water 55:45, 6–16 min acetonitrile/water 60:40, 16–24 min acetonitrile/water 90:10, 24–35 min acetonitrile/water 90:10, 35–40 min acetonitrile/water 35:65, 40–45 min acetonitrile/water 35:65, all at a flow rate 0.6 mL/min.

The amounts of amines were expressed as mg/kg with reference to a calibration curve obtained through aqueous dansyl-chloride-derivatized amine standards of concentrations ranging from 10 to 200 mg/L (Sigma-Aldrich, Milano, Italy). The detection limit for all the amines was 3 mg/kg of the sample under the adopted conditions. All the analyses were performed in triplicate.

2.7. Aroma Profile Analysis

Gas-chromatography-mass spectrometry coupled with the solid-phase microextraction (GC-MS-SPME) technique was employed for sausage volatile organic compound (VOCs) analysis. A total of 3 g of samples were combined with a known amount of 4-methyl-2-pentanol (Sigma-Aldrich, Steinheim, Germany) as internal standard and analysed according to the protocol reported by Montanari et al. [18]. Volatile peak identification was carried out using Agilent Hewlett–Packard NIST 2011 mass spectral library (Gaithersburg, MD, USA) [19].

The mass spectrum identification was confirmed in the same conditions by the injection of the pure standards (Sigma- Aldrich, St. Louis, MO, USA). Data are expressed as the ratio between each molecule's peak area and the peak area of internal standard and are the mean of six determinations for each sample.

2.8. Statistical Analysis

Principal component analysis (PCA) and cluster analysis (LDA) were carried out using Statistica 8.0 (StatSoft Inc., Tulsa, OK, USA).

3. Results and Discussion

3.1. Geographical Origin and Manufacturing Processes

The dry-fermented sausages in the study were produced according to traditional recipes with the addition of salt and different spices. The peculiar characteristics of the fifteen products are described in Table 1. The sausage diameter varied between 2.5 and 6 cm. All products were obtained using only pork meat and fat, apart from two Slovenian samples (SN and SWO) that also contained 20% of beef meat. The ripening conditions (temperature, time etc.) differed between the samples, and only three samples were smoked, i.e., two Slovenian samples (SN and SWO) and a Croatian sample (HS sample).

3.2. Physico-Chemical Characterization

In Table 2 the chemical–physical parameters of the fermented sausages at the end of ripening are shown. Wide differences in the final pH were observed, with values ranging from a minimum of 4.52 in the ECO samples to a maximum of 6.42 in IM1.

In general, Italian products have the highest pH values (between 5.88 and 6.42), together with Croatian ones (between 5.72 and 6.05). On the other hand, Slovenian sausages showed lower pH values (5.20 and 5.39). These values agree with those reported by Lešić et al. [20] for Croatian and Slovenian sausages and with the data reported by Cardinali et al. [21] for Italian Fabriano sausages.

Slovenian products are probably more subjected to northern European production influences. In fact, northern products are generally dried sausages with a pH around or even below 5 and often undergo a smoking phase (that inhibits molds), while Mediterranean sausages are usually long-ripened and with pH values up to 6.2–6.4, given the possible growth of desirable molds [22,23].

Table 1. Main characteristics of sausages collected in the four MCs (Italy, Slovenia, Spain and Croatia) in terms of ingredients, type of casing, presence of preservatives and smoking phase.

Characteristics of the Tested Sausages	Italy			Slovenia		Spain							Croatia		
	IM1	IM2	IAL	SN	SWO	ESA	ESB	ESE	ESO	ECB	ECE	ECO	HNS	HS	HZK
Section															
Diameter (cm)	5	3.5	6	5	5	4	5.5	4.5	5.5	4.5	4	5	3.5	3.5	2.5
Type of lean meat	pork	pork	pork	pork/bovine (3:1)	pork/bovine (3:1)	pork	pork	pork	pork	pork	pork	pork	pork	pork	pork
Fat in the meat batter (%)	8–12	8–12	20–30	19	19	20–25	20–30	20–30	20–25	25–30	25–30	15–20	30–35	30–35	20
Fat characteristics	cubes	cubes	minced	cubes	cubes	minced	minced	minced	minced	minced	minced	minced	cubes	cubes	cubes
Spices	pepper, white wine	pepper, white wine	pepper, cinnamon, nutmeg, cloves	pepper, garlic	pepper, garlic	pepper, nutmeg	pepper, nutmeg	pepper, nutmeg	pepper, nutmeg	paprika, oregano, nutmeg, coriander	paprika, oregano, nutmeg, coriander	paprika, oregano, nutmeg, coriander	pepper, garlic, mild paprika, hot paprika	pepper, garlic, mild paprika, hot paprika	pepper, garlic, wine
Nitrate/Nitrite	no	no	yes	yes	no	yes	yes	yes	yes	no	no	no	no	no	no
Type of casing	natural	natural	natural (pork or bovine)	collagen	collagen	collagen	collagen	collagen	collagen	collagen	collagen	collagen	natural (pork intestine)	natural (pork intestine)	natural (pork intestine)
Smoking	not smoked	not smoked	not smoked	smoked	smoked	not smoked	not smoked	not smoked	not smoked	not smoked	not smoked	not smoked	not smoked	smoked	not smoked

Table 2. Chemical-physical parameters of the fifteen dry fermented sausages at the end of ripening.

	Italy			Slovenia		Spain							Croatia		
	IM1	IM2	IAL	SN	SWO	ESA	ESB	ESE	ESO	ECB	ECE	ECO	HNS	HS	HZK
pH	6.42 ± 0.02	6.09 ± 0.01	5.88 ± 0.03	5.20 ± 0.02	5.39 ± 0.01	5.83 ± 0.03	5.63 ± 0.04	5.80 ± 0.02	5.13 ± 0.03	4.77 ± 0.02	5.04 ± 0.01	4.52 ± 0.03	5.81 ± 0.05	5.72 ± 0.04	6.05 ± 0.03
a_w	0.824 ± 0.003	0.760 ± 0.002	0.879 ± 0.002	0.823 ± 0.002	0.832 ± 0.003	0.917 ± 0.002	0.811 ± 0.002	0.848 ± 0.003	0.911 ± 0.004	0.895 ± 0.001	0.870 ± 0.003	0.908 ± 0.001	0.928 ± 0.001	0.903 ± 0.003	0.890 ± 0.001
Humidity (%)	26.15 ± 0.35	20.42 ± 0.38	30.11 ± 0.40	25.05 ± 0.27	26.07 ± 0.38	38.54 ± 0.19	25.45 ± 0.22	27.60 ± 0.18	39.04 ± 0.44	31.12 ± 0.30	30.25 ± 0.41	38.54 ± 0.29	28.75 ± 0.33	25.11 ± 0.38	32.69 ± 0.23
Fat (%)	34.13 ± 0.31	35.79 ± 0.33	34.18 ± 0.26	40.65 ± 0.17	41.33 ± 0.41	29.75 ± 0.28	42.15 ± 0.30	36.05 ± 0.35	29.42 ± 0.21	43.75 ± 0.25	40.38 ± 0.29	29.71 ± 0.37	50.37 ± 0.40	47.21 ± 0.46	30.21 ± 0.26
Proteins (%)	34.32 ± 0.21	37.73 ± 0.35	29.27 ± 0.17	27.61 ± 0.60	26.15 ± 0.44	26.47 ± 0.29	26.83 ± 0.33	28.48 ± 0.24	25.20 ± 0.48	18.44 ± 0.41	22.53 ± 0.26	23.71 ± 0.35	15.14 ± 0.40	20.08 ± 0.29	30.29 ± 0.34
Collagen (%)	1.12 ± 0.05	1.62 ± 0.06	2.61 ± 0.07	3.25 ± 0.03	3.23 ± 0.09	1.02 ± 0.10	1.28 ± 0.04	3.72 ± 0.06	2.10 ± 0.09	3.34 ± 0.08	2.69 ± 0.05	3.61 ± 0.04	2.81 ± 0.11	3.84 ± 0.07	2.08 ± 0.10
Salt (%)	4.31 ± 0.10	4.48 ± 0.05	3.84 ± 0.09	3.44 ± 0.07	3.24 ± 0.11	4.26 ± 0.02	4.24 ± 0.08	4.10 ± 0.06	4.28 ± 0.12	3.31 ± 0.05	4.12 ± 0.04	4.42 ± 0.08	2.94 ± 0.09	3.78 ± 0.05	4.69 ± 0.07

The Spanish fermented products were widely different in terms of pH, due to the heterogeneous traditional manufacturing process developed in the whole country. In fact, chorizo samples showed a very low pH (between 4.52 and 5.04), while the salchichón had a pH ranging from 5.13 to 5.83. There are different set of data reported in the literature for chorizo characteristics, and among them the pH can vary: results of Galician chorizo showed a pH value of about 5.6 [24], while Asturian products had pH between 5.0 and 5.1 [25].

Great variability was also observed in the final a_w of the products. In some case, this parameter reached extremely low values, such as in IM2 (0.760), ESB (0.811), IM1 (0.824) and SN (0.823). On the other hand, HNS showed the highest value (0.928). In any case, the aw values reflected the water content of the final products. This parameter ranged from 20.42% in IM2 to 39.04% in ESO. The salt content in the final products ranged from 2.94 (HNS) to 4.48 (IM2).

3.3. Microbiological Analysis

Microbial counts were performed to enumerate in the final product LAB, CNC, enterococci, enterobacteria (including *E. coli*) and yeasts (Table 3).

The largest microbial population was generally represented by LAB, whose counts ranged from 4.4 (ECB) to 8.7 (HNS) log cfu/g. The highest LAB counts were found in samples collected from Croatia and Italy. The lowest numbers were associated with sausages from Spain, particularly ECB and ECE. LAB have been described as the main bacterial population in the dominant microbiota of Croatian traditionally fermented sausages and in Italian spontaneously fermented salami, with loads of 7–8 log cfu/g in the final products [26–28].

The counts of CNC were higher in Italian Fabriano salami (7.1 log cfu/g) and in HZK (7.2 log cfu/g). This microbial group was below the detection limit in SN, ECB, ECE and ECO, while in SWO, its concentration was very low (1.4 log cfu/g). The presence of CNC was strongly related to pH, and in particular, they seem to be inhibited by lower pH.

Enterococci were not detected or were detected in sporadic quantities (<1 log cfu/g) in Italian and Slovenian samples and in three Spanish samples (ESA, ECB, ECO). The highest counts of this microbial group were found in Croatian samples (3.2–4.8 log cfu/g).

Yeasts ranged from 2.3 to 5.4 log cfu/g, except for SN sausage. This microbial population can have an important role in sausage-ripening, contributing to the formation of the aroma and to the evolution of product sensory features [29,30].

Enterobacteriaceae were present in detectable amounts only in Croatian products, with values ranging between 2.7 (HS) and 5.1 (HZK). In all other samples, this population was below the limit of determination. High levels of *Enterobacteriaceae* can indicate the low microbiological quality of the product, and high concentrations (>4 log cfu/g) of this population have been enumerated in several ripened traditional products [31–33].

The counts of *E. coli* were always below the detection limit. In addition, a search for pathogenic microorganisms such as *Listeria monocytogenes* and *Salmonella* was carried out, and no positive samples were found.

3.4. Metagenomic Analysis

A more detailed picture of the bacterial composition of spontaneously fermented sausage was obtained through amplicon sequencing and metagenomic analysis.

Only species and genera that reached a concentration higher than 0.5% of amplicon sequence variants (ASVs) in at least one of the samples were considered. A total of more than 500 ASVs were detected, indicating very high biodiversity in the composition of the microbiota of the European salamis taken into consideration. The relative abundance of the ASVs attributed to genera or species is given in Table 4.

Beyond a large number of microbial species, their relative composition was also very variable among the products, even for those collected in the same geographical area.

While in some samples most of the ASVs were attributed to a single group/species, some sausages were characterized by higher biodiversity, with an important diversification in the microbiota composition (e.g., HZK and ESO).

Among LAB, several genus and species were present. Members of the genus *Latilactobacillus* were found in all the sausages. *Latilactobacillus sakei* was the dominant species (>50% of ASVs in IM2, IAL, and SN). The *Latilactobacillus sakei* group (which included *Lat. curvatus*, *Lat. graminis* and *Lat. fuchuensis*) was found in all the sausages, even if in lower percentage, except for IM1 and ESA. *Lat. sakei*, and to a lesser extent *Lat. curvatus*, have been reported as the prevailing LAB species in fermented meat products originating from France, Italy and Spain [6,34,35]. In fact, LAB species diversity of fermented sausages is known to be limited, being *Lat. sakei*-predominant during the ripening process, due to the species' excellent adaptation, competitiveness and assertiveness in the meat matrix [36–41]. This predominance over other LAB can be attributed to its salt-tolerant and psychrotrophic nature and its specialization in metabolic pathways in the meat environment, including the arginine deiminase pathway and the utilization of nucleosides [36,42–46].

Other lactobacilli were sporadically detected in low amounts, such as *Lacticaseibacillus casei* group in ESB, *Lactiplantibacillus plantarum* group in ESO, *Ligilactobacillus* sp. and *Loigolactobacillus rennini* in HS and *Dellaglioa algida* in SWO. High levels of the member of the genus *Companilactobacillus* (*Com. alimentarius*, *Com. heilongjiangensis* and *Com. versmoldensis*) were present in many Spanish sausages, in particular ESE and ECB, in which they represented 55.3 and 45.0% of the total ASVs. The abundance of *Companilactobacillus* found only in Spanish sausages, is a regional peculiarity already reported in the literature [47,48]. Moreover, these species have been isolated as spoilage organisms in ready-to-eat meats and from other dry fermented sausages originating from Belgium and Italy [6,49,50] and have been proposed as autochthonous probiotic starter cultures for meat products [51].

Heterofermentative lactobacilli were present only in ESO and ECE (*Levilactobacillus yonginensis* group and *Limosilactobacillus mucosae*, respectively). In addition, ASVs belonging to some dairy LAB (*Lactobacillus helveticus*, *Streptococcus thermophilus* and *Lactococcus* sp.) were found in some Spanish products. These species can be related to the use of powdered milk or other dairy derivatives during manufacturing. The heterofermentative cocci were represented by *Leuconostoc carnosum*, present in relevant percentages in IM1 (15,8%), SN (19.1%), SWO (23.3%) and HZK (14.8%); *Leuconostoc* sp. (17.6% in IAL and 15.6% in SWO); and *Weissella* sp., found only in ESO.

Five species of staphylococci were detected (*Staph. epidermidis*, *Staph. equorum*, *Staph. saprophyticus*, *Staph. succinus* and *Staph. xylosus*). They were found only in some sausages, and they were dominant in IM1 (67.7% of ASVs belonging to *Staph. equorum* and 1 5% to *Staph. succinus*) and ESA (98.1 of ASVs belonging to *Staph. xylosus*).

Among CNS, *Staphylococcus xylosus* is the prevalent species associated with salamis, even if greater species diversity can be found, *Staphylococcus epidermidis*, *Staphylococcus equorum* and *Staphylococcus saprophyticus* also being reported in these products [52–54].

Staphylococci exert several important technological roles in sausage production, such as contribution to flavor and color [54,55], and since they are able to use alternative energy sources, such as arginine and nucleoside, their ecological persistence is assured [37]. These bacteria were detected as subdominant fractions in several fermented meat products, and their presence was associated with mammal skin, being competitive in the acidic and anaerobic conditions prevailing during meat fermentation [56,57]. It is noteworthy that many samples showed a very low pH (i.e., Spanish products) that could have hindered their growth. The results therefore underline the sensitivity of this genus to low pH, even in a potentially favorable environment, thanks to the low aw values. These populations probably suffered the strong initial acidification in these products, which was not followed by a significant increase in pH (scarcity of molds, etc.).

Carnobacterium sp. was highlighted only in some Croatian samples, i.e., HS and HNS, albeit with a relevant percentage (19.11% and 9%, respectively). This genus has been found in Fabriano-like products [21], and it has been associated with meat spoilage phenomena [58].

Among *Actinobacteria*, *Kocuria* sp. and *Corynebacterium* sp. were present only in HZK, while *Brevibacterium casei* and *Rothia* sp. were present in ECE.

The meat spoiler *Brochothrix thermosphacta* was detected in several sausages, and its concentration was particularly relevant in HZK (19.58%), IM2 (14.91%) and ESE (14.23%). This species has been reported to be part of the Fabriano-like sausage microbiota by Cardinali et al. [21], and it is commonly associated with spoiled meat-based products [59,60].

Among Gram-negative bacteria, *Pseudomonas* sp. appeared at a high concentration in the Slovenian samples, in the Spanish ESE (16.7% of ASV) and especially in HNS, in which this genus represented 26.8% of ASVs. Enterobacteria (*Klebsiella* sp. and *Escherichia*/*Shigella* group) were found in ESB, ESO and ECE.

To better evidence the differences in the ASV composition, a cluster analysis was carried out. The results are reported in Figure 1.

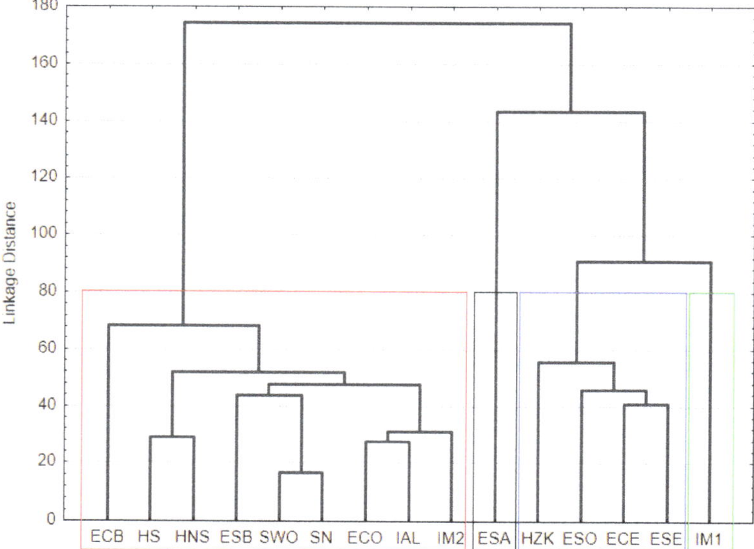

Figure 1. Tree diagram for the similarity of sausage metagenomic analysis obtained using Euclidean distances as distance metric and Ward's method for amalgamation rule.

Four different clusters were obtained. The first contained only the Italian sausage IM1. In the second cluster were grouped three Spanish sausages (ESE, ECE and ESO) and the Croatian sample HZK. The third cluster was represented only by ESA. The fourth cluster included nine samples, and in particular the Spanish sausages ECB, ECO and ESB; the Croatian HNS and HS; the Slovenian sausages; and the Italian samples IM2 and IAL.

To highlight the microorganisms responsible for the clustering of sausages, a PCA was applied whose first two components explained 59.76% of the variability (Figure 2).

The grouping of the sausages reflected the results of the tree diagram. In Figure 2, factor coordinates of the metagenomic variables are shown. Only the metagenomic variables with coordinates characterized by an absolute value higher than 1 are reported. The *Lat. sakei* and the *Lat. sakei* group were the main ones responsible for the description of the largest group located in the first quadrant, while *Staph. xylosus* determined the collocation of ESA in the left part of the second quadrant. The position of IM1 depends on the variable *Staph. equorum*, while the remaining group collocation is mainly influenced by the presence of *Companilactobacillus* (especially for the Spanish samples), *Corynebacterium* for the Croatian sausage and *Brochothrix thermosphacta* for all the samples grouped in this cluster.

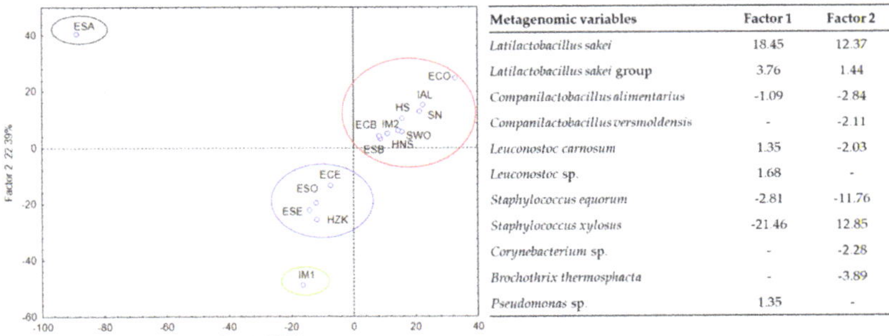

Metagenomic variables	Factor 1	Factor 2
Latilactobacillus sakei	18.45	12.37
Latilactobacillus sakei group	3.76	1.44
Companilactobacillus alimentarius	-1.09	-2.84
Companilactobacillus versmoldensis	-	-2.11
Leuconostoc carnosum	1.35	-2.03
Leuconostoc sp.	1.68	-
Staphylococcus equorum	-2.81	-11.76
Staphylococcus xylosus	-21.46	12.85
Corynebacterium sp.	-	-2.28
Brochothrix thermosphacta	-	-3.89
Pseudomonas sp.	1.35	-

Figure 2. PCA factor coordinates for the first two factors explaining the variability of the microbial populations of the sausages according to the metagenomic analysis. The factor coordinates of the most relevant metagenomic variables (characterized by values higher than |1|) for the first two factors are also reported, which represent the projection of the metagenomic variables on the factor-plane.

3.5. Biogenic Amine Concentrations

The biogenic amine concentrations in the sausages are reported in Table 5.

Tyramine was present in all the samples, with concentrations ranging from 47.7 mg/kg (IM2) to 366.8 mg/kg (HNS). The mean content of this biogenic amine was 165.5 mg/kg, with a standard deviation of 88.8 mg/kg, indicating a fair variability among the fifteen dry fermented sausages. In general, the Italian products showed lower tyramine concentrations, while the highest amounts were found in the Croatian salamis HNS and HS. These amounts are similar to those reported by EFSA [61], which indicated tyramine mean concentration of 136 mg/kg, with the 95th percentile of 397 mg/kg, in 400 European salami samples.

Histamine was found only in two Spanish samples (ESB and ECE) at concentrations of 195.8 and 174.7 mg/kg, respectively. This amine is considered the most dangerous for human health as it can cause various adverse effects known as "histamine poisoning" [62]. The quantities found in the two Spanish samples, although not very high, are nevertheless significant, especially when compared with the maximum amounts allowed in some fish products. These latter are the only ones regulated for histamine presence and generally admit a maximum quantity of 100 mg/kg in fresh fish and of 200 mg/kg for processed products [63].

The presence of putrescine was more variable: in four samples, this amine was not detected, while two Croatian samples (HNS and HS) showed the highest amounts (about 300 mg/kg). In the same samples, higher quantities of cadaverine were also observed.

Both Gram-negative and Gram-positive bacteria have been described as biogenic amine producers, with wide variability in aminobiogenetic potential between different strains of the same species. Spoilage microorganisms such as enterobacteria and pseudomonads can be strong histamine, cadaverine and putrescine accumulators, and biogenic amines produced by these microbial populations can also be found in fermented sausages [64,65]. On the other hand, decarboxylase activity has been found in Gram-positive strains, also belonging to the genus *Staphylococcus* and LAB. While *Lat. sakei* is usually known for its inability to produce BAs, in many other LAB species, strains with decarboxylase activity are present. *Lat. curvatus*, for example, can accumulate both tyramine and histamine as well as *Com. alimentarius* [66–68]. The ability to produce BAs has been found in other genera found in this investigation, such as *Leuconostoc* sp., *Weissella* sp. and *Carnobacterium* [68].

Table 3. Concentrations (log cfu/g) of the main microbial groups in the fifteen sausages at the end of ripening.

		Italy		Slovenia					Spain					Croatia	
	IM1	IM2	IAL	SN	SWO	ESA	ESB	ESE	ESO	ECB	ECE	ECO	HNS	HS	HZK
LAB	7.07 ± 0.48	8.52 ± 0.11	8.26 ± 0.16	6.57 ± 0.02	7.28 ± 0.28	7.85 ± 0.75	6.96 ± 0.21	6.32 ± 0.30	7.78 ± 0.14	4.41 ± 0.72	5.88 ± 1.02	7.73 ± 0.27	8.67 ± 0.11	8.43 ± 0.09	8.54 ± 0.03
CNS	7.12 ± 0.09	7.13 ± 0.12	5.22 ± 1.05	<1	1.44 ± 2.03	3.65 ± 0.22	3.05 ± 4.31	4.54 ± 0.16	5.40 ± 0.12	<1	<1	<1	5.34 ± 0.82	5.09 ± 0.01	7.24 ± 0.32
Enterococci	<1	0.60 ± 0.75	0.89 ± 0.71	<1	<1	<1	2.37 ± 0.10	2.19 ± 0.53	2.05 ± 0.38	<1	1.19 ± 0.35	0.89 ± 0.46	3.74 ± 0.12	3.18 ± 0.74	4.80 ± 0.16
Yeasts and molds	5.44 ± 0.05	5.01 ± 0.26	3.46 ± 0.49	0.95 ± 1.35	3.26 ± 0.37	5.34 ± 0.25	3.41 ± 0.80	4.10 ± 0.58	4.59 ± 0.59	3.10 ± 0.27	2.60 ± 0.25	3.70 ± 0.08	2.27 ± 0.31	3.85 ± 0.31	4.15 ± 0.10
Enterobacteriaceae	<1	<1	<1	<1	<1	<1	<1	<1	<1	<1	<1	<1	3.55 ± 0.50	2.71 ± 0.82	5.12 ± 0.49

Table 4. Relative abundance of amplicon sequence variants (ASVs) in the fifteen sausage samples analysed by metagenomic analysis. Only species and genera which reached a concentration higher than 0.5% in at least one of the samples are reported.

		Italy		Slovenia					Spain					Croatia	
	IM1	IM2	IAL	SN	SWO	ESA	ESB	ESE	ESO	ECB	ECE	ECO	HNS	HS	HZK
Latilactobacillus sakei	9.51	52.71	54.54	55.24	44.75	0.71	36.32	2.89	3.88	41.24	15.12	72.43	35.68	45.36	9.72
Latilactobacillus sakei group	- *	9.55	17.99	4.94	2.34	-	4.85	10.87	23.16	4.95	6.67	22.25	24.63	16.44	2.57
Lactiplantibacillus plantarum group	-	-	-	-	-	-	-	-	2.64	-	-	-	-	-	-
Lacticaseibacillus casei group	-	-	-	-	-	-	1.16	-	-	-	-	-	-	-	-
Companilactobacillus alimentarius	-	-	-	-	-	-	-	21.75	5.81	-	32.32	5.32	-	-	-
Companilactobacillus heilongjiangensis	-	-	-	-	-	-	-	-	1.32	-	-	-	-	-	-
Companilactobacillus versmoldensis	-	-	-	-	-	-	-	33.58	3.96	44.98	2.20	-	-	-	-
Ligilactobacillus sp.	-	-	-	-	-	-	-	-	-	-	-	-	-	4.80	-
Loigolactobacillus rennini	-	-	-	-	-	-	-	-	-	-	-	-	-	7.94	-
Dellaglioa algida	-	-	-	-	3.66	-	-	-	-	3.32	-	-	-	-	-
Levilactobacillus yonginensis group	-	-	-	-	-	-	-	-	3.36	-	-	-	-	-	-
Limosilactobacillus mucosae	-	-	-	-	-	-	-	-	-	-	4.44	-	-	-	-
Lactobacillus helveticus	-	3.07	-	-	-	-	25.54	-	19.45	0.94	-	-	-	-	-
Lactococcus sp.	-	-	1.12	-	-	-	-	-	-	-	2.72	-	-	-	-
Streptococcus sp.	-	-	-	-	-	-	3.68	-	-	2.09	4.61	-	-	-	-
Leuconostoc carnosum	15.84	-	8.75	19.06	23.26	-	6.70	-	1.70	0.09	-	-	-	-	14.80
Leuconostoc sp.	0.71	-	17.60	2.74	10.20	-	15.63	-	-	-	-	-	-	-	3.65
Weissella sp.	-	-	-	-	-	-	-	-	11.40	-	-	-	-	-	-
Carnobacterium sp.	-	-	-	-	-	-	-	-	-	-	-	-	9.00	19.11	-
Staphylococcus epidermidis	-	-	-	-	-	-	-	-	-	-	4.58	-	-	-	-
Staphylococcus equorum	67.74	11.65	-	-	-	-	-	-	4.47	-	-	-	-	-	14.04
Staphylococcus saprophyticus	-	4.20	-	-	-	-	-	-	-	-	-	-	-	-	-

Table 4. *Cont.*

	Italy			Slovenia		Spain							Croatia		
	IM1	IM2	IAL	SN	SWO	ESA	ESB	ESE	ESO	ECB	ECE	ECO	HNS	HS	HZK
Staphylococcus succinus	1.47	2.09	-	-	-	-	-	-	4.35	-	-	-	-	-	5.27
Staphylococcus xylosus	-	1.81	-	-	-	98.13	-	-	-	2.38	-	-	-	-	-
Kocuria sp.	-	-	-	-	-	-	-	-	-	-	-	-	-	-	1.25
Corynebacterium sp.	-	-	-	-	-	-	-	-	-	-	-	-	-	-	27.91
Corynebacterium variabile	-	-	-	-	-	-	-	-	-	-	-	-	-	-	1.21
Brevibacterium casei	-	-	-	-	-	-	-	-	-	-	1.59	-	-	-	-
Rothia sp.	-	-	-	-	-	-	-	-	-	-	4.10	-	-	-	-
Brochothrix thermosphacta	4.73	14.91	-	1.73	5.81	-	-	14.23	12.77	-	7.90	-	3.91	1.65	19.58
Escherichia/Shigella sp.	-	-	-	-	-	-	1.45	-	-	-	0.68	-	-	-	-
Klebsiella sp.	-	-	-	-	-	-	1.21	-	1.20	-	4.56	-	-	-	-
Pseudomonas sp.	-	-	-	14.64	7.57	-	-	16.68	-	-	1.42	-	26.79	4.70	-
Photobacterium piscicola	-	-	-	-	-	-	1.67	-	-	-	-	-	-	-	-
Acinetobacter sp.	-	-	-	1.64	2.42	1.16	-	-	-	-	2.26	-	-	-	-
Erysipelothrix sp.	-	-	-	-	-	-	1.80	-	0.52	-	4.84	-	-	-	-

*: not detected.

Table 5. Concentrations (mg/kg) of the main biogenic amines detected in the fifteen samples at the end of ripening.

	Italy			Slovenia		Spain							Croatia		
	IM1	IM2	IAL	SN	SWO	ESA	ESB	ESE	ESO	ECB	ECE	ECO	HNS	HS	HZK
Histamine	-*	-	-	-	-	-	195.79 ± 27.29	-	-	-	170.74 ± 28.54	-	-	-	-
Tyramine	73.87 ± 21.83	47.65 ± 12.93	78.66 ± 1.31	209.38 ± 0.71	180.52 ± 10.34	199.24 ± 30.75	171.35 ± 37.52	149.92 ± 31.19	67.96 ± 14.34	146.06 ± 48.04	173.72 ± 39.46	202.50 ± 8.04	366.78 ± 38.31	312.93 ± 26.46	105.31 ± 29.18
Putrescine	-	-	115.67 ± 3.78	59.03 ± 2.70	67.58 ± 3.64	-	108.07 ± 15.41	42.79 ± 8.54	110.54 ± 4.26	99.28 ± 13.57	79.30 ± 10.81	155.95 ± 15.52	256.59 ± 8.92	359.59 ± 80.64	-
Cadaverine	-	-	-	83.38 ± 2.61	100.87 ± 0.44	67.94 ± 1.91	136.90 ± 20.01	-	-	-	-	-	436.03 ± 29.87	252.52 ± 30.31	-
TOTAL	73.87	47.65	194.33	268.41	348.98	267.18	612.10	192.72	178.50	245.34	423.76	358.46	1059.40	925.04	105.31

*: under the detection level (3–5 mg/kg).

3.6. Sausage Aroma Profile

The aroma profile of the fifteen ripened sausages has been studied. The unidentified compounds accounted for less than 1% of the total peak area in each sample.

The analysis of the volatile organic compounds (VOCs) allowed the clear differentiation of the samples, reflecting the different formulations and production and ripening conditions traditionally adopted in the Mediterranean geographical areas. Indeed, it has been reported that these differences can influence sausage aroma, being dry fermented sausage flavor affected by many processing factors, i.e., raw materials, spices, microbiota composition, smoking, etc. [55].

Within this wide variability, some common characteristics can be found by grouping the identified molecules in homogeneous chemical groups: ketones, aldehydes, alcohols, acids, and esters (Table 6), as well as molecules deriving from spices and smoking, reported as Supplementary Materials (Tables S1 and S2, respectively).

Concerning the molecules associated with the spices included in the meat batter formulation (Table S1), they belong to terpenes and terpenoids, phenylpropenes and compounds deriving from garlic (dimethyl disulfide, diallyl sulfide, etc.) [69]. These latter compounds are particularly present in Slovenian salami, Spanish chorizo and Croatian products, except for the HNS sample. Among terpenes, D-limonene certainly is the prevalent molecule in products in which pepper has been used. Many other terpenes deriving from this spice, such as myrcene, linalool, copaene, carene, o-cymene, etc. [70], have been detected in these products, while they are absent in chorizo sausages, where oregano has been used. On the other hand, eugenol, safrole and methyl-eugenol are associated with the use of spices such as nutmeg, cinnamon and cloves [71,72].

The VOCs derived from smoke (Table S2) include furans and phenols, already reported for smoked products [73], and their presence is higher in the Slovenian sausages (SN and SWO) and HS, characterized by a smoking phase during their production. Furthermore, some of these VOCs were detected also in the Spanish chorizo samples, due to the traditional use of smoked paprika [74].

Table 6 reports VOCs derived from the microbial biochemical activities that occurred during sausage fermentation and ripening. The compounds are grouped into chemical families, whose total amounts are shown in Figure 3. Higher quantities of VOCs were found in some Spanish products and two Croatian samples.

Ketones were evenly distributed in the analysed samples, except for chorizo and HNS samples where their total amount was higher. On the other hand, lower values were found in salchichón ESA. These molecules mostly derive from fatty acid oxidation and, in particular, from β-oxidation. Some microbial groups such as staphylococci and fungi can have a role in these phenomena. It is interesting to observe that products characterized by high concentrations of fat (for example HNS) showed the higher presence of ketones. HS, having the same formulation of HNS but subjected to smoking, presented lower ketone amounts: this can be attributed to the antioxidant role of some of the compounds produced by smoking. It is also interesting to note that chorizo samples are characterized by higher ketone amounts among Spanish products. These fermented sausages did not contain nitrates or nitrites, which exert a well-known antioxidant activity, while these preservatives were present in salchichón formulations.

Among ketones, 2-butanone prevailed in some samples, i.e., ECB and HNS (Table 6). The presence of this molecule in fermented meat products is common but its contribution to aroma profile can be negative depending on its amounts and the balance with other VOCs [75]. Diacetyl (2,3-butanedione) and acetoin (3-hydroxy-2-butanone) are particularly present in some Spanish products (ECB and ECO), in Slovenian samples and in two Croatian sausages (HNS and HZK). Diacetyl and acetoin are mainly produced through the catabolism of pyruvic acid by LAB [10,76].

Table 6. Volatile organic compounds (VOCs) detected by SPME-GC-MS in the samples, expressed as a ratio between the peak area of each molecule and the peak area of the internal standard (4-methyl-2-pentanol). The standard deviation was always below 5%.

VOCs	Italy			Slovenia		Spain							Croatia		
	IM1	IM2	IAL	SN	SWO	ESB	ESE	ESA	ESO	ECB	ECE	ECO	HNS	HS	HZK
Acetone	5.85	3.00	2.61	0.87	1.31	3.70	3.20	- *	1.79	2.26	13.48	4.85	1.86	0.66	1.38
2-butanone	0.55	1.08	1.13	-	0.44	1.42	0.86	-	0.97	47.36	2.70	2.51	94.05	3.50	1.30
2,3-butanedione	1.87	1.54	0.23	0.41	0.48	0.50	2.44	0.18	0.52	5.93	3.68	1.32	2.14	3.42	2.03
2-pentanone	0.73	0.52	0.51	-	-	-	1.45	-	-	-	1.53	-	-	-	-
Methyl Isobutyl Ketone	0.89	0.81	0.70	0.62	0.50	0.64	0.52	0.61	2.87	0.79	2.46	3.00	2.97	3.94	3.90
4-methyl-3-penten-2-one	3.61	3.42	3.51	2.87	2.87	2.98	2.96	2.03	7.43	1.84	6.52	7.72	7.48	5.96	10.08
2,6-dimethyl-4-heptanone	-	-	-	-	-	-	-	-	-	0.65	5.39	6.87	3.20	3.42	-
2-heptanone	1.71	4.15	1.99	1.39	1.02	4.94	2.44	-	3.14	-	4.32	1.70	2.76	-	-
3-octanone	1.10	0.87	0.72	0.41	0.28	0.48	0.34	-	2.88	-	2.65	3.23	1.55	-	1.76
2-octanone	1.13	1.44	0.72	0.86	1.64	0.97	1.08	1.01	1.60	1.32	2.37	2.57	2.03	1.06	0.96
3-hydroxy-2-butanone	3.72	2.85	0.71	10.72	7.98	1.27	9.37	-	1.57	4.19	-	10.74	9.03	-	11.78
2,5-octanedione	0.39	4.64	0.72	1.34	0.78	-	-	-	2.84	-	-	-	14.68	-	-
2-nonanone	1.06	4.40	0.81	-	-	5.74	0.30	-	1.58	-	-	0.70	1.49	-	-
2-undecanone	0.84	0.65	-	0.78	0.71	3.06	0.40	-	-	-	-	-	1.47	2.88	-
Ketones	**23.46**	**29.36**	**14.35**	**20.26**	**18.02**	**25.70**	**25.37**	**3.83**	**27.18**	**64.33**	**45.09**	**45.21**	**144.71**	**24.84**	**33.19**
3-methyl-butanal	1.03	0.38	0.73	-	0.29	1.65	1.74	-	-	-	-	-	1.20	-	-
Pentanal	0.44	2.01	0.69	0.70	0.90	0.95	0.74	0.21	4.22	-	-	0.65	6.56	-	-
Hexanal	4.90	38.66	6.59	3.38	3.51	1.56	5.77	1.09	48.94	0.81	3.29	3.44	110.21	-	2.92
Heptanal	-	-	-	-	-	-	-	-	-	-	-	-	6.53	-	-
Octanal	-	-	-	-	-	-	-	-	-	-	2.46	-	5.10	1.47	-
2-heptenal	-	-	-	-	-	-	-	-	-	-	-	-	51.97	-	-
Nonanal	4.01	6.33	4.60	6.75	3.97	4.37	3.45	3.92	6.11	1.03	11.14	8.97	14.11	7.47	7.85
Decanal	1.72	2.59	1.57	-	-	2.63	0.91	1.27	2.01	0.78	-	-	5.16	-	-
Benzaldehyde	3.64	3.99	1.58	2.06	3.18	3.70	1.37	0.56	1.14	3.80	8.98	25.37	38.39	6.26	2.24
Benzeneacetaldehyde	5.47	4.03	14.61	5.30	13.70	6.28	5.66	6.27	4.85	1.30	4.82	2.49	30.31	127.99	23.42
Hexadecanal	1.31	2.93	2.63	1.40	1.61	1.39	1.03	1.10	1.30	3.49	5.58	3.26	-	-	-
Aldehydes	**22.52**	**60.92**	**33.00**	**19.59**	**27.17**	**22.54**	**20.67**	**14.42**	**68.57**	**11.21**	**36.27**	**44.18**	**269.54**	**143.19**	**36.43**
Ethyl alcohol	15.46	6.17	20.31	24.81	18.94	101.21	20.17	270.32	173.69	41.81	29.99	18.04	71.23	118.85	52.12
2-butanol	-	-	-	-	-	-	-	-	-	0.49	-	-	4.72	3.87	0.83
1-propanol	-	-	-	-	-	-	-	-	-	9.77	4.51	-	6.49	14.93	-
2-propen-1-ol	-	-	-	0.67	0.46	-	-	-	-	0.33	1.42	1.29	3.22	-	0.41

Table 6. *Cont.*

VOCs	Italy			Slovenia		Spain							Croatia		
	IM1	**IM2**	**IAL**	**SN**	**SWO**	**ESB**	**ESE**	**ESA**	**ESO**	**ECB**	**ECE**	**ECO**	**HNS**	**HS**	**HZK**
Isoamyl alcohol	0.66	0.44	0.35	0.49	0.40	2.10	3.45	6.19	2.66	0.51	2.13	1.03	3.72	-	1.73
1-pentanol	0.43	1.04	0.87	0.79	0.41	2.07	0.91	-	1.21	-	-	0.67	5.46	-	-
1-hexanol	2.08	3.88	2.00	1.69	2.26	1.89	5.01	2.68	8.71	6.98	5.48	3.96	-	3.00	-
1-octen-3-ol	2.85	5.01	1.30	0.62	1.08	0.49	1.41	-	2.81	0.44	2.70	1.19	19.92	1.52	0.84
1-octanol	0.69	0.88	0.67	0.81	0.62	0.55	0.94	0.52	0.61	0.41	1.36	0.93	3.09	1.26	1.03
Benzyl alcohol	0.94	-	0.37	0.85	0.93	1.06	0.91	0.82	1.08	3.43	2.12	4.85	43.88	11.04	-
Phenylethyl alcohol	1.16	1.17	0.59	2.26	3.40	1.62	3.01	7.85	1.70	-	4.11	2.84	5.05	28.23	1.45
Alcohols	**24.27**	**18.59**	**26.46**	**32.99**	**28.50**	**111.00**	**35.81**	**288.37**	**192.46**	**64.17**	**53.81**	**34.81**	**166.79**	**182.70**	**58.40**
Acetic acid	11.28	25.95	27.28	95.75	79.25	87.10	14.38	11.72	65.64	203.95	349.10	185.04	73.10	114.92	16.40
Propanoic acid	0.43	0.71	0.43	2.93	0.87	1.10	0.48	-	-	7.07	7.65	0.94	14.39	22.71	-
Butanoic acid	10.88	5.70	4.23	10.48	8.37	15.04	9.36	19.13	20.17	4.91	14.94	20.07	4.68	8.52	3.43
Isovaleric acid	8.56	3.06	0.76	1.39	1.43	4.11	2.89	1.03	3.99	2.48	4.35	2.74	-	-	3.38
Pentanoic acid	0.68	0.78	0.59	1.08	0.87	1.07	0.91	0.68	1.17	1.21	2.90	1.11	0.95	2.82	1.00
Hexanoic acid	2.98	3.25	2.62	5.04	4.12	5.30	6.03	2.74	13.22	6.21	16.35	6.68	16.21	5.94	2.60
4-hexenoic acid	-	-	-	-	-	-	-	-	-	54.84	-	-	-	-	-
Heptanoic acid	0.89	1.18	0.74	1.32	1.08	1.41	0.98	0.66	1.45	1.60	2.22	1.32	1.30	1.86	1.06
Octanoic acid	2.55	3.14	2.65	6.45	4.63	3.84	3.89	2.63	6.19	7.34	5.93	4.90	4.57	6.79	2.76
Nonanoic acid	1.99	2.01	1.75	2.64	1.91	2.45	1.22	1.94	2.27	1.81	2.28	3.03	3.58	3.62	2.01
n-decanoic acid	1.82	2.98	2.56	6.68	5.18	5.58	2.52	2.35	4.38	7.37	5.86	5.09	4.84	6.10	1.79
Dodecanoic acid	1.70	4.24	1.99	8.87	1.99	4.81	0.89	2.23	2.36	4.10	1.66	1.87	1.50	2.67	0.84
Acids	**43.75**	**52.98**	**45.61**	**142.64**	**109.69**	**131.80**	**43.55**	**45.12**	**120.82**	**302.91**	**413.24**	**232.80**	**125.11**	**175.96**	**35.27**
Acetic acid, methyl ester	-	-	-	-	-	-	-	-	-	2.74	9.55	2.73	-	1.72	-
Ethyl Acetate	1.30	1.02	2.93	2.78	2.42	23.96	1.60	12.85	22.98	14.30	6.52	3.31	2.90	13.65	1.07
Butanoic acid, ethyl ester	-	-	0.33	0.91	0.72	1.94	0.78	4.71	5.68	0.90	0.97	0.46	1.00	8.84	-
Hexanoic acid, ethyl ester	2.58	1.80	1.73	1.79	1.76	4.17	2.53	6.97	17.18	2.72	1.71	-	-	7.58	-
4-Hexenoic acid, ethyl ester	-	-	-	-	-	-	-	-	-	37.68	-	-	-	-	-
Octanoic acid, ethyl ester	-	-	1.14	1.60	1.07	7.61	1.12	14.60	13.55	6.24	4.20	-	2.89	7.18	1.00
Dodecanoic acid, methyl ester	-	-	-	-	-	-	-	-	-	1.45	2.28	0.61	-	-	-
Dodecanoic acid, ethyl ester	1.75	-	1.35	1.84	1.03	7.95	0.62	13.02	6.98	4.65	2.56	-	2.63	9.86	-
Benzoic acid, ethyl ester	-	-	-	-	-	-	-	-	2.82	11.28	-	-	-	-	-
Esters	**5.62**	**2.82**	**7.48**	**8.93**	**7.00**	**45.64**	**6.65**	**52.15**	**69.20**	**81.96**	**27.79**	**7.11**	**9.42**	**48.82**	**2.07**

* not detected under the adopted conditions.

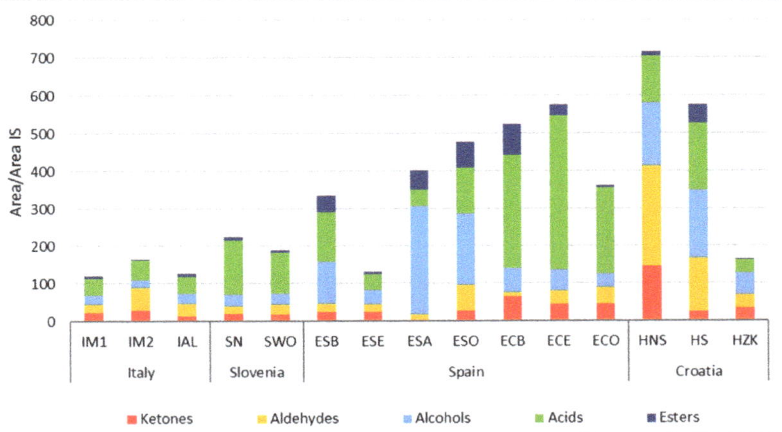

Figure 3. Presence of the different classes of volatile organic compounds (VOCs) in the fifteen fermented sausage samples considered. The values are expressed as the ratio between the peak area of the compound considered and the area of the internal standard.

Aldehydes are particularly relevant in the Italian sausage IM2 and a Spanish salchichón (ESO), but the maximum amounts were found in Croatian HNS and HS. Most of the detected aldehydes can derive from fatty acids oxidation. Within certain limits, these compounds can contribute to the typical sausage aroma profile. Nevertheless, given their strong aroma perception characterized by herbaceous notes, excessive quantities can lead to organoleptic defects (i.e., rancidity) [10]. Hexanal, together with nonanal, is certainly the most characteristic molecule of this VOC group. On the other hand, the methyl-branched aldehydes, such as isovaleric aldehyde (3-methyl-butanal), have often been associated with fermented sausage aroma and can derive from the bacterial metabolism of branched amino acids, in particular leucine [44,77]. Benzaldehyde and benzeneacetaldehyde are present in most of the samples. In particular, HS contained higher amounts of benzeneacetaldehyde, while benzaldehyde is predominant in HNS and ECO. These VOCs are the result of aromatic amino acid (phenylalanine and tyrosine) metabolism and can contribute to the sausage flavor imparting floral and almond notes [78].

Alcohols are present in all the samples, but their amounts were higher in HNS and HS and some Spanish samples, particularly in ESA and ESO samples (Table 6). Among alcohols, ethanol was the most abundant in the samples, with high amounts in some Spanish salchichón (ESB, ESA, ESO) and in Croatian sausages, while a lower presence was highlighted in Italian and Slovenian salamis. The presence of this compound can be influenced by the wine addition in meat batter formulation or it can be the result of several microbial pathways, i.e., pyruvate or amino acid metabolisms [77,79], being therefore strongly influenced by the natural microbiota composition of each product. Phenethyl alcohol, which is the results of benzenacetaldehyde reduction and that can give a rose odor, was present in significant amounts in HS and ESA, samples characterized by high amounts of its precursor.

Acids prevailed in chorizo samples but they were the most represented molecular class in Slovenian sausages (Figure 3). Acetic acid was found in low amounts in Italian salamis and in ESA, ESE and HZK, which showed high pH values. In contrast, significantly higher amounts of acetic acid were detected in the chorizo samples, particularly in ECE. These samples had the lowest pH, ranging from 4.52 to 5.04. Acetic acid can be produced, similarly to ethanol, with many bacterial metabolic pathways and different microbial groups able to be responsible for its accumulation, including LAB, staphylococci and fungi.

It is interesting to underline the presence of isovaleric acid (3-methyl, butanoic acid) (Table 6), whose occurrence in fermented sausages is well-documented and which can exert a very strong organoleptic impact even in low quantities [80,81]. Higher amounts of this VOC have been detected in IM1, while it was not present in very small quantities in HS and HNS.

Esters were present in lower amounts in comparison to other VOC classes and were mainly represented by ethyl acetate, ethyl hexanoate and ethyl octanoate (Table 6). Their levels showed dependence on sausage type. They were more abundant in some Spanish samples and in HS (Figure 3), being related to the presence of their precursors and to the esterase activities typical of the microbial communities forming the microbiome.

4. Conclusions

The analysis of the microbial communities associated with traditional spontaneous fermented meat products highlighted the high variability in the qualitative and quantitative composition of the microbiota involved in these natural fermentations. LAB and CNS were the most representative microorganisms in all the sausages. However, their relative ratio drastically changed. In addition, within each group, the relative presence of species and genus was extremely different. From this point of view, LAB was characterized by high biodiversity, and *Latilactobacillus* was the only genus found in all the products. The great LAB biodiversity can derive from both meat and the production environment, affecting the growth and survival of the different microbial groups.

These differences are reflected in the first instance in sausage safety characteristics, i.e., biogenic amine concentrations. Moreover, also the volatilome, and consequently the peculiar sensory traits of traditional products, is dependent on the complexity of the microbiota. In many sausages considered here, increased microbial biodiversity caused VOCs to be more complex, both qualitatively and quantitatively.

In other words, the biodiversity highlighted the need for these microbes to be exploited to find new strain candidates to be used as autochthonous starter cultures or bioprotective agents in meat products. Indeed, this work provided a basin of indigenous LAB strains, belonging to different species, that will be further studied for their features, considering both safety and technological aspects, in order to select the most promising for food applications. Being highly adapted to specific ecological niches, they can be successfully used in traditional meat products to control undesirable microbiota (also avoiding the accumulation of biogenic amines) and/or to endow the final product with peculiar aromatic characteristics. Moreover, they can become an important part of innovation in small-scale productions, enhancing the sustainability and competitiveness of these small industries.

Finally, a deep knowledge of the peculiar characteristics of traditional fermented sausages can valorize these niche productions, guaranteeing their recognizability.

Supplementary Materials: The following are available online at https://www.mdpi.com/article/10.3390/foods10112691/s1, Table S1: Volatile organic compounds (VOCs) detected by SPME-GC-MS deriving from spices or allium. Results are expressed as a ratio between the peak area of each molecule and the peak area of the internal standard (4-methyl-2-pentanol). The standard deviation was always below 5%. Table S2: Volatile organic compounds (VOCs) detected by SPME-GC-MS deriving from smoking. Results are expressed as a ratio between the peak area of each molecule and the peak area of the internal standard (4-methyl-2-pentanol). The standard deviation was always below 5%.

Author Contributions: Conceptualization, G.T., V.Š., F.Ö. and F.G.; Methodology, S.S.M., A.B.; Formal analyses, F.B., C.M. and N.D.; Investigation, C.M. and F.B.; Data curation: C.F.; Writing—original draft preparation, G.T. and D.B.; Writing—review and editing, C.M. and F.G.; Supervision, F.G. All authors have read and agreed to the published version of the manuscript.

Funding: This research was supported by the PRIMA programme, under BioProMedFood project (Reference Number: 2019-SECTION2-4; CUP: J34I19004820005). The PRIMA programme is supported by the European Union H2020 programme and innovation programme.

Institutional Review Board Statement: Not applicable.

Informed Consent Statement: Not applicable.

Data Availability Statement: Not applicable.

Conflicts of Interest: The authors confirm that they have no conflict of interest with respect to the work described in this manuscript.

References

1. Leroy, F.; Scholliers, P.; Amilien, V. Elements of innovation and tradition in meat fermentation: Conflicts and synergies. *Int. J. Food Sci.* **2015**, *212*, 2–8. [CrossRef] [PubMed]
2. Ojha, K.S.; Kerry, J.P.; Duffy, G.; Beresford, T.; Tiwari, B.K. Technological advances for enhancing quality and safety of fermented meat products. *Trends Food Sci. Technol.* **2015**, *44*, 105–116. [CrossRef]
3. Spitaels, F.; Wieme, A.D.; Janssens, M.; Aerts, M.; Daniel, H.-M.; Van Landschoot, A.; De Vuyst, L.; Vandamme, P. The microbial diversity of traditional spontaneously fermented lambic beer. *PLoS ONE* **2014**, *9*, e95384. [CrossRef]
4. Fenger, M.H.; Aschemann-Witzel, J.; Hansen, F.; Grunert, K.G. Delicious word. Assessing the impact of short storytelling messages on consumer preferences for variations of a new processed meat product. *Food Qual. Prefer.* **2015**, *41*, 237–244. [CrossRef]
5. Palla, M.; Cristani, C.; Giovannetti, M.; Agnolucci, M. Identification and characterization of lactic acid bacteria and yeasts of PDO Tuscan bread sourdough by culture dependent and independent methods. *Int. J. Food Microbiol.* **2017**, *250*, 19–26. [CrossRef] [PubMed]
6. Van Reckem, E.; Geeraerts, W.; Charmpi, C.; Van der Veken, D.; De Vuyst, L.; Leroy, F. Exploring the link between the geographical origin of European fermented foods and the diversity of their bacterial communities: The case of fermented meats. *Front. Microbiol.* **2019**, *10*, 2302. [CrossRef] [PubMed]
7. Vignolo, G.; Fontana, C.; Fadda, S. Semidry and dry fermented sausages. In *Handbook of Meat Processing*; Toldrá, F., Ed.; Blackwell Publishing: Oxford, UK, 2010; Chapter 22.
8. Cocconcelli, P.S.; Fontana, C. Starter cultures for meat fermentation. In *Handbook of Meat Processing*; Toldrá, F., Ed.; Blackwell Publishing: Oxford, UK, 2010; Chapter 10.
9. Franciosa, L.; Alessandria, V.; Dolci, P.; Rantsiou, K.; Cocolin, L. Sausage fermentation and starter cultures in the era of molecular biology methods. *Int. J. Food Microbiol.* **2018**, *279*, 26–32. [CrossRef]
10. Flores, M. Understanding the implications of current health trends on the aroma of wet and dry cured meat products. *Meat Sci.* **2018**, *144*, 53–61. [CrossRef] [PubMed]
11. Palavecino Prpich, N.Z.; Camprubí, G.E.; Cayré, M.E.; Castro, M.P. Indigenous microbiota to leverage traditional dry sausage production. *Int. J. Food Microbiol.* **2021**, *2021*, 15.
12. ISO. ISO 11290-1:2017. *Microbiology of the Food Chain—Horizontal Method for the Detection and Enumeration of Listeria Monocytogenes and of Listeria spp.; Part 1: Detection Method*; International Organization for Standardization: Geneva, Switzerland, 2017.
13. ISO. ISO 6579-1: 2017. *Microbiology of the Food Chain—Horizontal Method for the Detection, Enumeration and Serotyping of Salmonella. Part 1: Detection of Salmonella spp. Method*; International Organization for Standardization: Geneva, Switzerland, 2017.
14. Callahan, B.J.; McMurdie, P.J.; Rosen, M.J.; Han, A.W.; Johnson, A.J.; Holmes, S.P. DADA2: High-resolution sample inference from Illumina amplicon data. *Nat. Methods* **2016**, *13*, 581–583. [CrossRef]
15. Davis, N.M.; Proctor, D.M.; Holmes, S.P.; Relman, D.A.; Callahan, B.J. Simple statistical identification and removal of contaminant sequences in marker-gene and metagenomics data. *Microbiome* **2018**, *6*, 226. [CrossRef] [PubMed]
16. Pasini, F.; Soglia, F.; Petracci, M.; Caboni, M.F.; Marziali, S.; Montanari, C.; Gardini, F.; Grazia, L.; Tabanelli, G. Effect of fermentation with different lactic acid bacteria starter cultures on biogenic amine content and ripening patterns in dry fermented sausages. *Nutrients* **2018**, *10*, 1497. [CrossRef] [PubMed]
17. Martuscelli, M.; Crudele, M.A.; Gardini, F.; Suzzi, G. Biogenic amine formation and oxidation by *Staphylococcus xylosus* strains from artisanal fermented sausages. *Lett. Appl. Microbiol.* **2000**, *31*, 228–232. [CrossRef]
18. Montanari, C.; Bargossi, E.; Gardini, A.; Lanciotti, R.; Magnani, R.; Gardini, F.; Tabanelli, G. Correlation between volatile profiles of Italian fermented sausages and their size and starter culture. *Food Chem.* **2016**, *192*, 736–744. [CrossRef]
19. NIST. *NIST/NIH/EPA Mass Spectral Library, Standard Reference Database 1, NIST 11. Standard Reference Data Program*; National Institute of Standards and Technology: Gaithersburg, MD, USA, 2011.
20. Lešić, T.; Vahčić, N.; Kos, I.; Zadravec, M.; Sinčić Pulić, B.; Bogdanović, T.; Petričević, S.; Listeš, E.; Škrivanko, M.; Pleadin, J. Characterization of traditional Croatian household-produced dry-fermented sausages. *Foods* **2020**, *9*, 990. [CrossRef] [PubMed]
21. Cardinali, F.; Milanović, V.; Osimani, A.; Aquilanti, L.; Taccari, M.; Garofalo, C.; Polverigiani, S.; Clementi, F.; Franciosi, E.; Tuohy, K.; et al. Microbial dynamics of model Fabriano-like fermented sausages as affected by starter cultures, nitrates and nitrites. *Int. J. Food Microbiol.* **2018**, *278*, 61–72. [CrossRef]
22. Holck, A.; Heir, E.; Johannessen, T.C.; Axelsson, L. Northern European products. In *Handbook of Fermented Meat and Poultry*, 2nd ed.; Toldrà, F., Hui, Y.H., Astiasarán, I., Sebranek, J.G., Talon, R., Eds.; John Wiley and Sons: Chichester, UK, 2015; pp. 313–320.
23. Leroy, F.; Geyzen, A.; Janssens, M.; De Vuyst, L.; Scholliers, P. Meat fermentation at the crossroads of innovation and tradition: A historical outlook. *Trends Food Sci. Technol.* **2013**, *31*, 130–137. [CrossRef]

24. Rodríguez-González, M.; Fonseca, S.; Centeno, J.A.; Carballo, J. Biochemical changes during the manufacture of Galician Chorizo sausage as affected by the addition of autochthonous starter cultures. *Foods* **2020**, *9*, 1813. [CrossRef] [PubMed]
25. Prado, N.; Sampayo, M.; González, P.; Lombó, F.; Díaz, J. Physicochemical, sensory and microbiological characterization of Asturian Chorizo, a traditional fermented sausage manufactured in Northern Spain. *Meat Sci.* **2019**, *156*, 118–124. [CrossRef] [PubMed]
26. Kozačinski, L.; Zdolec, N.; Hadžiosmanović, M.; Cvrtila, Ž.; Filipović, I.; Majić, T. Microbial flora of the Croatian traditionally fermented sausage. *Archiv. Für. Lebensm.* **2006**, *5710*, 141–147.
27. Aquilanti, L.; Santarelli, S.; Silvestri, G.; Osimani, A.; Petruzzelli, A.; Clementi, F. The microbial ecology of a typical Italian salami during its natural fermentation. *Int. J. Food Microbiol.* **2007**, *120*, 136–145. [CrossRef]
28. Nikodinoska, I.; Tabanelli, G.; Baffoni, L.; Gardini, F.; Di Gioia, D. Characterization of Lactobacillus Strains Isolated form Italian Artisanal Salami: Safety and Techno-Functional Aspects. In Proceedings of the Food Micro 2018 (FM 2018), Berlin, Germany, 3–6 September 2018.
29. Flores, M.; Corral, S.; Cano-García, L.; Salvador, A.; Belloch, C. Yeast strains as potential aroma enhancers in dry fermented sausages. *Int. J. Food Microbiol.* **2015**, *212*, 16–24. [CrossRef] [PubMed]
30. Andrade, J.M.; Córdoba, J.J.; Casado, E.M.; Córdoba, M.G.; Rodríguez, M. Effect of selected strains of *Debaryomyces hansenii* on the volatile compound production of dry fermented sausage "salchichón". *Meat Sci.* **2010**, *85*, 256–264. [CrossRef]
31. Roccato, A.; Uyttendaele, M.; Barrucci, F.; Cibin, V.; Favretti, M.; Cereser, A.; Cin, M.D.; Pezzuto, A.; Piovesana, A.; Longo, A.; et al. Artisanal Italian salami and soppresse: Identification of control strategies to manage microbiological hazards. *Food Microbiol.* **2017**, *61*, 5–13. [CrossRef] [PubMed]
32. Gounadaki, A.S.; Skandamis, P.N.; Drosinos, E.H.; Nychas, G.-J.E. Microbial ecology of food contact surfaces and products of small-scale facilities producing traditional sausages. *Food Microbiol.* **2008**, *25*, 313–323. [CrossRef] [PubMed]
33. Talon, R.; Lebert, I.; Lebert, A.; Leroy, S.; Garriga, M.; Aymerich, T.; Drosinos, E.H.; Zanardi, E.; Ianieri, A.; Fraqueza, M.J.; et al. Traditional dry fermented sausages produced in small-scale processing units in Mediterranean countries and Slovakia. 1: Microbial ecosystems of processing environments. *Meat Sci.* **2007**, *77*, 570–579. [CrossRef] [PubMed]
34. Cocolin, L.; Dolci, P.; Rantsiou, K. Biodiversity and dynamics of meat fermentations: The contribution of molecular methods for a better comprehension of a complex ecosystem. *Meat Sci.* **2011**, *89*, 296–302. [CrossRef] [PubMed]
35. Talon, R.; Leroy, S. Diversity and safety hazards of bacteria involved in meat fermentations. *Meat Sci.* **2011**, *89*, 303–309. [CrossRef]
36. Chaillou, S.; Champomier-Vergès, M.C.; Cornet, M.; Crutz-Le Coq, A.M.; Dudez, A.M.; Martin, V.; Beaufils, S.; Darbon-Rongère, E.; Bossy, R.; Loux, V.; et al. The complete genome sequence of the meat-borne lactic acid bacterium *Lactobacillus sakei* 23K. *Nat. Biotechnol.* **2005**, *23*, 1527–1533. [CrossRef] [PubMed]
37. Janssens, M.; Myter, N.; De Vuyst, L.; Leroy, F. Community dynamics of coagulase-negative staphylococci during spontaneous artisan-type meat fermentations differ between smoking and moulding treatments. *Int. J. Food Microbiol.* **2013**, *166*, 168–175. [CrossRef] [PubMed]
38. Aquilanti, L.; Garofalo, C.; Osimani, A.; Clementi, F. Ecology of lactic acid bacteria and coagulase negative cocci in fermented dry sausages manufactured in Italy and other Mediterranean countries: An overview. *Int. Food Res. J.* **2016**, *23*, 429–445.
39. Montanari, C.; Gatto, V.; Torriani, S.; Barbieri, F.; Bargossi, E.; Lanciotti, R.; Grazia, L.; Magnani, R.; Tabanellia, G.; Gardini, F. Effects of the diameter on physico-chemical, microbiological and volatile profile in dry fermented sausages produced with two different starter cultures. *Food Biosci.* **2018**, *22*, 9–18. [CrossRef]
40. Stavropoulou, D.A.; Filippou, P.; De Smet, S.; De Vuyst, L.; Leroy, F. Effect of temperature and pH on the community dynamics of coagulase-negative staphylococci during spontaneous meat fermentation in a model system. *Food Microbiol.* **2018**, *76*, 180–188. [CrossRef] [PubMed]
41. Janßen, D.; Eisenbach, L.; Ehrmann, M.A.; Vogel, R.F. Assertiveness of *Lactobacillus sakei* and *Lactobacillus curvatus* in a fermented sausage model. *Int. J. Food Microbiol.* **2018**, *285*, 188–197. [CrossRef]
42. Rimaux, T.; Vrancken, G.; Pothakos, V.; Maes, D.; De Vuyst, L.; Leroy, F. The kinetics of the arginine deiminase pathway in the meat starter culture *Lactobacillus sakei* CTC 494 are pH-dependent. *Food Microbiol.* **2011**, *28*, 597–604. [CrossRef] [PubMed]
43. Rimaux, T.; Vrancken, G.; Vuylsteke, B.; De Vuyst, L.; Leroy, F. The pentose moiety of adenosine and inosine is an important energy source for the fermented-meat starter culture *Lactobacillus sakei* CTC 494. *Appl. Environ. Microbiol.* **2011**, *77*, 6539–6550. [CrossRef] [PubMed]
44. Ravyts, F.; Vuyst, L.D.; Leroy, F. Bacterial diversity and functionalities in food fermentations. *Eng. Life Sci.* **2012**, *12*, 356–367. [CrossRef]
45. Montanari, C.; Barbieri, F.; Magnani, M.; Grazia, L.; Gardini, F.; Tabanelli, G. Phenotypic diversity of *Lactobacillus sakei* strains. *Front. Microbiol.* **2018**, *9*, 2003. [CrossRef] [PubMed]
46. Fontana, C.; Bassi, D.; López, C.; Pisacane, V.; Otero, M.C.; Puglisi, E.; Rebecchi, A.; Cocconcelli, P.S.; Vignolo, G. Microbial ecology involved in the ripening of naturally fermented llama meat sausages. A focus on lactobacilli diversity. *Int. J. Food Microbiol.* **2016**, *236*, 17–25. [CrossRef]
47. Fontán, M.C.G.; Lorenzo, J.M.; Parada, A.; Franco, I.; Carballo, J. Microbiological characteristics of "androlla", a Spanish traditional pork sausage. *Food Microbiol.* **2007**, *24*, 52–58.
48. Fontán, M.C.G.; Lorenzo, J.M.; Martínez, S.; Franco, I.; Carballo, J. Microbiological characteristics of Botillo, a Spanish traditional pork sausage. *LWT* **2007**, *40*, 1610–1622. [CrossRef]

49. Zheng, J.; Wittouck, S.; Salvetti, E.; Franz, C.M.A.P.; Harris, H.M.B.; Mattarelli, P.; O'Toole, P.W.; Pot, B.; Vandamme, P.; Walter, J.; et al. A taxonomic note on the genus *Lactobacillus*: Description of 23 novel genera, emended description of the genus *Lactobacillus Beijerinck* 1901, and union of *Lactobacillaceae* and *Leuconostocaceae*. *Int. J. Syst. Evol.* **2020**, *70*, 2782–2858. [CrossRef] [PubMed]
50. Coppola, S.; Mauriello, G.; Aponte, M.; Moschetti, G.; Villani, F. Microbial succession during ripening of Naples-type salami, a southern Italian fermented sausage. *Meat Sci.* **2000**, *56*, 321–329. [CrossRef]
51. Munekata, P.E.S.; Pateiro, M.; Zhang, W.; Domínguez, R.; Xing, L.; Fierro, E.M.; Lorenzo, J.M. Autochthonous probiotics in meat products: Selection, identification, and their use as starter culture. *Microorganisms* **2020**, *8*, 1833. [CrossRef] [PubMed]
52. Bermúdez, R.; Lorenzo, J.M.; Fonseca, S.; Franco, I.; Carballo, J. Strains of *Staphylococcus* and *Bacillus* isolated from traditional sausages as producers of biogenic amines. *Front. Microbiol.* **2012**, *3*, 151. [CrossRef] [PubMed]
53. Greppi, A.; Ferrocino, I.; La Storia, A.; Rantsiou, K.; Ercolini, D.; Cocolin, L. Monitoring of the microbiota of fermented sausages by culture independent rRNA based approaches. *Int. J. Food Microbiol.* **2015**, *212*, 67–75. [CrossRef] [PubMed]
54. Sánchez Mainar, M.; Stavropoulou, D.; Leroy, F. Exploring the metabolic heterogeneity of coagulase-negative staphylococci to improve the quality and safety of fermented meats: A review. *Int. J. Food Microbiol.* **2017**, *247*, 24–37. [CrossRef]
55. Leroy, F.; Verluyten, J.; De Vuyst, L. Functional meat starter cultures for improved sausage fermentation. *Int. J. Food Microbiol.* **2006**, *106*, 270–285. [CrossRef]
56. Drosinos, E.H.; Mataragas, M.; Xiraphi, N.; Moschonas, G.; Gaitis, F.; Metaxopoulos, J. Characterization of the microbial flora from a traditional Greek fermented sausage. *Meat Sci.* **2005**, *69*, 307–317. [CrossRef] [PubMed]
57. Ravyts, F.; Steen, L.; Goemaere, O.; Paelinck, H.; De Vuyst, L.; Leroy, F. The application of staphylococci with flavour-generating potential is affected by acidification in fermented dry sausages. *Food Microbiol.* **2010**, *27*, 945–954. [CrossRef]
58. Casaburi, A.; Nasi, A.; Ferrocino, I.; Di Monaco, R.; Mauriello, G.; Villani, F.; Ercolini, D. Spoilage-related activity of *Carnobacterium maltaromaticum* strains in air-stored and vacuum-packed meat. *Appl. Environ. Microbiol.* **2011**, *77*, 7382–7393. [CrossRef] [PubMed]
59. Ercolini, D.; Federica, R.; Torrieri, E.; Masi, P.; Villani, F. Changes in the spoilagerelated microbiota of beef during refrigerated storage under different packaging conditions. *Appl. Environ. Microbiol.* **2006**, *72*, 4663–4671. [CrossRef] [PubMed]
60. Raimondi, S.; Nappi, M.R.; Sirangelo, T.M.; Leonardi, A.; Amaretti, A.; Ulrici, A.; Magnani, R.; Montanari, C.; Tabanelli, G.; Gardini, F. Bacterial community of industrial raw sausage packaged in modified atmosphere throughout the shelf life. *Int. J. Food Microbiol.* **2018**, *280*, 78–86. [CrossRef] [PubMed]
61. EFSA. Scientific opinion on risk based control of biogenic amine formation in fermented foods. *EFSA J.* **2011**, *9*, 2393–2486. [CrossRef]
62. Landete, J.M.; Pardo, I.; Ferrer, S. Regulation of hdc expression and HDC activity by enological factors in lactic acid bacteria. *J. Appl. Microbiol.* **2008**, *105*, 1544–1551. [CrossRef]
63. European Commission. Commission Regulation (EC) No. 2073/2005 of 15 November 2005 on microbiological criteria for foodstuffs. *Off. J. Eur. Union* **2005**, *L338*, 1–26.
64. Ruiz-Capillas, C.; Jiménez-Colmenero, F. Biogenic amines in meat and meat products. *Crit. Rev. Food Sci. Nutr.* **2004**, *44*, 489–499. [CrossRef] [PubMed]
65. Gardini, F.; Özogul, Y.; Suzzi, G.; Tabanelli, G.; Özogul, F. Technological factors affecting biogenic amine content in foods: A review. *Front. Microbiol.* **2016**, *7*, 1218. [CrossRef] [PubMed]
66. Yalçınkaya, S.; Kılıç, G.B. Isolation, identification and determination of technological properties of the halophilic lactic acid bacteria isolated from table olives. *J. Food Sci. Technol.* **2019**, *56*, 2027–2037. [CrossRef]
67. Straub, B.W.; Kicherer, M.; Schilcher, S.M.; Hammes, W.P. The formation of biogenic amines by fermentation organisms. *Z. Lebensm. Unters. Forch.* **1995**, *201*, 79–82. [CrossRef] [PubMed]
68. Barbieri, F.; Montanari, C.; Gardini, F.; Tabanelli, G. Biogenic amine production by lactic acid bacteria: A review. *Foods* **2019**, *8*, 17. [CrossRef] [PubMed]
69. Schmidt, S.; Berger, R.G. Aroma compounds in fermented sausages of different origins. *Lebensm. Wiss. Technol.* **1998**, *31*, 559–567. [CrossRef]
70. Menon, A.N.; Padmakumari, K.P. Studies on essential oil composition of cultivars of black pepper (*Piper nigrum L.*)-V. *J. Essent. Oil Res.* **2005**, *17*, 153–155. [CrossRef]
71. Piras, A.; Rosa, A.; Marongiu, B.; Atzeri, A.; Dessì, M.A.; Falconieri, D.; Porcedda, S. Extraction and separation of volatile and fixed oils from seeds of Myristica fragrans by supercritical CO_2: Chemical composition and cytotoxic activity on Caco-2 cancer cells. *J. Food Sci.* **2012**, *77*, C448–C453. [CrossRef] [PubMed]
72. Batiha, G.E.S.; Alkazmi, L.M.; Wasef, L.G.; Beshbishy, A.M.; Nadwa, E.H.; Rashwan, E.K. *Syzygium aromaticum* L. (Myrtaceae): Traditional uses, bioactive chemical constituents, pharmacological and toxicological activities. *Biomolecules* **2020**, *10*, 202. [CrossRef]
73. Sikorski, Z.E.; Sinkiewicz, I. Principles of smoking. In *Handbook of Fermented Meat and Poultry*, 2nd ed.; Toldrà, F., Ed.; John Wiley and Sons: Chichester, UK, 2015; Chapter 6.
74. Pereira, C.; de Guía Córdoba, M.; Aranda, E.; Hernández, A.; Velázquez, R.; Bartolomé, T.; Martín, A. Type of paprika as a critical quality factor in Iberian chorizo sausage manufacture. *CyTA-J. Food* **2019**, *17*, 907–916. [CrossRef]

75. Tabanelli, G.; Montanari, C.; Grazia, L.; Lanciotti, R.; Gardini, F. Effects of aw at packaging time and atmosphere composition on aroma profile, biogenic amine content and microbiological features of dry fermented sausages. *Meat Sci.* **2013**, *94*, 177–186. [CrossRef]
76. Liu, S.Q. Practical implications of lactate and pyruvate metabolism by lactic acid bacteria in food and beverage fermentations. *Int. J. Food Microbiol.* **2003**, *83*, 115–131. [CrossRef]
77. Carballo, J. The role of fermentation reactions in the generation of flavor and aroma of foods. In *Fermentation, Effects on Food Properties*; Mehta, B.M., Kamal-Eldin, A., Iwanski, R.Z., Eds.; CRC Press: Boca Raton, FL, USA, 2012; pp. 51–83.
78. Smid, E.J.; Kleerebezem, M. Production of aroma compounds in lactic fermentations. *Annu. Rev. Food Sci. Technol.* **2014**, *5*, 313–326. [CrossRef]
79. von Wright, A.; Axelsson, L. Lactic acid bacteria: An introduction. In *Lactic Acid Bacteria, Microbial and Functional Aspects*, 4th ed.; Lahtinen, S., Ouwehand, A.C., Salminen, S., von Wright, A., Eds.; CRC Press: Boca Raton, FL, USA, 2011; pp. 1–16.
80. Olivares, A.; Navarro, J.L.; Flores, M. Establishment of the contribution of volatile compounds to the aroma of fermented sausages at different stages of processing and storage. *Food Chem.* **2009**, *115*, 1464–1472. [CrossRef]
81. Gianelli, M.P.; Olivares, A.; Flores, M. Key aroma components of a dry-cured sausage with high fat content (Sobrassada). *Food Sci. Technol. Int.* **2011**, *17*, 63–71. [CrossRef] [PubMed]

Article

Contribution of Microorganisms to Biogenic Amine Accumulation during Fish Sauce Fermentation and Screening of Novel Starters

Xinxiu Ma [1,2], Jingran Bi [1,2], Xinyu Li [1,2], Gongliang Zhang [1,2], Hongshun Hao [1,2] and Hongman Hou [1,2,*]

1 School of Food Science and Technology, Dalian Polytechnic University, No. 1, Qinggongyuan, Ganjingzi District, Dalian 116034, China; maxinxiu@hotmail.com (X.M.); bijr@dlpu.edu.cn (J.B.); lixinyu990519@163.com (X.L.); zgl_mp@163.com (G.Z.); beike1952@163.com (H.H.)
2 Liaoning Key Lab for Aquatic Processing Quality and Safety, No. 1, Qinggongyuan, Ganjingzi District, Dalian 116034, China
* Correspondence: houhongman@dlpu.edu.cn; Tel.: +86-411-8632-2020

Abstract: In this study, high-throughput sequencing and culture-dependent and HPLC methods were used to investigate the contribution and regulation of biogenic amines (BAs) by dominant microorganisms during fish sauce fermentation. The results showed that the microbial composition constantly changed with the fermentation of fish sauce. *Tetragenococcus* (40.65%), *Lentibacillus* (9.23%), *Vagococcus* (2.20%), *Psychrobacter* (1.80%), *Pseudomonas* (0.98%), *Halomonas* (0.94%) and *Staphylococcus* (0.16%) were the dominant microflora in fish sauce. The content of BAs gradually increased as the fermentation progressed. After 12 months of fermentation, the histamine content (55.59 mg/kg) exceeded the toxic dose recommended by the Food and Drug Administration (FDA). Correlation analysis showed that dominant microorganisms have a great contribution to the accumulation of BAs. By analyzing the BA production capacity of dominant isolates, the accumulation of BAs in fish sauce might be promoted by *Tetragenococcus* and *Halomonas*. Moreover, four strains with high BA reduction ability were screened out of 44 low BA-producing dominant strains, and their influence on BA accumulation in fermented foods was determined. Results demonstrated that *Staphylococcus nepalensis* 5-5 and *Staphylococcus xylosus* JCM 2418 might be the potential starters for BA control. The present study provided a new idea for the control of BAs in fermented foods.

Keywords: fish sauce; biogenic amines; microbial community dynamics; starter; correlation analysis

Citation: Ma, X.; Bi, J.; Li, X.; Zhang, G.; Hao, H.; Hou, H. Contribution of Microorganisms to Biogenic Amine Accumulation during Fish Sauce Fermentation and Screening of Novel Starters. *Foods* **2021**, *10*, 2572. https://doi.org/10.3390/foods10112572

Academic Editors: Valentina Bernini and Juliano De Dea Lindner

Received: 16 September 2021
Accepted: 19 October 2021
Published: 25 October 2021

1. Introduction

Fish sauce, a seasoning with a unique flavor, is spontaneously fermented from anchovies and salt in a certain proportion, and the fermentation time was more than 6 months [1,2]. Fish sauce, with a particular aroma and rich in amino acids, is often used as a seasoning and nutritional supplement in Southeast Asian countries. In China, fish sauce is mainly popular in eastern coastal areas such as Shandong, Guangdong and Fujian, with an annual output of over 100,000 tons [3,4]. In the process of fish sauce fermentation, a large number of microorganisms are introduced, which may produce biogenic amines from amino acids. Due to increasing occurrences of food-borne disease outbreaks caused by biogenic amines, food safety has received more concern in the production of fish sauce [5].

BAs are small molecule nitrogen-containing compounds which are formed by the decarboxylation or transamination of amino acids by microorganisms [6]. Although BAs have certain physiological functions in the human body, excessive intake may bring safety risks [7]. Due to the hazardous effects of BAs on humans, such as rash, migraine, hypertension and hypotension, the determination of BAs in food products has attracted worldwide attention [5]. Fish products are typical foods containing high contents of BAs [8]. The common BAs in fish sauce are Tryptamine, β-phenylethylamine, putrescine, cadaverine, histamine, tyramine, spermine and spermidine, among which putrescine, cadaverine, histamine and tyramine are the most common BAs [6,9,10]. In particular, some fish sauce products have been reported to contain histamine of 349 to 5487 mg/kg, which exceeded the FDA limit (50 mg/kg) for histamine [11–14]. Therefore, it is very important to demonstrate the formation mechanism and control of BAs in the fermentation of fish sauce.

The microbial community in naturally fermented fish sauce is complex, but it is a significant factor in determining product quality and safety [15]. Therefore, by defining the succession of bacterial communities and their correlation with BAs, the contribution of microorganisms on the accumulation of BAs in fish sauce fermentation could be explained. Previous studies have shown that only a small number of bacteria may contribute to the formation of biogenic amines [16]. For example, Sang et al. [17] showed that the 30 high BA-producing strains detected in the shrimp paste might be the key factor causing the high BA content in shrimp paste. Microorganisms not only have the ability to produce BAs, but can also reduce BAs by producing amine oxidase [18,19]. Mah et al. [20] found that the content of BAs in Myeolchi-jeot inoculated with *S. xylosus* NO.0538 was 16.0% lower than that in the control group. From this perspective, evaluating bacterial BA production and reduction ability is critical in understanding the BAs accumulation mechanism. Therefore, compared with previous studies, our study not only analyzed the role of microorganisms in BA accumulation, but also identified the strains that can reduce BAs on this basis, which were successfully applied to fish sauce and other fermented foods.

The fish sauce factories in northern China are concentrated in Shandong, and the fish sauce factory that provided samples for us is the largest factory in Shandong, with a complete and representative fermentation process. In this study, the fish sauce samples at different fermentation stages from Rongcheng (Shandong province, PR China) were characterized by BAs assays, high-throughput sequencing analysis of bacterial diversity and cultural isolation of dominant bacteria. The microbial community information was used to clarify the contribution of microorganisms to the accumulation of BAs. Candidate starters isolated from fish sauce to control BA accumulation were screened. This research provides new ideas for improving the quality of fish sauce and reducing the risk of eating fish sauce.

2. Materials and Methods

2.1. Sample Collection

2.1.1. Fish Sauce Samples Collection

The fish sauce samples were collected from the Rongcheng Baozhiyuan fish sauce factory (Shandong province, PR China). Fish sauce is made by the natural fermentation of anchovies and salt at a ratio of 7:3 for more than 12 months (Figure 1). The annual factory output of the fish sauce is 2000–2500 tons, and it is a large-scale factory in northern fish sauce factories. Samples were collected at 3, 6, 9, 12 and 18 months of fermentation. Three groups of samples were taken in parallel from three fermentation tanks. In order to ensure the uniformity and representativeness of samples, equal amounts (0.25 kg) of samples were taken from the upper, middle and lower parts of each fermentation tank, and the final samples were mixed from the three parts. We placed the sample in the sterile sampling bags and quickly transported them back to the laboratory. Parts of the samples were immediately used for microbial analysis, and the remaining samples were stored at -80 °C.

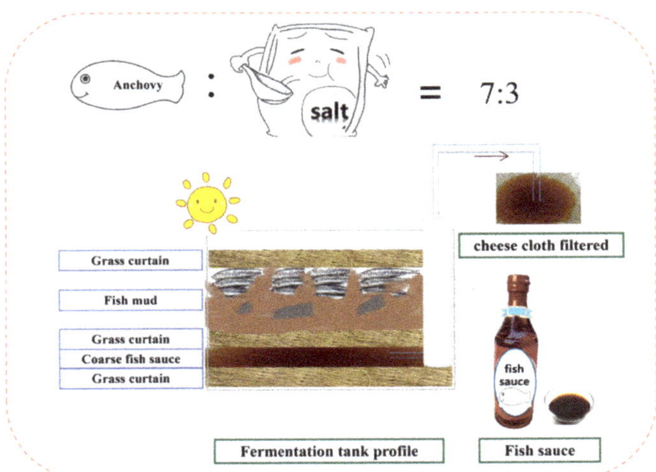

Figure 1. Schematic diagram of fish sauce natural fermentation.

2.1.2. Fermentation Experiments

Fish sauce samples fermented for 9 months and shrimp paste, Doujiang and Sufu, purchased at the market, were selected as the raw materials to determine the effect of strains on the accumulation of BAs in different fermented foods. The isolated strain (OD_{600} = 0.4) was inoculated into the above-mentioned fermented foods (50 g) with an inoculum of 5% (*v*/*v*), and the fermentation was continued for 7 days at 37 °C. The control group was inoculated with 5% (*v*/*v*) sterilized water in different fermented foods. All experiments were performed in triplicate.

2.2. Microbiological Analyses

2.2.1. Microbial Community Analysis

To further analyze the changes in microbial composition in the fish sauce during fermentation, the fish sauce samples were subjected to high-throughput sequencing analysis. Primer pairs 338F and 806R were used to amplify the hypervariable V3–V4 region of bacterial *16S rRNA* gene. The amplified product was purified using the AxyPrep DNA gel extraction kit (Axygen Biosciences, Union City, CA, USA) according to the manufacturer's instructions, and was run on the Miseq Illumina platform at Majorbio Bio-Pharm Technology Co., Ltd. (Shanghai, China) [21].

2.2.2. Strains Identification

Luria-Bertani (LB) medium with 15% NaCl, de Man Rogosa and Sharp (MRS) medium with 15% NaCl and Plate Count Agar are the main media used in this study. A total of 90 mL of sterile normal saline (0.85%, *v*/*v*) was added to 10 g fish sauce sample, and the mixture was shaken at 37 °C at 150 rpm for 30 min. The mixtures were serially diluted (10^{-1} to 10^{-5}) with the sterile normal saline, and 100 µL of each dilution was plated onto the above medium. The single colony was isolated and purified on the corresponding medium at least 4 times to obtain the pure isolates. For the identification of bacteria, the Gen-EluteTM Kit (Tiangen Biotechnology Co., Ltd., Beijing, China) was used to extract the genomic DNA of the bacteria. The bacterial *16S rDNA* gene was amplified with the universal primer pairs 27F and 1492R. The PCR programswere carried out as follows: pre-denaturation at 94 °C for 5 min, followed by 30 cycles at 94 °C for 30 s, 56 °C for 30 s, 72 °C for 1 min and elongation at 72 °C for 10 min. The amplified fragments were then sent to sequencing (Beijing BGI Huada Biotech Co., Ltd., Beijing, China). The pure isolates were stored at −80 °C in the corresponding liquid culture medium containing 32% (*v*/*v*) glycerin [17].

2.3. Biogenic Amines Determination

2.3.1. Determination of BAs

Eight BAs or amine hydrochlorides, namely Tryptamine, β-phenylethylamine, putrescine, cadaverine, histamine, tyramine, spermine and spermidine (Aladdin, Shanghai, China), and derivating agent (dansyl chloride (Meilunbio, Dalian, China)) were purchased from Sangon Biotech (Shanghai, China) and Meilun biotech (Dalian, China).

The BA content in each sample was assessed according to the method developed by Sang et al. [21]. In detail, add 80 μL 2 M NaOH, 120 μL saturated $NaHCO_3$, and 800 μL dansyl chloride solution (10 mg/mL in acetone) to 400 μL pretreated sample. The mixtures were incubated in water at 45 °C for 40 min. A total of 550 uL of acetonitrile was added to the mixture to dissolve the residue and then the mixture was centrifuged at 3000× *g* for 5 min at 4 °C. The supernatant was filtered through the 0.22 filter for three times. The quantification of BA was carried out using a Huapu S6000 HPLC unit (ACCHROM, Beijing, China) which consisted of an ACCHROM Tnature C18 (4.6 × 250 mm) column coupled to a quaternary pump and a diode array detector. Two elution solutions are (A) ultrapure water and (B) acetonitrile. A 10 μL sample was injected into the column at a flow rate of 1.0 mL/min. The gradient elution program was carried out as follows: 0–10 min, 45% A; 10–15 min, 35–45% A; 15–20 min, 20–35% A; 20–25 min, 20% A; 25–30 min, 10–20% A; 30–33 min, 10% A; 33–35 min, 10–45% A; 35–40 min, 45% A. During analysis, the column temperature was lower than 35 °C, and the absorbance was monitored with a UV detector at 254 nm. Each BA was quantified with a calibration curve generated by analyzing standard BA mixed solution. The experiment was carried out in triplicate.

2.3.2. Pretreatment of Samples of Fish Sauce and Other Fermented Foods

After adding 20 mL of 10% trichloroacetic acid (TCA) (SCR, Shanghai, China) to 5 g of sample to extract biogenic amines, the mixture was left at 4 °C for 2 h. The mixture was centrifuged at 8000× *g* for 10 min at 4 °C. An equal volume of 10% TCA was used to extract the remaining BA in the supernatant again. The obtained mixture was centrifuged at 3000× *g* for 10 min at 4 °C. The BA in the samples was determined according to the method described in Section 2.3.1.

2.3.3. Determination of BA Production and Reduction Properties of Strains

The strains were cultured in a 5 mL medium containing 0.5% (*w*/*v*) histidine, tyrosine, ornithine hydrochloride and lysine (BBI, Shanghai, China). The BA concentrations after 48 h of incubation were determined to evaluate the BA production ability of the strains. The strain was incubated at 37 °C for 12–24 h and then centrifuged at 6000× *g* for 5 min to collect bacterial pellets. The bacterial pellets were washed twice with 0.05 mol/L phosphate buffer (PBS, pH = 7), and then centrifuged at 6000× *g* for 5 min. We added 0.05 mol/L PBS containing putrescine, cadaverine and histamine (100 mg/L) to the bacterial pellet until the OD_{600} of the mixture reached 0.4. The mixture was then cultured at 37 ○C for 48 h and the residual BAs in the suspension were determined. Sterile PBS (including 100 mg/L putrescine, cadaverine and histamine) was used as a control. The BA reduction efficiency was calculated as:

$$A = [(C - S) \div C] \times 100\% \quad (1)$$

A is the BA reduction percentage and *C* and *S* are the BA concentrations of the control and strain specimens, respectively [22]. To extract the biogenic amine in the bacterial solution, 9 mL TCA was added to 1 mL bacterial solution, and the mixture was left at 4 °C for 2 h. BAs were determined by the method described in Section 2.3.1.

2.4. Statistical Analysis

The statistical software SPSS 26 was used to analyze the significance of the difference by one-way ANOVA method. The high-throughput sequencing analysis was performed using the free online platform of the Majorbio I-Sanger Cloud Platform (www.i-sanger.com, accessed on 2 February 2021). For the data of high-throughput sequencing, raw data were quality-filtered by Fastp (version 0.20.0). Operational taxonomic units (OTUs) were clustered with 97% similarity cutoff using Uparse (version 7.1) (http://www.drive5.com/uparse/, accessed on 2 February 2021). The taxonomy of each 16S rRNA gene sequence was analyzed by RDP Classifier (version 2.2) algorithm (https://sourceforge.net/projects/rdp-classifier/, accessed on 16 July 2021) against the Silva 128/16s_bacteria, using a confidence threshold of 0.7.

2.5. Nucleotide Sequence Accession Numbers

The raw reads of bacterial 16S rRNA gene sequencing were deposited into the NCBI Sequence Read Archive (SRA) database (Accession Number: SRR15851749).

3. Results

3.1. BA Contents in Fish Sauce Samples at Different Fermentation Stages

The content of BA in fish sauce samples at different fermentation stages was shown in Table 1. In the fish sauce samples of five fermentation stages, six main BAs were detected, including histamine (18.39–74.50 mg/kg), tyramine (13.23–20.88 mg/kg), tryptamine (9.59–20.41 mg/kg), β-phenylethylamine (7.63–27.26 mg/kg), putrescine (14.71–60.41 mg/kg), and cadaverine (23.44–90.61 mg/kg), and the content of total BAs ranged from 89.38 to 296.82 mg/kg. Spermidine was not detected in the early period (3–6 months) and was formed during the medium (6–12 months) and late period (12–18 months), while spermine was only detected in the early period of fermentation. During the fermentation process, the concentration of BAs gradually increased. The content of histamine and cadaverine increased to the maximum at the 18th month of fermentation, which was 74.50 mg/kg and 90.61 mg/kg, respectively.

Table 1. Content of BAs in fish sauce samples at different fermentation stages.

BAs (mg/kg)	R3M	R6M	R9M	R12M	R18M
TRY	9.59 ± 0.40 ^{e}	10.78 ± 0.24 ^{d}	12.77 ± 0.34 ^{c}	17.76 ± 0.26 ^{b}	20.41 ± 0.42 ^{a}
PHE	7.63 ± 0.25 ^{e}	9.14 ± 0.12 ^{d}	13.56 ± 0.62 ^{c}	21.24 ± 0.25 ^{b}	27.26 ± 0.06 ^{a}
PUT	14.71 ± 0.92 ^{e}	19.77 ± 0.10 ^{d}	28.15 ± 1.25 ^{c}	47.67 ± 0.42 ^{b}	60.41 ± 0.27 ^{a}
CAD	23.44 ± 1.21 ^{e}	29.75 ± 0.20 ^{d}	42.08 ± 1.90 ^{c}	71.03 ± 0.55 ^{b}	90.61 ± 0.28 ^{a}
HIS	18.39 ± 0.91 ^{e}	22.99 ± 0.17 ^{d}	32.75 ± 1.48 ^{c}	55.59 ± 0.45 ^{b}	74.5 ± 0.17 ^{a}
TYR	13.23 ± 0.50 ^{e}	15.18 ± 0.28 ^{d}	17.86 ± 0.59 ^{c}	19.07 ± 0.16 ^{b}	20.88 ± 0.06 ^{a}
SPD	ND	ND	2.41 ± 0.01 ^{c}	2.7 ± 0.01 ^{b}	2.75 ± 0.002 ^{a}
SPM	2.39 ± 0.04 ^{a}	2.41 ± 0.01 ^{a}	1.98 ± 0.04 ^{b}	ND	ND
Total	89.38 ± 3.97 ^{e}	110.01 ± 1.01 ^{d}	151.55 ± 7.24 ^{c}	235.05 ± 1.79 ^{b}	296.82 ± 0.71 ^{a}

Data are expressed as mean ± SDs (n = 3).; ND: not detected. Different letters (a, b, c, d, e) indicate the mean value of significant difference at $p < 0.05$. (R3M, R6M, R9M, R12M, R18M indicated for 3, 6, 9, 12, 18 months, respectively. TRY: tryptamine, TYR: tyramine, PHE: β-phenylethylamine, HIS: histamine, PUT: putrescine, CAD: cadaverine, SPM: spermine, SPD: spermidine).

3.2. Dynamic Changes of Microorganisms in Fish Sauce Samples during Fermentation

We analyzed the change in microbial diversity during fish sauce fermentation by high throughput sequencing. At the phylum level, *Firmicutes* predominated during the entire fermentation, particularly in 18-month samples with up to 94.29% of the total sequences. The species abundance of *Cyanobacteria* was high during the process from 3 months to 12 months of fermentation, especially in 12-month samples (21.10%), while after 18 months of fermentation, the abundance of *Cyanobacteria* dropped sharply to 0.01%. *Proteobacteria* and *Actinobacter* rose rapidly between 3 and 12 months of fermentation, and the abundance reached 19.14% and 15.13%, respectively. The abundance of these two phyla decreased significantly after fermentation to 18 months. (Figure 2A). At the genus level, *Lentibacillus* (approximately 45.90%) was the most dominant bacteria in the early phase of fish sauce fermentation. Then, *Tetragenococcus* (9.15–45.78%) rapidly replaced *Lentibacillus* as the dominant genus at the 6th month of fermentation. In the 18th month of fermentation, *Tetragenococcus* became the most dominant bacteria (approximately 91.99%), and *Halomonas* became the second most dominant genus with an abundance of 2.74%. The genus *Pseudomonas* appeared as the dominant genus during the late fermentation period (9–18 month), especially in the 12-month samples, and the abundance reached 3.69% (Figure 2B).

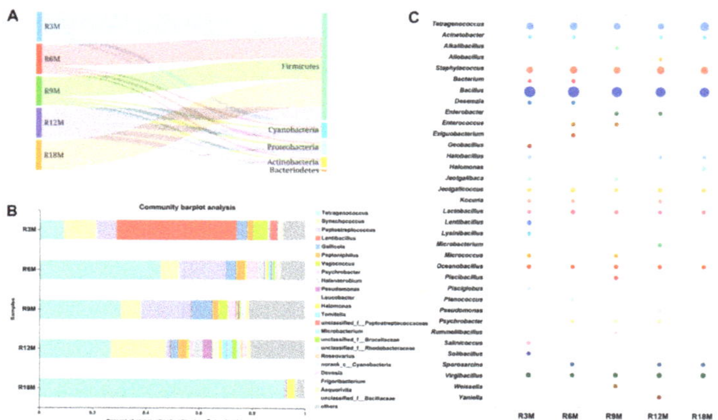

Figure 2. Analysis of bacterial community dynamics in the fish sauce fermentation process. (**A**) Bacterial community dynamics at the phylum level. (**B**) Bacterial community dynamics at the genus level. (**C**) Bacterial community dynamics analysis by traditional separation and screening methods. (R3M, R6M, R9M, R12M, R18M indicated for 3, 6, 9, 12, 18 months, respectively).

To further determine the microbial diversity of fish sauce, we used traditional separation and screening methods to isolate and identify the microorganisms in fish sauce. According to *16S rDNA* gene sequence, a total of 284 strains were classified into 35 genera and 77 species (Figure 2C). Some microorganisms contributing to fermentation were isolated, such as *Tetragenococcus*, *Bacillus*, *Staphylococcus*, *Lentibacillus*, *Psychrobacter*, *Halomonas* and *Pseudomonas*. Species level analysis showed that two species of *Tetragenococcus* were found. Among them, the isolation frequency of *Tetragenococcus halophilus* strains was high during the fermentation process. Seven species of *staphylococcus* were also found. *Staphylococcus nepalensis* and *Staphylococcus saprophyticus* were more frequently found in the early period of fermentation, while *Staphylococcus epidermidis*, *Staphylococcus captis* and *Staphylococcus lentus* were in the later fermentation period. *Lentibacillus* and *Pseudomonas* strains were only isolated in the 3- and 12-month samples, respectively. These results also verified the results of high-throughput sequencing.

The effect of fermentation time on the microbial composition of fish sauce was analyzed. Bray–Curtis principal coordinate analysis (PCoA; beta-diversity) was used to qualitatively examine the differences in bacterial composition, and the results showed the clear cluster pattern of the five samples, and the microbiological composition of samples fermented for 6 months was more similar to that at 9 months (Figure 3A). ANOSIM (analysis of similarities) analyzed the significant differences between fish sauce samples in five fermentation stages, indicating that the bacterial composition varied greatly in time distribution (Figure 3B). The communities which had significant differences in the grouping of samples were found by LEfSe analysis. *Lentibacillus*, *Psychrobacter*, *Gallicola*, *Pseudomonas* and *Tetragenococcus* were the representative bacteria in the fish sauce samples fermented for 3, 6, 9, 12 and 18 months, respectively (Figure 3C).

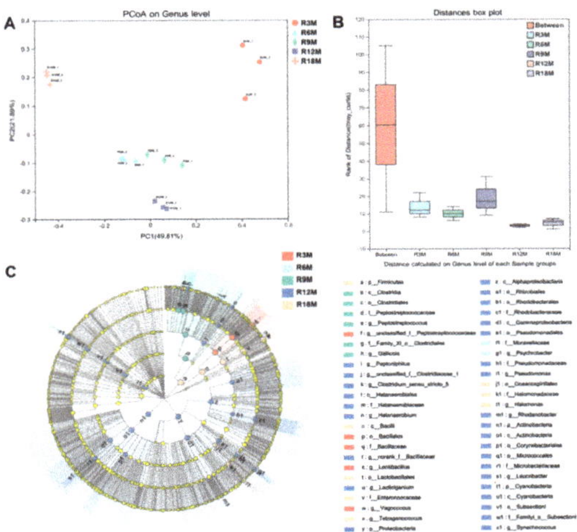

Figure 3. The influence of fermentation time on the microbial composition in fish sauce. (**A**) PCoA (Bray–Curtis). Different samples are represented by points of different colors or shapes. The closer the points are, the more similar the species composition of the samples. (**B**) ANOSIM. The "Between" boxes refer to differences between groups, the others represent differences within each group. (**C**) LEfSe. Microorganisms that are significantly enriched in the corresponding group are represented by nodes with different colors and have a significant impact on the differences between groups. (R3M, R6M, R9M, R12M, R18M indicated for 3, 6, 9, 12, 18 months, respectively).

3.3. Microbial Contribution to BA Contents in Fish Sauce

The RDA (redundancy analysis) analysis among BAs, fermentation time and dominant microorganisms are shown in Figure 4A. This result showed that all BAs are positively correlated with fermentation time, especially β-phenylethylamine. Among the dominant microorganisms, *Tetragenococcus* and *Lentibacillus* had the strongest correlation with BAs. *Tetragenococcus* is positively correlated with BAs, while *Lentibacillus* is negatively correlated with BAs. The correlation network analysis between species and BAs shows that most genera were negatively related to BAs. However, some bacterial genera still exhibited as being positively related to BAs (Figure 4B). For example, *Tetragenococcus* is positively correlated with all BAs, *Halomonas* is positively correlated with histamine, β-phenylethylamine, putrescine and tryptamine, and *unidentified genus from family Clostridiaceae* is positively correlated with tyramine, cadaverine, tryptamine and histamine. Pearson correlation analysis showed the correlation between BAs and the top 50 genera. Except for *Tetragenococcus* and *Lentibacillus,* which are strongly correlated with BAs, we found that the dominant genus *Pseudomonas* was weakly positively correlated with BAs, while *Staphylococcus* and *Psychrobacter* were weakly negatively correlated with BAs (Figure 4C).

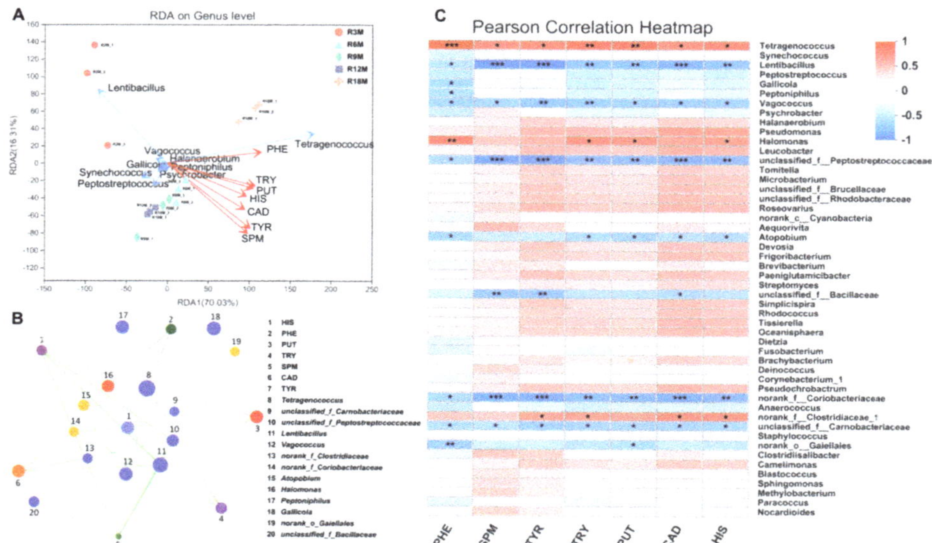

Figure 4. Correlation analysis between microorganisms and BAs during fish sauce fermentation. (**A**) RDA. Different samples are represented by points of different colors or shapes. The length of the arrows represents the degree of impact of environmental factors on the species data. The correlations are represented by the angle between the arrows (acute angle: positive correlation; obtuse angle: negative correlation; right angle: no correlation). (**B**) Correlation network analysis. The size of the nodes indicates the abundance of species, and different colors indicate different species. Red line: positive correlation, green line: negative correlation. (**C**) Pearson correlation heatmap diagram. * $0.01 < p \leq 0.05$, ** $0.001 < p \leq 0.01$, *** $p \leq 0.001$. (R3M, R6M, R9M, R12M, R18M indicated for 3, 6, 9, 12, 18 months, respectively. TRY: tryptamine, TYR: tyramine, PHE: β−phenylethylamine, HIS: histamine, PUT: putrescine, CAD: cadaverine, SPM: spermine).

In order to better illustrate the role of dominant bacteria on the accumulation of BAs, we evaluated the ability of producing the BAs of 100 dominant strains isolated from fish sauce (Table 2). A total of 56.00% of the dominant strains can produce a large amount of BAs. About 65.96% *Tetragenococcus* strains, 45.83% *Staphylococcus* strains, 33.33% *Lentibacilluis* strains and all strains of *Psychrobacter* and *Pseudomonas* had a high ability to produce BAs. These strains may be the main factor causing the large accumulation of BAs in fish sauce. The analysis of the correlation between dominant bacteria with BAs and the determination of BA production ability of the dominant isolates showed that *Tetragenococcus*, *Lentibacillus*, *Psychrobacter*, *Pseudomonas* and *Staphylococcus* and other bacteria had a regulatory effect on the accumulation of BAs in fish sauce fermentation. For example, *Tetragenococcus* positively regulated BAs and promoted the accumulation of BAs, while *Lentibacillus* negatively regulated BAs and inhibited the accumulation of BAs.

Table 2. Percentage of strains with BA production activities, shown in vitro by the different bacteria species isolated from fermented fish sauce.

Genus	Number of Total Strains	The Percent of BA-Production Strains
Tetragenococcus	47 (16)	65.96%
Staphylococcus	48 (26)	45.83%
Lentibacillus	3 (2)	33.33%
Psychrobacter	1 (0)	100.00%
Pseudomonas	1 (0)	100.00%

For each genus, the number of isolates with low production of BAs is shown in brackets.

3.4. Screening of Novel Starter Cultures for Reducing Biogenic Amines

Due to the regulation of microbes on the accumulation of BAs, it is feasible to select microorganisms to reduce the content of BAs in fish sauce. Therefore, eight strains with the lowest BA production capacity (especially histamine) were screened from the genera and their BA reduction ability was determined (Tables 3 and S1). Staphylococcus strains, particularly *S. nepalensis*, had the best ability to reduce BAs. The reduction efficiencies of *S. nepalensis* 5-5 on putrescine, cadaverine and histamine were significantly higher than other strains, which were 17.64%, 19.80% and 16.77%, respectively. The isolated Lentibacillus strains also showed a high ability to reduce BAs. The ability to reduce the histamine of Lentibacillus salicamp SF-20 was higher than that of Lentibacillus amyloliquefaciens LAM0015. Tetragenococcus halophilus NBRC 12172 and Tetragenococcus muriaticus LMG 18498 had a low ability to reduce BAs, especially histamine. Strains with high BA reduction ability can be used as candidates for reducing BAs in fermented foods.

Table 3. Percentage of strains with BA reduction activities.

Strains	PUT (%)	CAD (%)	HIS (%)
Staphylococcus nepalensis 5-5	17.64 ± 0.44 [a]	19.80 ± 0.93 [a]	16.77 ± 1.04 [a]
Staphylococcus xylosus JCM 2418	11.96 ± 0.71 [b]	6.74 ± 0.31 [c]	5.04 ± 0.27 [d]
Staphylococcus hominis ICC_10-1_SCI_contig_1	9.32 ± 0.34 [c]	8.68 ± 0.69 [b]	7.81 ± 0.73 [b]
Staphylococcus capitis +Y36	9.78 ± 0.37 [c]	7.26 ± 0.51 [c]	5.64 ± 0.79 [cd]
Lentibacillus salicamp SF-20	8.62 ± 0.12 [d]	8.21 ± 0.62 [bc]	6.87 ± 0.47 [c]
Lentibacillus amyloliquefaciens LAM0015	8.78 ± 0.89 [cd]	5.26 ± 0.68 [d]	4.54 ± 0.65 [d]
Tetragenococcus halophilus NBRC 12172	5.34 ± 0.64 [e]	3.50 ± 0.61 [e]	2.59 ± 0.63 [e]
Tetragenococcus muriaticus LMG 18498	4.68 ± 0.78 [e]	3.91 ± 0.24 [e]	2.00 ± 0.15 [e]

Data are expressed as mean ± SDs ($n = 3$).; Different letters (a, b, c, d, e.) indicate the mean value of significant difference at $p < 0.05$. (PUT: putrescine; CAD: cadaverine; HIS: histamine)

S. nepalensis 5-5, *S. xylosus* JCM 2418, *S. hominis* ICC_10-1_SCI_contig_1 and *L. salicampi* SF-20 might be potential candidates for controlling the BAs due to their high BA reduction ability as well as low BA production activity. In order to ascertain the BA reduction ability of the strains in the food matrix, we inoculated the strains into different fermented foods and determined the BA content in the samples after 7 days of continuous fermentation. All four strains can reduce the BAs in fish sauce, and *S. nepalensis* 5-5 and *S. xylosus* JCM 2418 especially showed the best BA reduction ability. In the shrimp paste, only *S. nepalensis* 5-5 and *S. xylosus* JCM 2418 can reduce the BAs, while the remaining strains increase the BA content. *S. xylosus* JCM 2418 has the highest reducing ability on HIS in Doujiang, while *S. hominis* ICC_10-1_SCI_contig_1 and *L. salicampi* SF-20 increased the content of HIS in Doujiang. *S. nepalensis* 5-5 had the best BA reduction ability in Sufu among all the strains (Figure 5). The above results indicate that *S. nepalensis* 5-5 and *S. xylosus* JCM 2418 had excellent capacities for controlling BAs in fermented food.

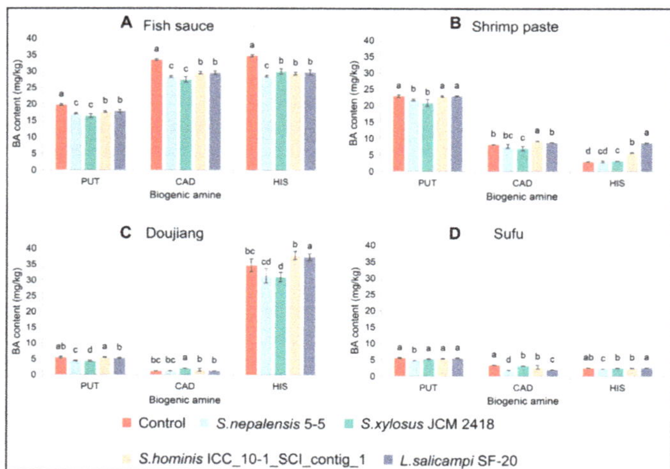

Figure 5. The content of BAs in fermented foods inoculated with different strains (**A**: Fish sauce. **B**: Shrimp paste. **C**: Doujiang. **D**: Sufu.). Different letters (a, b, c, etc.) indicate the mean value of significant difference at $p < 0.05$. (PUT: Putrescine; CAD: Cadaverine; HIS: Histamine).

4. Discussion

Excessive consumption of BAs can cause a variety of harmful effects [11]. Natural fermentation gives the fish sauce a rich microbial community, some of which can produce decarboxylase for the decarboxylation of amino acids to produce BAs, which is detrimental to food safety [2,9]. On the contrary, BAs can also be degraded by the amine oxidase of some bacteria. In this study, six common BAs were detected in the fish sauce samples at five fermentation stages, and putrescine, cadaverine and histamine existed in all samples. The histamine content of the fish sauce sample at the end of fermentation was much higher than the toxic level (50 mg/kg) suggested by the United States FDA [14] (Table 1). BAs could be recognized as signs of the quality and safety of products. Therefore, the contents of BAs in food should be effectively reduced and legitimately controlled.

The microbial community structure changed constantly during the fermentation of fish sauce. This may be due to the introduction of a large number of microorganisms in the natural fermentation mode, which made the microbial composition in the early fermentation samples significantly different from those in other groups. However, as the fermentation progressed, the stability of the microbial composition gradually improved, and the advantages of the fermented genus gradually appeared (Figures 2B and 3A,B). Traditional isolation and screening methods have obtained a large number of *Staphylococcus* strains. The high-throughput sequencing results showed that the abundance of *Staphylococcus* was not very high, but it could be detected at every fermentation stage, indicating that it was also a very important genus for fermentation (Figure 2C).

There are differences in the representative bacterial genera in the samples at different fermentation stages (Figure 3C). *Lentibacillus* is a representative microorganism in fish sauce samples fermented for 3 months, which is considered as a common microorganism in fish sauce [22]. *Psychrobacter* as a genus, often detected in fermented food, was a representative microorganism in fish sauce samples fermented for 6 months. It is reported that the presence of *Psychrobacter celer* can lead to a high production of pleasant volatile aroma compounds such as aldehydes, ketones and sulfur compounds [23]. However, some reports stated that *Psychrobacter* inoculated into cheese might produce HIS [24]. *Pseudomonas* as a representative microorganism in fish sauce samples fermented for 12 months was a common spoilage bacterium, which could produce a large amount of hydrogen sulfide in fish juice and has the ability to produce BAs [25]. *Tetragenococcus* is the dominant genus in each fermentation stage, but the abundance in the samples fermented for 18 months is significantly different from other fermentation stages (the abundance in samples of 3, 6, 9, 12 and 18 months of fermentation was 9.15%, 45.78%, 30.00%, 26.34% and 91.99%, respectively). Therefore, it might be a key genus that determines the quality and safety of the final products. *Tetragenococcus* exhibits certain halotolerant characteristics and plays a major role in sauce flavor [26,27]. In addition, the previous report showed that *Tetragenococcus halophilus* MJ4 clearly repressed the formation of cadaverine during fermentation to improve the safety of fish sauce products [28]. However, it was reported that *Tetragenococcus* had a high BAs production ability, which had a great effect on the accumulation of BAs in sausage, shrimp paste, etc. [17,29]. Some genera detected throughout the fermentation process, such as *Staphylococcus*, also played an important role in the quality and safety of fish sauce products, which is often used as a starter in fermented foods and can be used to control BAs during food fermentation [30].

The microbial composition of fish sauce samples had significant differences in time distribution, and BAs were also related to fermentation time (Figure 4A) [5], indicating that the characteristic microorganisms in samples of specific fermentation time might have a great contribution to the formation of BAs. Therefore, it is necessary to analyze the correlation between the characteristic microorganisms and BAs. The accumulation of BAs in the samples was low at the early fermentation period (3–6 months), and the early representative microorganism, *Lentibacillus* and *Psychrobacter,* were negatively correlated with BAs, suggesting the BA accumulation might be inhibited by *Lentibacillus* and *Psychrobacter* during this period. As the microbial composition of the samples fermented for 6 months was similar to that of 9 months, BA accumulation was also low during the fermentation of fish sauce from 6 to 9 months. *Pseudomonas*, positively related to BAs, was a representative microorganism in fish sauce samples fermented for 12 months. During the fermentation period of 9–12 months, the content of BAs in fish sauce increased greatly, indicating that *Pseudomonas* might be promote the accumulation of BAs. *Tetragenococcus* was dominant during the whole fermentation process, and it was significantly positively correlated with BAs, indicating that *Tetragenococcus* might play a key role in the accumulation of BAs (Figure 4B,C).

To verify the results of the correlation analysis, we tested the BA production capacity of the isolated dominant strains (Table 2). The BA production ability of isolated strains could prove the results of the correlation analysis, indicating that *Tetragenococcus* and *Pseudomonas* may promote the accumulation of BAs, while *Lentibacillus* and *Staphylococcus* may inhibit the accumulation of BAs. Isolated *Psychrobacter* strains produced high BAs, while the results of high-throughput sequencing showed a negative correlation between *Psychrobacter* and BAs, possibly due to the limitations of isolation and screening methods, and the ability to produce BAs was strain specific. The accumulation of BAs in fish sauce fermentation could be affected by changing the species and abundance of dominant microorganisms. The method of adjusting the microbial community structure by changing the fermentation process is very common. The difference of salt concentration in fermentation affected the composition of microorganisms in the samples, and the content of BAs in the products with different salt concentrations was obviously different [16]. In addition, temperature can also affect the accumulation of BAs during fermentation. Delgado-Ospina et al. [31] demonstrated that temperature helped to simulate the off and on of the amine metabolism during cocoa fermentation, thus regulating BA accumulation and reduction. Recently, microbial fermentation as a means to improve the safety and quality of fermented food has been widely used. Xia et al. [32] used *Lactobacillus plantarum* and *Staphylococcus xylosus* as starters to co-inoculate Chinese rice wine. Co-inoculation induced a significant reduction in total BAs (43.7%), putrescine (43.0%), tyramine (42.8%), and histamine (42.6%) content. Therefore, it is feasible to select suitable starters to reduce the accumulation of biogenic amines during the fermentation of fish sauce.

About 44.00% of the dominant strains in this study produce very little Bas; therefore, they may have BA reduction ability and potential as fish sauce starter cultures. Eight isolated strains were selected from the dominant species of fish sauce samples, including four *Staphylococcus* strains, two *Lentibacillus* strains and two *Tetragenococcus* strains, and their BA reduction abilities were evaluated (Table 3). Among them, the *Staphylococcus* and *Lentibacillus* isolates showed high BA reduction ability. The results show that microbial-based solutions can be designed to reduce the BA content in fermented foods [30]. In previous studies, *Staphylococcus* is often used as a starter to reduce BAs in fermented foods. For example, Muhammad et al. [33] found that HIS concentration was reduced by 27.7% by *Staphylococcus carnosus* FS19 and could also reduce other amines during fermentation. Li et al.'s [30] study showed that the *Staphylococcus* isolates including *Staphylococcus pasteuri*, *Staphylococcus epidermidis*, *Staphylococcus carnosus*, and *Staphylococcus simulans* could significantly reduce BAs and have the potential to control BAs in fermented meat products. *Lentibacillus* is a common genus of bacteria in fish sauce [34]. Thus, the effects of Lentibacillus and Staphylococcus isolates on BA accumulation were assessed in different fermented foods to screen the potential starters for BA control (Figure 5). The content of BAs in fermented foods inoculated with *S. nepalensis* 5-5 and *S. xylosus* JCM 2418 were significantly reduced, indicating that the two strains could be utilized as potential candidates for controlling BAs in fermented foods.

5. Conclusions

In summary, the bacterial community structure and BA profiles in fish sauce at different fermentation stages were comprehensively evaluated. The dominant bacteria of fish sauce were found to be *Tetragenococcus*, *Lentibacillus*, *Pseudomonas*, *Halomonas*, *Vagococcus* and *Staphylococcus* by high-throughput sequencing. The correlation analysis between microorganisms and BAs indicated that *Tetragenococcus*, *Halomonas* and *Pseudomonas* were positively correlated with BAs, while *Lentibacillus*, *Vagococcus* and *Staphylococcus* were negatively correlated with BAs. By analyzing the BA production capacity of dominant isolates, the accumulation of BAs in fish sauce might be promoted by *Tetragenococcus* and *Halomonas*. Additionally, *S. nepalensis* 5-5 and *S. xylosus* JCM 2418 isolated from fish sauce were applied as starters to fish sauce and other fermented foods for reducing health risk. Our research assessed the BA reducing application of *S. nepalensis* 5-5 and *S. xylosus* JCM 2418 during fish sauce fermentation for the first time, and provided a theory to support the standardized and secure production of fish sauce.

Supplementary Materials: The following are available online at https://www.mdpi.com/article/10.3390/foods10112572/s1, Table S1: Determination of BA production capacity of dominant isolates.

Author Contributions: Conceptualization, X.M.; methodology, X.M. and G.Z.; Investigation, X.M. and X.L.; Data curation, H.H. (Hongshun Hao); Writing—original draft preparation, X.M.; writing—review and editing, H.H. (Hongman Hou) and J.B.; Supervision, H.H. (Hongman Hou); Project administration, H.H. (Hongman Hou); Funding acquisition, H.H. (Hongman Hou). All authors have read and agreed to the published version of the manuscript.

Funding: This research was funded by the National Key R & D Program of China (2017YFC1600403), the Liaoning Province's Program for Promoting Liaoning Talents (XLYC1808034).

Institutional Review Board Statement: Not applicable.

Informed Consent Statement: Not applicable.

Data Availability Statement: The datasets generated for this study are available on request to the corresponding author.

Conflicts of Interest: All authors declare no conflict of interest.

References

1. Fukui, Y.; Yoshida, M.; Shozen, K.-I.; Funatsu, Y.; Takano, T.; Oikawa, H.; Yano, Y.; Satomi, M. Bacterial communities in fish sauce mash using culture-dependent and -independent methods. *J. Gen. Appl. Microbiol.* **2012**, *58*, 273–281. [CrossRef]
2. Lee, S.H.; Jung, J.Y.; Jeon, C.O. Bacterial community dynamics and metabolite changes in myeolchi-aekjeot, a Korean traditional fermented fish sauce, during fermentation. *Int. J. Food Microbiol.* **2015**, *203*, 15–22. [CrossRef]

3. Du, F.; Zhang, X.; Gu, H.; Song, J.; Gao, X. Dynamic Changes in the Bacterial Community During the Fermentation of Traditional Chinese Fish Sauce (TCFS) and Their Correlation with TCFS Quality. *Microorganisms* **2019**, *7*, 371. [CrossRef]
4. Jiang, W.; Xu, Y.; Li, C.; Dong, X.; Wang, D. Biogenic amines in commercially produced Yulu, a Chinese fermented fish sauce. *Food Addit. Contam. Part B Surveill.* **2014**, *7*, 25–29. [CrossRef]
5. Zaman, M.Z.; Abu Bakar, F.; Jinap, S.; Bakar, J. Novel starter cultures to inhibit biogenic amines accumulation during fish sauce fermentation. *Int. J. Food Microbiol.* **2011**, *145*, 84–91. [CrossRef]
6. Biji, K.B.; Ravishankar, C.N.; Venkateswarlu, R.; Mohan, C.O.; Gopal, T.K. Biogenic amines in seafood: A review. *J. Food Sci. Technol.* **2016**, *53*, 2210–2218. [CrossRef]
7. Onal, A.; Tekkeli, S.E.; Onal, C. A review of the liquid chromatographic methods for the determination of biogenic amines in foods. *Food Chem.* **2013**, *138*, 509–515. [CrossRef]
8. Spano, G.; Russo, P.; Lonvaud-Funel, A.; Lucas, P.; Alexandre, H.; Grandvalet, C.; Coton, E.; Coton, M.; Barnavon, L.; Bach, B.; et al. Biogenic amines in fermented foods. *Eur. J. Clin. Nutr.* **2010**, *64*, S95–S100. [CrossRef]
9. Yongsawatdigul, J.; Choi, Y.J.; Udomporn, S. Biogenic Amines Formation in Fish Sauce Prepared from Fresh and Temperature-abused Indian Anchovy (*Stolephorus indicus*). *JFS Food Chem. Toxicol.* **2004**, *69*, 312–319. [CrossRef]
10. Mooraki, N.; Sedaghati, M. Reduction of biogenic amines in fermented fish sauces by using Lactic acid bacteria. *J. Surv. Fish. Sci.* **2019**, *5*, 99–110. [CrossRef]
11. Kuda, T.; Miyawaki, M. Reduction of histamine in fish sauces by rice bran nuka. *Food Control* **2010**, *21*, 1322–1326. [CrossRef]
12. Naila, A.; Flint, S.; Fletcher, G.C.; Bremer, P.J.; Meerdink, G. Chemistry and microbiology of traditional Rihaakuru (fish paste) from the Maldives. *Int. J. Food Sci. Nutr.* **2011**, *62*, 139–147. [CrossRef]
13. Tsai, Y.; Lin, C.; Chien, L.; Lee, T.; Wei, C.; Hwang, D. Histamine contents of fermented fish products in Taiwan and isolation of histamine-forming bacteria. *Food Chem.* **2006**, *98*, 64–70. [CrossRef]
14. FDA. *Scombrotoxin (Histamine) Formation*; Department of Health and Human Services, Public Health Service, Food and Drug Administration, Center for Food Safety and Applied Nutrition, Office of Seafood: Washington, DC, USA, 2001.
15. Wang, Y.; Li, C.; Li, L.; Yang, X.; Chen, S.; Wu, Y.; Zhao, Y.; Wang, J.; Wei, Y.; Yang, D. Application of UHPLC-Q/TOF-MS-based metabolomics in the evaluation of metabolites and taste quality of Chinese fish sauce (Yu-lu) during fermentation. *Food Chem.* **2019**, *296*, 132–141. [CrossRef]
16. Bao, X.; Xiang, S.; Chen, J.; Shi, Y.; Chen, Y.; Wang, H.; Zhu, X. Effect of *Lactobacillus reuteri* on vitamin B12 content and microbiota composition of furu fermentation. *LWT Food Sci. Technol.* **2019**, *100*, 138–143. [CrossRef]
17. Sang, X.; Ma, X.; Hao, H.; Bi, J.; Zhang, G.; Hou, H. Evaluation of biogenic amines and microbial composition in the Chinese traditional fermented food grasshopper sub shrimp paste. *LWT Food Sci. Technol.* **2020**, *134*, 109979. [CrossRef]
18. Li, L.; Wen, X.; Wen, Z.; Chen, S.; Wang, L.; Wei, X. Evaluation of the Biogenic Amines Formation and Degradation Abilities of *Lactobacillus curvatus* From Chinese Bacon. *Front. Microbiol.* **2018**, *9*, 1015. [CrossRef]
19. Xia, X.; Zhang, Q.; Zhang, B.; Zhang, W.; Wang, W. Insights into the Biogenic Amine Metabolic Landscape during Industrial Semidry Chinese Rice Wine Fermentation. *J. Agric. Food Chem.* **2016**, *64*, 7385–7393. [CrossRef]
20. Mah, J.-H.; Hwang, H.-J. Inhibition of biogenic amine formation in a salted and fermented anchovy by *Staphylococcus xylosus* as a protective culture. *Food Control* **2009**, *20*, 796–801. [CrossRef]
21. Sang, X.; Li, K.; Zhu, Y.; Ma, X.; Hao, H.; Bi, J.; Zhang, G.; Hou, H. The Impact of Microbial Diversity on Biogenic Amines Formation in Grasshopper Sub Shrimp Paste During the Fermentation. *Front. Microbiol.* **2020**, *11*, 782. [CrossRef]
22. Jung, W.Y.; Lee, S.H.; Jin, H.M.; Jeon, C.O. *Lentibacillus garicola* sp. nov., isolated from myeolchi-aekjeot, a Korean fermented anchovy sauce. *Antonie Van Leeuwenhoek* **2015**, *107*, 1569–1576. [CrossRef]
23. Irlinger, F.; Yung, S.A.; Sarthou, A.S.; Delbes-Paus, C.; Montel, M.C.; Coton, E.; Coton, M.; Helinck, S. Ecological and aromatic impact of two Gram-negative bacteria (*Psychrobacter celer* and *Hafnia alvei*) inoculated as part of the whole microbial community of an experimental smear soft cheese. *Int. J. Food Microbiol.* **2012**, *153*, 332–338. [CrossRef]
24. Helinck, S.; Perello, M.-C.; Deetae, P.; de Revel, G.; Spinnler, H.-E. Debaryomyces hansenii, Proteus vulgaris, *Psychrobacter* sp. and *Microbacterium foliorum* are able to produce biogenic amines. *Dairy Sci. Technol.* **2013**, *93*, 191–200. [CrossRef]
25. Ge, Y.; Zhu, J.; Ye, X.; Yang, Y. Spoilage potential characterization of *Shewanella* and *Pseudomonas* isolated from spoiled large yellow croaker (*Pseudosciaena crocea*). *Lett. Appl. Microbiol.* **2017**, *64*, 86–93. [CrossRef]
26. Amadoro, C.; Rossi, F.; Piccirilli, M.; Colavita, G. *Tetragenococcus koreensis* is part of the microbiota in a traditional Italian raw fermented sausage. *Food Microbiol.* **2015**, *50*, 78–82. [CrossRef]
27. He, G.; Wu, C.; Huang, J.; Zhou, R. Metabolic response of *Tetragenococcus halophilus* under salt stress. *Biotechnol. Bioprocess Eng.* **2017**, *22*, 366–375. [CrossRef]
28. Kim, K.H.; Lee, S.H.; Chun, B.H.; Jeong, S.E.; Jeon, C.O. *Tetragenococcus halophilus* MJ4 as a starter culture for repressing biogenic amine (cadaverine) formation during saeu-jeot (salted shrimp) fermentation. *Food Microbiol.* **2019**, *82*, 465–473. [CrossRef]
29. Kimura, B.; Konagaya, Y.; Fujii, T. Histamine formation by *Tetragenococcus muriaticus*, a halophilic lactic acid bacterium isolated from fish sauce. *Int. J. Food Microbiol.* **2001**, *70*, 71–77. [CrossRef]
30. Li, L.; Zou, D.; Ruan, L.; Wen, Z.; Chen, S.; Xu, L.; Wei, X. Evaluation of the Biogenic Amines and Microbial Contribution in Traditional Chinese Sausages. *Front. Microbiol.* **2019**, *10*, 872. [CrossRef]

31. Delgado-Ospina, J.; Acquaticci, L.; Molina-Hernandez, J.B.; Rantsiou, K.; Martuscelli, M.; Kamgang-Nzekoue, A.F.; Vittori, S.; Paparella, A.; Chaves-Lopez, C. Exploring the Capability of Yeasts Isolated from Colombian Fermented Cocoa Beans to Form and Degrade Biogenic Amines in a Lab-Scale Model System for Cocoa Fermentation. *Microorganisms* **2020**, *9*, 28. [CrossRef]
32. Xia, X.; Luo, Y.; Zhang, Q.; Huang, Y.; Zhang, B. Mixed Starter Culture Regulate Biogenic Amines Formation via Decarboxylation and Transamination during Chinese Rice Wine Fermentation. *J. Agric. Food Chem.* **2018**, *66*, 1–32. [CrossRef]
33. Zaman, M.Z.; Bakar, F.A.; Selamat, J.; Bakar, J.; Ang, S.S.; Chong, C.Y. Degradation of histamine by the halotolerant *Staphylococcus carnosus FS19* isolate obtained from fish sauce. *Food Control* **2014**, *40*, 58–63. [CrossRef]
34. Namwong, S.; Tanasupawat, S.; Smitinont, T.; Visessanguan, W.; Kudo, T.; Itoh, T. Isolation of *Lentibacillus salicampi* strains and *Lentibacillus juripiscarius* sp. nov. from fish sauce in Thailand. *Int. J. Syst. Evol. Microbiol.* **2005**, *55*, 315–320. [CrossRef]

 foods

MDPI

Article

Impact of Thyme Microcapsules on Histamine Production by *Proteus bacillus* in Xinjiang Smoked Horsemeat Sausage

Honghong Yu [1,†], Yali Huang [1,†], Liliang Lu [2], Yuhan Liu [2], Zonggui Tang [2] and Shiling Lu [1,*]

1 Laboratory of Meat Processing and Quality Control, College of Food Science and Technology, Shihezi University, Shihezi 832000, China; Yhh195812@163.com (H.Y.); hyl2026@163.com (Y.H.)
2 Analysis and Testing Center, Xinjiang Academy of Agriculture and Reclamation Science, Shihezi 832000, China; lubeyond@126.com (L.L.); yuhanliu0815@163.com (Y.L.); zongguitang@163.com (Z.T.)
* Correspondence: lushiling_76@163.com; Tel.: +86-131-7993-3069; Fax: +86-0993-205-7399
† These authors contributed equally to this work.

Abstract: Here, we explored the influences of thyme microcapsules on the growth, gene expression, and histamine accumulation by *Proteus bacillus* isolated from smoked horsemeat sausage. RT-qPCR was employed to evaluate the gene expression level of histidine decarboxylase (HDC) cascade-associated genes. We used HPLC to monitor histamine concentration both in pure culture as well as in the processing of smoked horsemeat sausage. Results showed that histamine accumulation was suppressed by thyme microcapsule inhibitory effect on the histamine-producing bacteria and the reduction in the transcription of hdcA and hdcP genes. Besides, compared with thyme essential oil (EO), thyme microcapsules exhibited higher antibacterial activity and had a higher score for overall acceptance. Therefore, the addition of thyme microcapsules in Xinjiang smoked horsemeat sausage inhibits histamine accumulation.

Keywords: thyme microcapsules; *Proteus bacillus*; histamine; histidine decarboxylation pathway; smoked horsemeat sausage

Citation: Yu, H.; Huang, Y.; Lu, L.; Liu, Y.; Tang, Z.; Lu, S. Impact of Thyme Microcapsules on Histamine Production by *Proteus bacillus* in Xinjiang Smoked Horsemeat Sausage. *Foods* **2021**, *10*, 2491. https://doi.org/10.3390/foods10102491

Academic Editor: Maria Martuscelli

Received: 16 September 2021
Accepted: 10 October 2021
Published: 18 October 2021

1. Introduction

Histamine is a low molecular weight nitrogenous organic compound that plays a pivotal role in human learning as well as memory, body temperature, and immune responses [1]. Histamine oxidases quickly degrade the exogenous histamine that is ingested in food under normal conditions; however, when the process of detoxification is insufficient, or the concentrations of histamines present in food are very high, the histamine might lead to histamine intolerance or intoxication [2,3]. A food-poisoning incidence reported in January 2009 that was associated with histamine toxicity caused illness in 53 persons [4]. The Xinjiang smoked horsemeat sausage made via spontaneous fermentation is popular for its appetizing sensory features as well as excellent nutritional properties [5]. Nevertheless, the traditional sausage is often produced in small-scale plants after a process of spontaneous fermentation [6], which more easily leads to the accumulation of high levels of biogenic amines (BAs), e.g., putrescine, tyramine, as well as histamine [7]. In fermented meat products, histamine is formed through histidine decarboxylation mediated by the histidine decarboxylase enzymes originating from the bacteria present in food [8]. The bacterial histidine decarboxylases have been broadly studied and characterized in various organisms [9], and two distinct enzyme classes have been differentiated, namely pyridoxal phosphate dependent as well as the pyruvoyl dependent. The diverse species of gram-negative bacteria, especially the lactic acid bacterial species involved in food spoilage or fermentation possess, the pyridoxal phosphate-dependent decarboxylases, whereas the gram-positive bacteria possess the pyruvoyl-dependent decarboxylases [10]. Studies have shown that the histidine genes clusters related to histidine decarboxylase (HDC) consist of 4 ORFs, hdcA, hdcRS, hdcP, as well as hdcB, and their sequences are highly conserved in numerous lactic acid bacteria. The hdcA OFR codes for the histidine decarboxylase, hdcP encodes a histidine/histamine protein, whereas hdcRS encodes a histidyl-tRNA synthetase; nonetheless, the role of the hdcB gene is still unclear [8,11].

The histamine content in fermented sausages depends on many factors, including the existence of microorganisms, which decarboxylate the amino acids, as well as a conducive environment for the growth of decarboxylase-producing microbes [12]. Furthermore, due to its heat stability, once histamine is formed, it is difficult to destroy with high-temperature treatment [13]. Hence, microbial food contamination should be prevented to minimize the possibility of histamine production [5,14]. Therefore, varieties of antibacterial agents have been applied in food to prevent the histamine aggregation. Thyme oil obtained from *Thymus vulgaris* L. has been opined to possess elevated antimicrobial, insecticidal, as well as phytotoxic features owing to the phenolic constituents, particularly carvacrol (2-methyl-5-methyl ethyl)-phenol) as well as thymol (5-methyl-2 (1methylethyl) phenol) [15,16]. However, the application of thyme essential oil (EO) in food preservation is limited because it is likely to change the original flavor of food products. Moreover, when the oil is exposed to heat, oxygen, or light, it is highly susceptible to oxidative deterioration [17]. The technology of encapsulation provides an efficient approach to stabilize EOs as well as impede volatile ingredients loss [18]. Encapsulation of cinnamon EO is achieved through the inclusion complexation approach using the β-cyclodextrin and has high antifungal bioactivity against *Botrytis* spp. [17]. Ginger EO microcapsules, with a particle size of 8.2–15.3 μm, were encapsulated by gum arabic (GA) and/or maltodextrin (MD) through spray drying and possess effective antioxidant activity, extending the shelf life of the food products [19]. Hence, EO microcapsules confer antibacterial and antioxidants effects, with prospective application in meat products

However, to the best of our knowledge, there is a knowledge gap in the application of thyme microcapsules in sausages. Hence, the purpose of the research was to assess if microcapsules of thyme affect the accumulation of histamine, the growth of microorganisms, transcription of histidine decarboxylase (HDC) cascade-associated genes, and the sensory quality of smoked horsemeat sausage.

2. Materials and Methods

2.1. Microorganisms and Growth Conditions

A previously isolated histamine-producing microorganism, *Proteus bacillus,* from naturally smoked horsemeat sausage in our laboratory, was used in this study (Animal Product Processing Laboratory, Shihezi, China). The sequences were deposited in GenBank databases under accession number MN483275. *Proteus bacillus* was grown on Brain-Heart Infusion Broth (BHI) media, enriched with 100 Mm histidine (Sigma, Santa Clara, CA, USA) at 37 °C. The strain used in this work had been previously identified by molecular methods [20]. Overnight cultures of *P. bacillus* strains were used as inoculum for all fermentation assays. Each fermentation experiment was performed using the same stock culture medium and overnight culture inoculants to ensure the same CFU/mL (10^5–10^6). Thyme microcapsules and thyme essential oil were supplied by the Boao Extension Technology Co. Ltd. in Beijing. Five batches were prepared (10 mL): a batch without histidine (control); histidine and 0% thyme microcapsules (0% microcapsules); histidine and minimum inhibitory concentration (MIC) thyme microcapsules (MIC microcapsules); histidine and 1/2 MIC thyme microcapsules (1/2MIC microcapsules); and histidine and the same amount of essential oil as microcapsules (essential oil). Brain-Heart Infusion Broth (BHI) media (10 mL) required the following ingredients: beef brain (20%), beef heart infusion juice (25%), peptone (1%), glucose (0.2%), NaCl (0.5%), and agar (2%). Brain-Heart Infusion Broth (BHI) media was obtained from AoBoXing Company (Beijing, China). Samples were collected (2 mL) every four hours for 48 h. To check the microbial growth in all cultures, we measured the absorbance at 600 nm (OD600) using a multi-modal reader (Bio Tek, Unalaska, AK, USA).

2.2. Determination of Relevant Indicators in a Pure Culture System

The pH of the samples was measured using a Sartorius UB pH-meter (Sartorius, Gogenting, Germany). We calibrated the pH meter with phosphate buffer and potassium hydrogen phthalate buffer. After the calibration was completed, we washed the composite electrode with distilled water and dried the filter paper and inserted it into the sample solution to ensure that the glass bulb at the front of the electrode was in uniform contact with the standard buffer solution. After the pH meter stabilized, showing the pH value of the unknown solution, we pressed the "OK" button to measure the solution again [2]. The measurement accuracy of the pH meter is 0.01. *P bacillus* MN483275 was cultured in BHI medium and supplemented with 100 Mm histidine to activate the transcription of the hdc gene [21]. To harvest the cells, we centrifuged the broth culture after 24 h of growth at 10,000× *g* for 3 min at 4 °C and resuspended in 1 mL of TRIzol reagent. TRIzol reagent (Sigma, Santa Clara, CA, USA) was used to extract total RNA, as previously described by del Rio et al. [22]. Sample supernatants were obtained by centrifugation at 5000× *g* for 10 min at 4 °C. High-performance liquid chromatography (HPLC) was used to determine histamine concentration [5]. Three biological replicates were performed for each experiment.

2.3. Sausage Sample Preparation

The horse meat was cut into cubes about 1 cm^3 in size. Four batches of smoked horsemeat sausage were processed: a control batch without *P. bacillus* or thyme microcapsules (batch CK); a batch inoculated with *P. bacillus* (batch P); a batch inoculated with *P. bacillus* and thyme microcapsules (batch PMT); and a batch inoculated with *P. bacillus* and thyme essential oil (batch PO). Sausage preparation (10 kg/batch) required the following ingredients: basic raw materials of lean horsemeat (80%) and fat horsemeat (20%); and auxiliary raw materials of pepper (0.1%), white sugar (2%), sodium chloride (2.5%), ginger powder (0.2%), monosodium glutamate (0.1%), spiced powder (0.1%), star anise (0.1%), and a smoke solution (1%). Thyme microcapsules and thyme oil were added at a concentration of 0.156%. Each batch was kept at 4 °C for 24 h and then stuffed (300–350 g/sausage) into a natural casing (horse's small intestine) previously soaked in salt water and with a diameter of 5–6 cm [3]. The culture of *P. bacillus* was suspended in sterile water (100 mL) to attain a final level of 10^4 CFU/g of sausage. Fermentation and ripening of the sausages was performed at a steady humidity as well as temperature in an incubator (DNP-9272, Jinghong Company, Shanghai, China) at 90–95% relative humidity (RH) and 18–20 °C first for 2 days and then at 75–80% RH for 26 days at 10–12 °C [23]. We collected the samples at 0 (initial sausages), 3 (after fermentation), 7, 14, 21, as well as 28 days.

2.4. Microbial Counts

The bacterial numbers were enumerated during the 28 days fermentation period using a previously document protocol [5] with slight adjustments. Using aseptic techniques, 10 g of each sample was homogenized in 90 mL of sterile saline (containing 0.85% NaCl), mixed in a stomacher for 10 min at 250 rpm, and the suspension was diluted serially (1:10) using sterilized saline in triplicates [24]. This was followed by inoculation of the diluted suspensions in different agars: gram-positive catalase-positive cocci (GCC+) were examined using mannitol salt agar (MSA) medium at 37 °C for 48 h; *Enterobacteriaceae* were analyzed using Violet Red Bile Glucose Agar (VRBGA) for 48 h at 37 °C, whereas lactic acid bacteria (LAB) were determined on de Man Rogosa Sharpe agar (MRS) medium at 30 °C for 2 days [5]. AoBoXing Company, Beijing, China provided all the growth media.

2.5. RNA Extraction

The RNA from experimental sausages was isolated, as per protocol documented by del Rio et al. [22] with slight adjustments. In brief, 1 mL of Trizol was added to 100 mg of shredded sausages in a 1.5-mL centrifuge tube, followed by homogenization, and left to stand for 5 min at room temperature. Chloroform (0.2 mL) was introduced into the homogenate and then mixed through, shaking for 15 s, then left standing for another 2 min. Next, we spun the mixture for 10 min at 12,000× *g* at 4 °C and then aliquoted the supernatant into clean tubes. We added 0.5 mL isopropanol into the mixtures, gently mixed, and maintained standing for 10 min at room temperature. After that, 10-min centrifugation of the mixture was performed at 12,000× *g* at 4 °C, and the supernatant was discarded. The resulting precipitate was rinsed using 1 Ml 75% ethanol to precipitate it further. We air-dried the precipitate and re-suspended in DEPC-treated H_2O. The RNA was inoculated with DNase (ABM, Vancouver, BC, Canada) to degrade any contaminating DNA. The concentration of the RNA was estimated by measuring its absorbance at 260 nm as well as 280 nm on a NanoDrop ND-2000c UV/vis Spectrophotometer (Thermo Scientific, Shanghai, China).

2.6. Assessment of RNA Levels by RT-qPCR

The 5X All-In-One RT Master Mix (AccuRT Genomic DNA Removal Kit) (ABM, Vancouver, BC, Canada) was employed to convert the RNA into cDNA via reverse transcription. Thereafter, qPCR was performed using the Stratagene MX 3000 sequence detection system (Agilent Technologies, Santa Clara, CA, USA) in 25 μL reaction mixture, as documented previously [25]. The reaction mixture contained the primers as well as the EvaGreen 2X qPCR Master Mix, which uses low ROX as a passive reference (ABM, Vancouver, BC, Canada). The specific primers utilized are listed in Table 1. The hdcA-F/hdcA-R as well as hdcP-F/hdcP-R primer pairs targeted the hdcA (the first gene of the HDC cluster) and hdcP (encoding for histidine/histamine exchanger) [11]. The hdcRS-F/hdcRS-R as well as hdcB-F/hdcB-R primer pairs [9] targeted the hdcRS gene (encoding the histidyl-tRNA synthetase) and hdcB (whose role is unclear), respectively [8]. The tufF/tufR and recA-F/recA-R primer pairs [9] targeted the thermo-unstable elongation factor (tuf) as well as the RNA polymerase alpha-subunit (recA) genes, respectively, which served as the reference genes. Negative controls were included as samples without DNA in each run. The $2^{-\Delta\Delta Ct}$ approach was employed to determine the relative gene-expression level [26]. RT-qPCR was conducted on RNA samples extracted from 3 different cultures for each condition.

Table 1. Primers used in this work.

Gene	Primer	Sequence (5′→3′)
hdcA	hdcA-F	GATGGTATTGTTTCKTATGA
	hdcA-R	CCAAACACCAGCATCTTC
hdcP	hdcP-F	GTCTGATCCATGGACACGGCTGAAC
	hdcP-R	GTTGCCGCGAATCTAGAATC
hdcB	hdcB-F	TACCGTTAGAGGCGAGTTCC
	hdcB-R	GGCAGCACAGGATTAGCATC
hisRS	hisRS-F	CACACAGATTGGTTGTGAGGC
	hisRS-R	CGTCCCGTGTTTCTTTGTCAC
tuf	tuf-F	TCTTCATCATCAACAAGGTCTGCTT
	tuf-R	GAACACATCTTGCTTTCACGTCAA
recA	recA-F	CAAGGCTTAGAGATTGCCGATG
	recA-R	ACGAGGAACTAACGCAGCAAC

2.7. Histamine Determination from Sausage Samples

The aggregated histamine in the experimental sausages was isolated using a previously described protocol [5] with slight adjustments. We homogenized 5 g of the sausages using an ULTRA-TURRAX T25 basic ZKA (WERKE, Sasel, Hamburg, German) in 20 mL of 0.4 M perchloric acid. The mixture was centrifuged for 10 min at 5000× *g* at 4 °C and collected the supernatant. (AllegraX-22, Santa Clara, CA, USA). The volume of the filtrate was adjusted to 50 mL using 0.4 M perchloric acid. Derivatization of biogenic amines was conducted using dansyl chloride according to the protocol documented by Lu et al. and

Sun et al. [5,27]. We put the sample extract (1 mL) into a 5-mL volumetric flask and then added sodium hydroxide (2 N, 200 mL), saturated sodium bicarbonate (300 mL), and dansyl chloride (10 mg/mL; Sigma, CA, USA) to the volumetric flask and incubated at 40 °C in the dark for 45 min. To remove residual dansyl chloride, we added 100 mL of ammonia, incubated at room temperature for 30 min, adjusted the volume of the reaction mixture to 5 mL with acetonitrile, and centrifuged at 3000× g for 5 min [5]. For HPLC analysis, a 0.45-μm membrane syringe filter was employed to filter the supernatant. Histamine content was detected by HPLC (LC-2010AHT, Shimadzu Corporation, Beijing, China) by a C18 column (Spherisob, 2.5 μm octadecylsilane, 250-mm 94.6-mm internal diameter), an injection volume of 10 μL, a flow rate of 0.8 mL/min, as well as column temperature of 35 °C. The mobile phase was composed of ultrapure water (eluent A) and acetonitrile (eluent B), and the gradient program was 40% A + 60% B at 0 min; 30% A + 70% B at 5 min; 10% A + 90% B at the 10th minute; 100% B at the 15th minute; and 40% A + 60% B at the 25th minute. The 1.00 mg/L histamine standard solution was appropriately diluted with acetonitrile and then derivatized and determined according to the method of Park et al. [28]. The detection limit (RSN = 3) and the limit of quantification (RSN = 10) were determined by the signal-to-noise ratio. Histamine was assayed at 254 nm. Moreover, all analytical determinations were performed in triplicate for each sausage sample.

2.8. Sensory Evaluation

The sensory quality of the smoked horsemeat sausages was assessed at the end of the fermentation, as previously documented, with slight modifications [27]. The sensory analysis was carried out using a sensory panel of 10 investigators, comprising of 5 females as well as 5 males. The assessors were selected based on their sensory potential and prior experience in performing a sensory assessment of meat products (Table 2). Each panel investigator was tasked with rating the appearance (color, gloss, dry), flavor (sourness, odor), and texture (hardness, organizational structures) of the sausage samples. Attributes were quantified with a numerical intensity scale from 0 to 7, where 0 = attribute not detected, and 7 = attribute very intense [29]. The total scores for each sample were calculated by adding the average score given by each panel member for each of the seven attributes.

Table 2. Sensory standard of smoked horse sausage.

Attribute		Definitions (Developed by the Panel)
Appearance	color	The actual hue of the color, pink brown to dark brown
	gloss	Shiny, attractive marinated color and uniform color
	dry	The absence of moistness, resembling dried meat
Flavor	sourness	The smell of oranges
	odor	Scented with horse meat and marinated flavors
Texture	hardness	Tough and hard to chew
	organizational structure	Tissue is delicate and elastic

2.9. Statistical Analyses

The Origin 8.5 software (Origin Lab) was utilized for data analysis. Means ± standard deviations were computed from three independent replicates. The SPSS software 23 package (IBM, Armonk, NY, USA) was employed to evaluate the differences between groups. $p < 0.05$ signified statistical significance.

3. Results and Discussion

3.1. The Thyme Microcapsules Effect on P. bacillus Growth and Histamine Generation

As indicated in (Figure 1a), the absorbance values (at 600 nm) of the microbial cultures treated with thyme oil was consistently lower than that of control batch at 48 h, suggesting that thyme essential oil (EO) repressed *P. bacillus* growth. Moreover, compared with the thyme oil group, thyme microcapsules exhibited higher antibacterial bioactivity [18,30].

However, thyme microcapsules only prolonged the life of the strain but did not change the growth pattern. Thyme microcapsule at varied levels had different suppressive influences on the strain ($p < 0.05$), and microbial growth decreased with escalating thyme microcapsules.

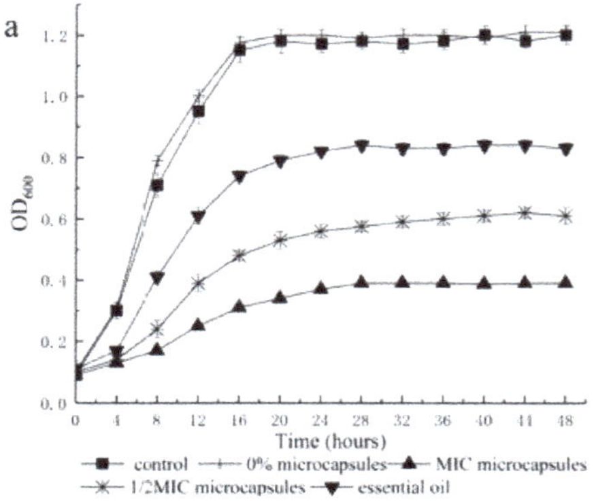

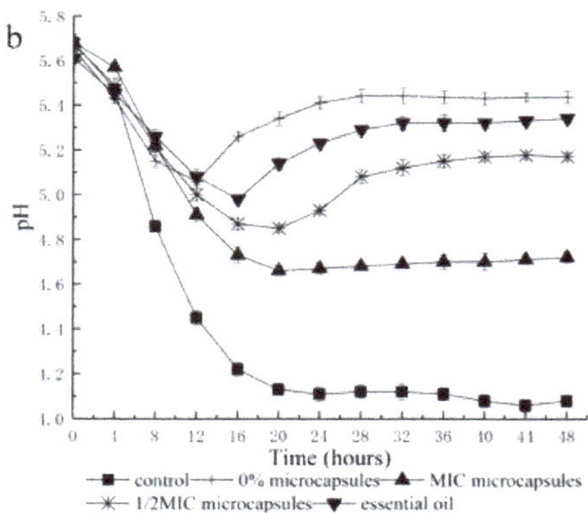

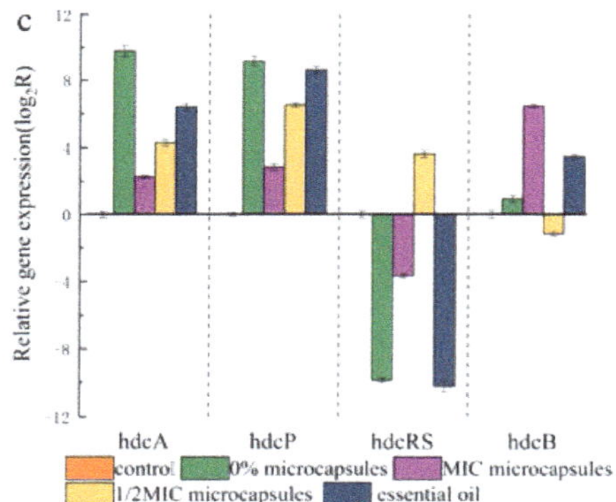

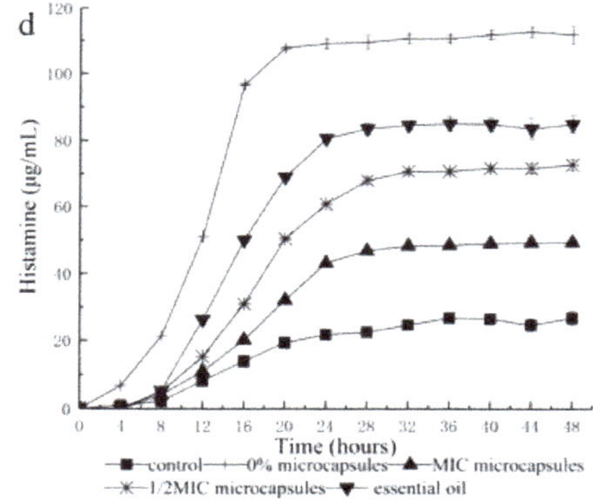

Figure 1. Effect of thyme microcapsules on *P. bacillus* growth (**a**), pH (**b**), gene expression of HDC cluster (**c**) and histamine accumulation (**d**) during 48 h. Batch Control: without histidine (control); 0% microcapsules: histidine + 0% microcapsules; MIC microcapsules: histidine + MIC microcapsules; 1/2MIC microcapsules: histidine +1/2MIC microcapsules; essential oil: histidine + essential oil.

The changes in pH over 48 h in all batches are shown in Figure 1b. The pH values decreased in the first 12 h and stabilized gradually after 20 h in the control batch, indicating that the microbes started to grow rapidly, and the products of fermentation were accumulating slowly. In the histidine-treatment batches, the microbial culture pH values (0% microcapsules, MIC microcapsules, 1/2 MIC microcapsules and essential oil) were higher relative to control batch, attributable to the aggregation of alkaline substances, such as histamine.

The thyme microcapsules group's pH was lower compared to the other groups, which revealed that thyme microcapsules inhibit *P. bacillus* growth as well as the HDC gene cluster expressions, thereby reducing the histamine accumulation.

The histamine standard solution with a mass concentration of 1.00 mg/L was diluted into different low-mass concentration solutions, and samples were injected for determination after derivatization. The results showed that when the RSN was 3, the mass concentration of histamine standard solution was 0.03 mg/L; when the RSN was 10, the mass concentration of histamine standard solution was 0.12 mg/L. Therefore, the detection limit of the instrument was 0.03 mg/L, and the limit of quantification was 0.12 mg/L. The accumulation of histamine over 48 h in all batches is shown in Figure 1d. Histamine was detected at 4 h and was found to gradually stabilized after 28 h in the control batch. In the histidine treatments, the histamine levels in the thyme microcapsules batches (MIC microcapsules and 1/2 MIC microcapsules) were significantly lower ($p < 0.05$) in contrast with the other batches (control and 0% microcapsules). Additionally, the thyme microcapsules influence on histamine accumulation was higher compared to that of thyme EO, suggesting that thyme microcapsule significantly decreased the accumulation of histamine. Moreover, the biosynthesis of histamine decreased with increased concentration of thyme microcapsules. The accumulation of histamine may confer resistance to acid stress acquired via the consumption of intracellular protons through a decarboxylation process, as has been documented in other bacterial species [31–33]. Thyme microcapsules suppressed histamine accumulation by inhibiting the histamine-producing bacteria and the expression of hdcA as well as hdcP genes.

3.2. Gene Expression in Pure Culture

The transcription of the HDC gene cluster is shown in Figure 1c. The HDC gene cluster includes the hdcA gene, histidyl-tRNA synthetase (hdcRS), a transporter (hdcP), as well as the hdcB gene, with the four genes oriented in a similar direction [34]. The comparative assessment showed that the expression of hdcB as well as hdcRS gene was not modulated by thyme microcapsule levels in the growth medium. However, the hdcA and hdcP gene expressions were remarkably repressed by increased thyme microcapsule levels. The highest expression activity of the hdcA gene was reported at 0% thyme microcapsules (181.02 times the expression in MIC thyme microcapsules), and the highest transcriptional activity of the hdcP gene was reported at 0% thyme microcapsules (78.25 times the expression in MIC thyme microcapsules). The hdcA gene encodes a 377 amino acid polypeptide and is believed to be a pyridoxal-P-dependent histidine decarboxylase. On the other hand, the hdcP gene is considered to be a histidine/histamine antiporter [31]. The lack of hdcB and hdcRS indicates that they are not important to the histidine decarboxylation cascade and are likely to encode the accessory roles [34]. The integration of the histidine/histamine exchanger (hdcP) with a histidine decarboxylase (hdcA) forms a classical decarboxylation cascade in bacteria [8].

3.3. Microbiological Analyses of Smoked Horsemeat Sausages

The alterations in bacterial numbers are indicated in Table 3. LAB as well as GCC+ comprised the primary microbes in the fermentation as well as ripening given the elevated numbers of indigenous LAB and GCC+ in raw horsemeat coupled with their rapid growth. The *Enterobacteriaceae* initial numbers were more remarkably elevated in batches PMT and PO than in batch CK ($p < 0.05$), associated with the richness of *P. bacillus* in the initial population. In addition, the number of LAB and GCC+ in all batches rose sharply, reached the maximum on the seventh day, and then decreased slightly. Besides, there was no significant difference between the four smoked horsemeat sausage batches in the later period ($p > 0.05$). Thyme microcapsules have no significant bactericidal biological activity against LAB and GCC+ in all batches, which may be attributed to the fact that the sensitivity of microorganisms to thyme microcapsules varied from species to species [24]. This is consistent to findings documented by Lu et al. [5] on a similar pattern in LAB and GCC+ numbers in smoked horsemeat sausages augmented with plant extracts. The numbers of *Enterobacteriaceae* remarkably reduced in the fermentation as well as the ripening ($p < 0.05$). On day 28, *Enterobacteriaceae* was completely depleted in the PMT batch, while the count in the CK, P, and PO batches were 1.01 ± 0.03, 3.36 ± 0.01, 1.85 ± 0.029 log10 CFU g^{-1}, respectively, revealing that thyme microcapsules repressed *Enterobacteriaceae* growth. Similarly, Scacchetti et al. [18] reported that thyme microcapsules showed high antimicrobial properties. The thyme microcapsule antibacterial mechanism is attributed to thyme EO slow release, which contains carvacrol, thymol, and other phenolic compounds, which are considered to be fungicides or antibacterial agents [35]. These components can attack the phospholipids in the cell membrane, increasing cell permeability, cytoplasmic leakage, or interaction with enzymes located on the cell wall to extend the shelf life of sausages in thyme microcapsule samples [17,18,35].

Table 3. Microbial counts (log 10 CFU g-1) during ripening of smoked horsemeat sausage.

Microbiological Counts	Batch	Days (d)				
		0	3	7	14	28
LAB	CK	3.86 ± 0.04 d	6.68 ± 0.01 c	7.19 ± 0.01 a	7.03 ± 0.02 b	7.11 ± 0.04 ab
	P	3.93 ± 0.07 d	6.81 ± 0.06 c	7.13 ± 0.01 ab	7.15 ± 0.02 a	7.09 ± 0.03 b
	PMT	3.91 ± 0.02 d	6.59 ± 0.02 c	7.09 ± 0.02 a	6.89 ± 0.01 b	7.01 ± 0.05 a
	PO	3.91 ± 0.03 e	6.66 ± 0.03 d	7.13 ± 0.02 b	6.92 ± 0.04 c	7.17 ± 0.08 a
GCC+	CK	3.97 ± 0.02 c	6.87 ± 0.01 b	7.79 ± 0.04 a	6.63 ± 0.08 b	6.54 ± 0.02 b
	P	3.95 ± 0.01 c	6.75 ± 0.04 b	7.46 ± 0.02 a	6.47 ± 0.03 b	6.45 ± 0.03 b
	PMT	3.95 ± 0.05 c	6.59 ± 0.06 b	7.79 ± 0.02 a	6.44 ± 0.04 b	6.26 ± 0.02 b
	PO	3.96 ± 0.03 c	6.61 ± 0.02 b	7.56 ± 0.02 a	6.43 ± 0.04 b	6.27 ± 0.07 b
Enterobacteria	CK	3.09 ± 0.07 a	3.16 ± 0.04 a	2.92 ± 0.02 b	2.11 ± 0.04 c	1.01 ± 0.01 d
	P	5.64 ± 0.05 a	4.92 ± 0.04 b	4.18 ± 0.06 c	3.97 ± 0.03 d	3.36 ± 0.01 e
	PMT	5.47 ± 0.02 a	4.31 ± 0.01 b	3.53 ± 0.02 c	3.09 ± 0.01 d	nd
	PO	5.59 ± 0.07 a	4.68 ± 0.02 b	3.96 ± 0.03 c	3.12 ± 0.01 d	1.85 ± 0.029 e

Batch CK: the spontaneously fermented as the control; batch P: inoculated with *P. bacillus*; batch PMT: inoculated with *P. bacillus* and thyme microcapsules; batch PO: inoculated with *P. bacillus* and essential oil; LAB, lactic acid bacteria; GCC+, gram-positive catalase-positive cocci. Data are expressed as mean ± standard deviation (n = 3). a–e: values in the same row and batch not followed by a common letter are significantly different ($p < 0.05$), differences between days. nd, not detected.

3.4. Gene Expression of hdcA and hdcP in Smoked Horsemeat Sausages

The hdcA and hdcP gene expression is involved in the production of histamine, which is found in our previous pure-culture system. The transcription of hdcA and hdcP genes in smoked horsemeat sausage is shown in Figure 2. The expression of hdcA and hdcP genes in all batches increased over the first seven days, then reached a maximum. There was a remarkable difference in the expression of hdcA and hdcP genes among the four batches ($p < 0.05$). The expression of the hdcA gene in batches CK, PO, and P was 9.58-, 18.64-, and 537.45-fold, respectively, which was higher compared to batch PMT. Similarly, the expression of the hdcp gene in batches CK, PO, and P was 10.06-, 1.89-, and 72.00-fold, respectively, which was higher than in batch PMT. Thyme microcapsules had a significantly higher effect on the expression of hdcA and hdcP compared to thyme oil. Studies have previously reported that the activity of the histamine synthesis gene cluster (HDC) affects the accumulation of histamine [31]. The factors affecting its activity include pH, temperature, salinity, and oxygen content, with pH being the main influencing factor. The acidic environment inhibits the growth of the test bacteria to a certain extent and affects the production and activity of enzymes. The expression of hdcA and hdcP genes increased under low pH conditions and the presence of extracellular histidine. Histamine accumulation by *P. bacillus* might confer resistance to the prevailing acidic stress via the consumption of intracellular protons through a decarboxylation reaction, as documented in other bacterial species [9,31,32]. Herein, the sausage processing conditions in all the samples were similar. Besides, the lack of significant difference in pH among the four batches showed that thyme microcapsules affected the activity of histidine decarboxylase and suppressed the expression of hdcA and hdcP genes.

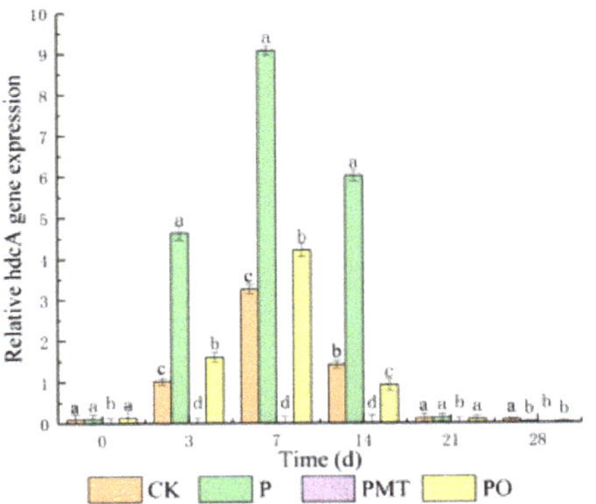

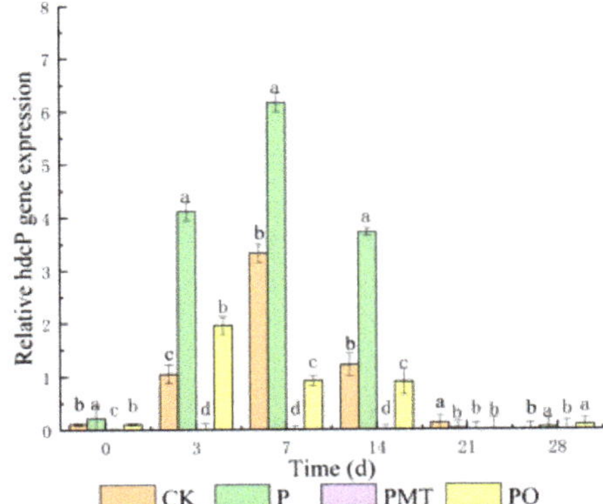

Figure 2. Effect of thyme microcapsules on gene expression of hdcA and hdcP in smoked horsemeat sausage during fermentation and ripening (average ± standard deviation, $n = 3$). (a–d): Values in the same column and batch not followed by a common letter are significantly different ($p < 0.05$) between days. Batch CK: the spontaneously fermented as the control; batch P: inoculated with *P. bacillus*; batch PMT: inoculated with *P. bacillus* and thyme microcapsules; batch PO: inoculated with *P. bacillus* and essential oil.

3.5. Histamine Accumulation in Smoked Horsemeat Sausages

The changes in histamine concentration in smoked horsemeat sausage are indicated in Figure 3. The histamine quantities in the four groups were measurable three days post fermentation and were found to increase during the first seven days, congruent with previous data in fermented sausages [24]. Higher levels of histamine (76.17, 35.18, and 44.09 mg/kg) were reported in the P, PO, and CK batches on day 7. On day 28, the histamine concentration in batches PMT and PO was 20.34 and 30.52 mg/kg, respectively, corresponding to a 66.15% and 49.22% decrease relative to batch P ($p < 0.05$) and suggesting that the presence of thyme oil prevented the accumulation of histamine. Compared with batch (PO), there was a 33.35% reduction in the accumulation of histamine in batch PMT, indicating that thyme microcapsule had a stronger inhibitory effect than thyme oil ($p < 0.05$). Additionally, thyme microcapsules decreased histamine aggregation by suppressing the expression of histidine decarboxylase (hdcA) as well as histidine/histamine antiporter (hdcP) genes ($p < 0.05$). Our data demonstrated that the encapsulation with thyme microcapsules reduced the accumulation of histamine, thus suggesting that the thyme microcapsule application can promote the safety of smoked horsemeat sausage as well as other meat products.

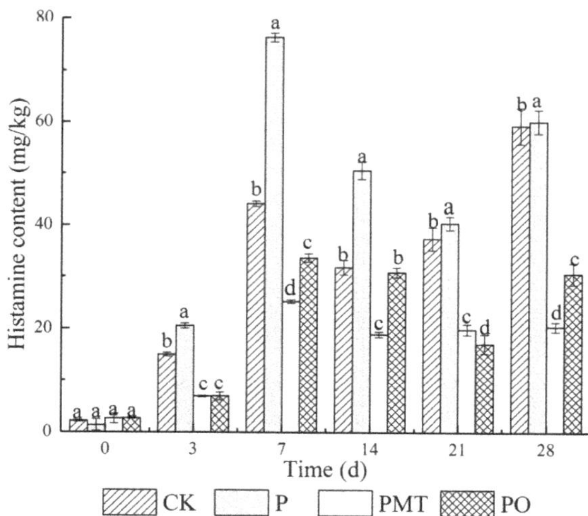

Figure 3. Effect of thyme microcapsules on histamine accumulation in smoked horsemeat sausage during fermentation and ripening (average ± standard deviation, $n = 3$). (a–d): Values in the same column and batch not followed by a common letter are significantly different ($p < 0.05$) between days. Batch CK: the spontaneously fermented as the control; batch P: inoculated with *P. bacillus;* batch PMT: inoculated with *P. bacillus* and thyme microcapsules; batch PO: inoculated with *P. bacillus* and essential oil.

3.6. Sensory Quality

The sensory qualities of the smoked horsemeat sausages were assessed at day 28 and are shown in Figure 4. The sausages with thyme microcapsules had significantly higher score for color than other batches (CK, P, PO) ($p < 0.05$), which may be attributed to the presence of nitrite. Thyme microcapsules can lower the pH value, and under acidic conditions, more nitrosomyoglobin will be produced, which plays an important role in color development [36]. The gloss of fermented meat products is an important sensory attribute. There was lower score for gloss in the batch P, which may be attributed to the presence of a large number of microorganisms than other batches. Numerous factors impact the sensory quality of smoked horsemeat sausage, such as raw meat, microbial activity, pH value, and any other factors in sausage [37]. There was no remarkable difference ($p > 0.05$) in the sourness and dry scores among all sausage batches. Compared with the thyme EO group, the thyme microcapsules group had a better odor. Besides, thyme microcapsule as an encapsulated plant extract covered up the bad smell of thyme EO and also slowed down its release. This may also be attributed to the fact that thyme microcapsules reduce the accumulation of histamine. Ardö et al. [38] reported that though lipid oxidation is the primary cascade to the generation of flavor compounds, amino acid metabolism also plays a pivotal role in development of flavor. Additionally, compared to the batches P and PO, there were higher score for hardness and structure in the batch PMT. Some reports showed that the addition of EO microcapsule to fermented products can maintain tissue hardness and prolong the food shelf-life [35,39]. Therefore, adding thyme microcapsules can effectively inhibit the accumulation of histamine in smoked horsemeat sausage and can also improve its sensory quality.

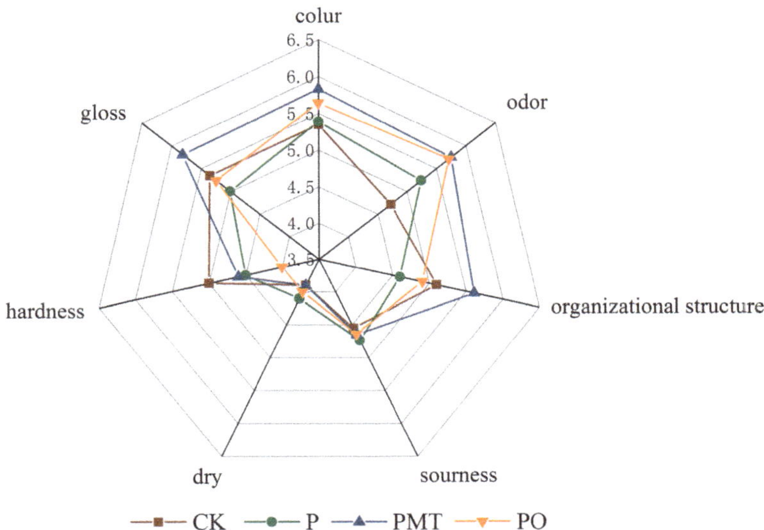

Figure 4. Sensory evaluation of smoked horsemeat sausage. Batch CK: the spontaneously fermented as the control; batch P: inoculated with *P. bacillus*; batch PMT: inoculated with *P. bacillus* and thyme microcapsules; batch PO: inoculated with *P. bacillus* and essential oil.

4. Conclusions

Herein, we demonstrated that both thyme microcapsules and thyme essential oil (EO) could diminish histamine levels, down-regulate the expression of HDC gene cluster, and repress the microbial growth, e.g., *Enterobacteria* in smoked horsemeat sausage. Accumulation of histamine is not only linked to histamine-generating microbes but also associated with the histidine decarboxylase (HDC) gene cluster expression. Thyme microcapsules and thyme EO had insignificant influence on the expression of hdcRS and hdcB; however, they remarkably suppressed the transcriptional activation of hdcA as well as hdcP genes ($p < 0.05$). The suppressive influence of thyme microcapsules was remarkably higher ($p < 0.05$) relative to that in thyme EO. Additionally, the addition of thyme microcapsules enhanced the sensory quality of the smoked horsemeat sausage. Hence, the use of thyme microcapsules not only improves the smoked horsemeat sausage sensory quality but also enhances its safety.

Author Contributions: Data curation, H.Y.; formal analysis, Y.H.; supervision, L.L., Y.L. and Z.T.; writing—original draft, H.Y.; writing—review and editing, S.L. All authors have read and agreed to the published version of the manuscript.

Funding: This work was awarded by the National Natural Science Foundation of China (Nos. 32060547), the Postgraduate Research and Innovation Project of the Autonomous Region (Nos. XJ2019G093), and the Xinjiang Production and Construction Corps Key Science and Technology Project (Nos. 2020AB012).

Institutional Review Board Statement: Not applicable.

Informed Consent Statement: Not applicable.

Data Availability Statement: The data presented in this study are available on request from the corresponding author. The data are not publicly available due to institutional privacy.

Acknowledgments: We thank the staff at the Animal Product Laboratory, College of Food, Shihezi University, for their technical as sistance.

Conflicts of Interest: The authors declare no conflict of interest.

References

1. Ladero, V.; Calles-Enríquez, M.; Fernández, M.; Álvarez, M.A. Toxicological effects of dietary biogenic amines. *Curr. Nutr. Food Sci.* **2010**, *6*, 145–156. [CrossRef]
2. Linares, D.M.; Del Río, B.; Ladero, V.; Martínez, N.; Fernández, M.; Martín, M.C.; Álvarez, M.A. Factors influencing biogenic amines accumulation in dairy products. *Front. Microbiol.* **2012**, *3*, 180. [CrossRef]
3. Zhao, L.; Xue, L.; Li, B.; Wang, Q.; Li, B.; Lu, S.; Fan, Q. Ferulic acid reduced histamine levels in the smoked horsemeat sausage. *Int. J. Food Sci. Tech.* **2018**, *53*, 2256–2264. [CrossRef]
4. Lin, C.; Tsai, H.; Lin, C.; Huang, C.; Kung, H.; Tsai, Y. Histamine content and histamine-forming bacteria in mahi-mahi (*Coryphaena hippurus*) fillets and dried products. *Food Control* **2014**, *42*, 165–171. [CrossRef]
5. Lu, S.; Ji, H.; Wang, Q.; Li, B.; Li, K.; Xu, C.; Jiang, C. The effects of starter cultures and plant extracts on the biogenic amine accumulation in traditional Chinese smoked horsemeat sausages. *Food Control* **2015**, *50*, 869–875. [CrossRef]
6. Lu, S.; Xu, X.; Shu, R.; Zhou, G.; Meng, Y.; Sun, Y.; Chen, Y.; Wang, P. Characterization of biogenic amines and factors influencing their formation in traditional Chinese sausages. *J. Food Sci.* **2010**, *75*, M366–M372. [CrossRef]
7. Bartkiene, E.; Bartkevics, V.; Mozuriene, E.; Krungleviciute, V.; Novoslavskij, A.; Santini, A.; Rozentale, I.; Juodeikiene, G.; Cizeikiene, D. The impact of lactic acid bacteria with antimicrobial properties on biodegradation of polycyclic aromatic hydrocarbons and biogenic amines in cold smoked pork sausages. *Food Control* **2017**, *71*, 285–292. [CrossRef]
8. Lucas, P.M.W.W. Histamine-Producing Pathway Encoded on an Unstable Plasmid in Lactobacillus hilgardii 0006. *Appl. Envirvron. Microb.* **2005**, *71*, 14178. [CrossRef]
9. Satomi, M.; Mori-Koyanagi, M.; Shozen, K.; Furushita, M.; Oikawa, H.; Yano, Y. Analysis of plasmids encoding the histidine decarboxylase gene in Tetragenococcus muriaticus isolated from Japanese fermented seafoods. *Fisheries Sci.* **2012**, *78*, 935–945. [CrossRef]
10. Landete, J.M.; Pardo, I.; Ferrer, S. Regulation of hdc expression and HDC activity by enological factors in lactic acid bacteria. *J. Appl. Microbiol.* **2008**, *105*, 1544–1551. [CrossRef]
11. Satomi, M.; Furushita, M.; Oikawa, H.; Yoshikawatakahashi, M.; Yano, Y. Analysis of a 30 kbp plasmid encoding histidine decarboxylase gene in Tetragenococcus halophilus isolated from fish sauce. *Int. J. Food Microbiol.* **2008**, *126*, 202–209. [CrossRef]
12. Pereira, C.I.; Crespo, M.T.; Romao, M.V. Evidence for proteolytic activity and biogenic amines production in Lactobacillus curvatus and *L. Homohiochii*. *Int. J. Food Microbiol.* **2001**, *68*, 211–216. [CrossRef]
13. Tapingkae, W.; Tanasupawat, S.; Parkin, K.L.; Benjakul, S.; Visessanguan, W. Degradation of histamine by extremely halophilic archaea isolated from high salt-fermented fishery products. *Enzyme Microb. Tech.* **2010**, *46*, 92–99. [CrossRef]
14. Wang, Y.; Li, F.; Zhuang, H.; Chen, X.; Li, L.; Qiao, W.; Zhang, J. Effects of plant polyphenols and α-tocopherol on lipid oxidation, residual nitrites, biogenic amines, and N-nitrosamines formation during ripening and storage of dry-cured bacon. *LWT* **2015**, *60*, 199–206. [CrossRef]
15. Burt, S. Essential oils: Their antibacterial properties and potential applications in foods—A review. *Int. J. Food Microbiol.* **2004**, *94*, 223–253. [CrossRef]
16. Van Haute, S.; Raes, K.; Van der Meeren, P.; Sampers, I. The effect of cinnamon, oregano and thyme essential oils in marinade on the microbial shelf life of fish and meat products. *Food Control* **2016**, *68*, 30–39. [CrossRef]
17. Hu, J.; Zhang, Y.; Xiao, Z.; Wang, X. Preparation and properties of cinnamon-thyme-ginger composite essential oil nanocapsules. *Ind. Crop. Prod.* **2018**, *122*, 85–92. [CrossRef]
18. Scacchetti, F.A.P.; Pinto, E.; Soares, G.M.B. Functionalization and characterization of cotton with phase change materials and thyme oil encapsulated in beta-cyclodextrins. *Prog. Org. Coat.* **2017**, *107*, 64–74. [CrossRef]
19. Simon-Brown, K.; Solval, K.M.; Chotiko, A.; Alfaro, L.; Reyes, V.; Liu, C.; Dzandu, B.; Kyereh, E.; Goldson Barnaby, A.; Thompson, I.; et al. Microencapsulation of ginger (*Zingiber officinale*) extract by spray drying technology. *LWT* **2016**, *70*, 119–125. [CrossRef]
20. Fernández, E.; Alegría, Á.; Delgado, S.; Martín, M.C.; Mayo, B. Comparative Phenotypic and Molecular Genetic Profiling of Wild *Lactococcus lactis* subsp.lactis Strains of the *L. lactis* subsp. lactis and *L. Lactis* subsp. cremoris Genotypes, Isolated from Starter-Free Cheeses Made of Raw Milk. *Appl. Environ. Microb.* **2011**, *77*, 5324–5335. [CrossRef]
21. Linares, D.M.; Alvarez-Sieiro, P.; Del Rio, B.; Ladero, V.; Redruello, B.; Martin, M.C.; Fernandez, M.; Alvarez, M.A. Implementation of the agmatine-controlled expression system for inducible gene expression in *Lactococcus lactis*. *Microb. Cell Fact.* **2015**, *14*, 1–12. [CrossRef]
22. Del Rio, B.; Linares, D.M.; Ladero, V.; Redruello, B.; Fernández, M.; Martin, M.C.; Alvarez, M.A. Putrescine production via the agmatine deiminase pathway increases the growth of *Lactococcus lactis* and causes the alkalinization of the culture medium. *Appl. Microbiol. Biot.* **2015**, *99*, 897–905. [CrossRef]
23. Lu, S.; Xu, X.; Zhou, G.; Zhu, Z.; Meng, Y.; Sun, Y. Effect of starter cultures on microbial ecosystem and biogenic amines in fermented sausage. *Food Control* **2010**, *21*, 444–449. [CrossRef]
24. Laranjo, M.; Gomes, A.; Agulheiro-Santos, A.C.; Potes, M.E.; Cabrita, M.J.; Garcia, R.; Rocha, J.M.; Roseiro, L.C.; Fernandes, M.J.; Fraqueza, M.J.; et al. Impact of salt reduction on biogenic amines, fatty acids, microbiota, texture and sensory profile in traditional blood dry-cured sausages. *Food Chem.* **2017**, *218*, 129–136. [CrossRef]
25. Del Rio, B.; Redruello, B.; Ladero, V.; Fernandez, M.; Martin, M.C.; Alvarez, M.A. Putrescine production by *Lactococcus lactis* subsp. Cremoris CECT 8666 is reduced by NaCl via a decrease in bacterial growth and the repression of the genes involved in putrescine production. *Int. J. Food Microbiol.* **2016**, *232*, 1–6. [CrossRef]

26. Livak, K.J.; Schmittgen, T.D. Analysis of relative gene expression data using real-time quantitative PCR and the 2(-Delta Delta C(T)) Method. *Methods* **2001**, *25*, 402–408. [CrossRef]
27. Sun, Q.; Zhao, X.; Chen, H.; Zhang, C.; Kong, B. Impact of spice extracts on the formation of biogenic amines and the physicochemical, microbiological and sensory quality of dry sausage. *Food Control* **2018**, *92*, 190–200. [CrossRef]
28. Park, J.S.; Lee, C.H.; Kwon, E.Y.; Lee, H.J.; Kim, J.Y.; Kim, S.H. Monitoring the contents of biogenic amines in fish and fish products consumed in Korea. *Food Control* **2010**, *21*, 1219–1226. [CrossRef]
29. Dansby, M.A.; Bovell-Benjamin, A.C. Sensory characterization of a ready-to-eat sweetpotato breakfast cereal by descriptive analysis. *J. Food Sci.* **2003**, *68*, 706–709. [CrossRef]
30. Bel Hadj Salah-Fatnassi, K.; Hassayoun, F.; Cheraif, I.; Khan, S.; Jannet, H.B.; Hammami, M.; Aouni, M.; Harzallah-Skhiri, F. Chemical composition, antibacterial and antifungal activities of flowerhead and root essential oils of *Santolina chamaecyparissus* L., Growing wild in Tunisia. *Saudi J. Biol. Sci.* **2017**, *24*, 875–882. [CrossRef]
31. Hsu, H.; Chuang, T.; Lin, H.; Huang, Y.; Lin, C.; Kung, H.; Tsai, Y. Histamine content and histamine-forming bacteria in dried milkfish (Chanos chanos) products. *Food Chem.* **2009**, *114*, 933–938. [CrossRef]
32. Kimura, B.; Takahashi, H.; Hokimoto, S.; Tanaka, Y.; Fujii, T. Induction of the histidine decarboxylase genes of Photobacterium damselae subsp. Damselae (formally *P. histaminum*) at low pH. *J. Appl. Microbiol.* **2009**, *107*, 485–497. [CrossRef]
33. Satomi, M.; Furushita, M.; Oikawa, H.; Yano, Y. Diversity of plasmids encoding histidine decarboxylase gene in *Tetragenococcus* spp. Isolated from Japanese fish sauce. *Int. J. Food Microbiol.* **2011**, *148*, 60–65. [CrossRef]
34. Calles-Enriquez, M.; Eriksen, B.H.; Andersen, P.S.; Rattray, F.P.; Johansen, A.H.; Fernandez, M.; Ladero, V.; Alvarez, M.A. Sequencing and transcriptional analysis of the streptococcus thermophilus histamine biosynthesis gene cluster: Factors that affect differential hdcA expression. *Appl. Environ. Microb.* **2010**, *76*, 6231–6238. [CrossRef]
35. Xiao, Z.; Kang, Y.; Hou, W.; Niu, Y.; Kou, X. Microcapsules based on octenyl succinic anhydride (OSA)-modified starch and maltodextrins changing the composition and release property of rose essential oil. *Int. J. Biol. Macromol.* **2019**, *137*, 132–138. [CrossRef]
36. Flores, M.; Toldrá, F. Microbial enzymatic activities for improved fermented meats. *Trends Food Sci. Tech.* **2011**, *22*, 81–90. [CrossRef]
37. Coloretti, F.; Chiavari, C.; Poeta, A.; Succi, M.; Tremonte, P.; Grazia, L. Hidden sugars in the mixture: Effects on microbiota and the sensory characteristics of horse meat sausage. *LWT* **2019**, *106*, 22–28. [CrossRef]
38. Ardö, Y. Flavour formation by amino acid catabolism. *Biotechnol. Adv.* **2006**, *24*, 238–242. [CrossRef]
39. Saelao, S.; Maneerat, S.; Thongruck, K.; Watthanasakphuban, N.; Wiriyagulopas, S.; Chobert, J.; Haertlé, T. Reduction of tyramine accumulation in Thai fermented shrimp (kung-som) by nisin Z-producing *Lactococcus lactis* KTH0-1S as starter culture. *Food Control* **2018**, *90*, 249–258. [CrossRef]

Article

Lipase Addition Promoted the Growth of *Proteus* and the Formation of Volatile Compounds in *Suanzhayu*, a Traditional Fermented Fish Product

Cuicui Jiang, Mengyang Liu, Xu Yan, Ruiqi Bao , Aoxue Liu, Wenqing Wang, Zuoli Zhang, Huipeng Liang, Chaofan Ji, Sufang Zhang and Xinping Lin *

National Engineering Research Center of Seafood, Collaborative Innovation Center of Provincial and Ministerial Co-Construction for Seafood Deep Processing, Liaoning Province Collaborative Innovation Center for Marine Food Deep Processing, School of Food Science and Technology, Dalian Polytechnic University, Dalian 116034, China; jcc07061013@foxmail.com (C.J.); morisaliu@hotmail.com (M.L.); yanxu8088@gmail.com (X.Y.); rickibao@sohu.com (R.B.); liuaoxue.lax@gmail.com (A.L.); wangwenqing1104@gmail.com (W.W.); zzl2632098660@gmail.com (Z.Z.); lianghp@dlpu.edu.cn (H.L.); jichaofan@outlook.com (C.J.); zhangsf@dlpu.edu.cn (S.Z.)

* Correspondence: yingchaer@1163.com or linxinping19851005@foxmail.com; Tel.: +86-0411-8631-8675

Abstract: This work investigated the effect of lipase addition on a Chinese traditional fermented fish product, *Suanzhayu*. The accumulation of lactic acid and the decrease of pH during the fermentation were mainly caused by the metabolism of *Lactobacillus*. The addition of lipase had little effect on pH and the bacterial community structure but promoted the growth of *Proteus*. The addition of lipase promotes the formation of volatile compounds, especially aldehydes and esters. The formation of volatile compounds is mainly divided into three stages, and lipase had accelerated the fermentation process. *Lactobacillus*, *Enterococcus* and *Proteus* played an important role not only in inhibition of the growth of *Escherichia-Shigella*, but also in the formation of flavor. This study provides a rapid fermentation method for the *Suanzhayu* process.

Keywords: fermented fish; *Proteus*; lipase; volatile compounds; aldehydes; esters

Citation: Jiang, C.; Liu, M.; Yan, X.; Bao, R.; Liu, A.; Wang, W.; Zhang, Z.; Liang, H.; Ji, C.; Zhang, S.; et al. Lipase Addition Promoted the Growth of *Proteus* and the Formation of Volatile Compounds in *Suanzhayu*, a Traditional Fermented Fish Product. *Foods* **2021**, *10*, 2529. https://doi.org/10.3390/foods10112529

Academic Editors: Valentina Bernini and Juliano De Dea Lindner

Received: 9 September 2021
Accepted: 15 October 2021
Published: 21 October 2021

Publisher's Note: MDPI stays neutral with regard to jurisdictional claims in published maps and institutional affiliations.

1. Introduction

Fermentation is a traditional preservation technique that provides a unique aroma and nutritional values with the action of microorganisms or endogenous enzymes. *Suanzhayu* is a type of traditional solid fermented fish in China, which is produced by mixing rice powders with seasonings and fresh fish meat in a sealed fermentation condition. Due to the metabolism of microorganisms during the fermentation process, the product develops a unique sour aroma; thus, it is popular among the consumers [1]. In recent years, the effects of a starter culture on *Suanzhayu's* quality, safety, bacterial community structure, and base flavor formation have been extensively studied [1]. However, the process still involves a long fermentation period. Moreover, it is difficult to control the changes of microbial community and flavor during fermentation, which often leads to unstable product quality. Therefore, the addition of enzymes, such as protease and lipase to shorten the fermentation maturation process has attracted the attention of many researchers [2,3].

Lipase increases the content of free fatty acids through hydrolysis reactions. Free fatty acids can be converted to volatiles such as aldehydes and ketones by microbial enzymatic or non-enzymatic reactions in the system [4], which increase the flavor of the product. Many papers have reported that lipases are associated with the formation of characteristic food flavors. For example, the lipolysis of lipase is essential for the characteristic flavor of food, such as that in the mature white cheese flavor [5]. Researchers found out that lipase addition improved the desirable dairy volatile compounds, such as butyric acid, hexanoic acid, 2-nonanone, and hexanal resulting from lipolysis in milk [6]. In addition, it has been reported that lipase addition can promote the formation of antibacterial flavor substances, such as octanal and nonanal, thereby inhibiting the growth of citrus *Geotrichum citri-aurantii* mycelium [7]. However, thus far, there are no studies on the relationship between lipase on bacterial community succession, physicochemical indicators and changes in aroma properties of *Suanzhayu*.

Based on the previous studies, we hypothesized that lipase addition might improve the aromas, such as aldehydes and esters during *Suanzhayu* fermentation. Therefore, in this study, the effect of lipase addition on *Suanzhayu* fermentation was investigated. The correlation between bacterial community succession and the changes of *Suanzhayu* product properties, such as pH, lactic acid content, and volatile components, were analyzed and discussed.

2. Materials and Methods

2.1. Preparation of Suanzhayu Samples

Suanzhayu samples were purchased from a local Qianhe Market (Dalian, Liaoning, China) using fresh carp (*Cyprinus carpio* L.) with an average weight of 2 ± 0.2 kg. The fish were gutted, cleaned, and cut into cubes of 2.5 × 2.5 × 2.5 cm.

The lipase addition is based on the following reasons. First, the lipase activity (PANGBO ENZYME, Nanning, China) is 100,000 U/g, and the recommended dose is 0.1 to 0.3%, i.e., 100 U/g to 300 U/g. Second, 100 to 300 U/g is a common addition in fermented products [8]. For example, Rani and Jagtap reported that 200 U/g of lipase was added to cheese, and the cheese maturation time was shortened from 90 days to 60 days while maintaining its quality [8]. Finally, different lipase addition at 100, 200 and 300 U/g were prepared, and the pre-experiment showed that the effect of adding 100 U/g fish meat of lipase was of good acceptance. Therefore, we decided to add 100 U/g lipase in *Suanzhayu*. Two groups were prepared: (i) group U100, in which 100 U/g fish meat of lipase was added; and (ii) group U0, in which the same volume of sterile water was added. Each fermented jar was added with 200 ± 5 g of fish meat, mixed with 3% salt, 30% rice flour and the enzyme solution. Then, the mixtures were placed in completely sealed jars and kept at 25 °C for 14 d. Samples was performed in triplicate from independent jars incubated for 0, 3, 5, 7, and 14 d. Samples for volatile compounds and microbiological analysis were kept at −80 °C and samples for other analysis were stored at −20 °C. Subsequent analysis was carried out as soon as possible within two months.

2.2. Determination of pH and Lactic Acid Contents

Sterile water (20 mL) was added into 2 g of sample, and the mixture was homogenized (4 × 15 s at 8000 rpm), (T25 digital ULTRA TURRAX®, IKA, Staufen, Germany). The supernatant was then subjected to pH measurement using a pH meter (FE28, Mettler Toledo, Greifensee, Switzerland). The lactic acid content was determined according to the previous method [9]. Briefly, 0.1 g sample was thoroughly minced with 10 mL sterile water. Lactic acid was extracted by ultrasonic for 10 min. The supernatant was filtered through a 0.45 μm membrane filter, following by a Cleaner SC18 SPE column. The lactic acid content was determined with high-performance ion chromatography (Dionex ICS-5000 + DC, Thermo Scientific, Waltham, MA, USA) with the same chromatographic condition described by Lv et al. [9]. All the experiments were carried out in triplet and expressed as the mean ± standard error.

2.3. Analysis of Bacterial Community

Suanzhayu was aseptically sampled from the fermented jar and stored at −80 °C, and then samples were sent to the Biomarker Bioinformatics Technology Co., Ltd. (Beijing, China) for 16S rRNA gene amplicon. The total genomic DNA was extracted from a 0.25 g sample with EZNA® DNA kit (Omega, Norcross, GA, USA). The hypervariable regions V3 and V4 of 16S rRNA were amplified using special primers 338F (5′-ACTCCTACGGGAGGCAGCA-3′) and 806R (5′-GGACTACHVGGGTWTCTAAT-3′). After amplification and purification of the fragments, the gene library was established using DNA Library Preparation Kit (Ade Technology Co., LTD, Beijing, China) and was subjected to sequencing using an Illumina HiSeq 2500 platform.

Paired-end reads were assembled using FLASH version 1.2.7 to obtain raw tags, which were then filtered based on quality using Trimmomatic version 0.33. The chimera tags were then removed using UCHIME version 4.2, and the effective tags were obtained. The effective tags with a similarity of more than 97% were clustered, and operational taxonomic units (OTUs) were determined using UPARSE software. For each OTU, a representative sequence with a high frequency of occurrence is classified by searching the Silva database using the Ribosomal Database Project (RDP) classifier version 2.2. The raw data have been submitted to the National Centre for Biotechnology Information (NCBI) website (Accession number: SRP230170).

2.4. Volatile Compounds Analysis by HS-SPME-GC-MS

Volatile compounds in the sample were determined by HS-SPME-GC-MS (Agilent Technologies, Santa Clara, CA, USA) equipped with HP-5MS capillary column (30 m $\times$ 250 μm $\times$ 0.25 μm) (Agilent, USA). *Suanzhayu* was steamed for 20 min and then minced thoroughly. Samples (2 g) were placed in a headspace extraction vial (20 mL, 18 mm) and cyclohexanone (Aladdin, American) (50 mg/L) were added to each sample (0–3 d 20 μL, 5–7 d 40 μL, 14 d 60 μL) as the internal standard. The volatile compounds were extracted with a solid-phase micro-extraction needle with divinylbenzene/Carboxen/polydimethylsiloxane (PDMS) fiber for 40 min. The GC-MS parameters were set according to Bao's study. The volatile compounds were thermally desorbed at 250 °C for 5 min. The initial oven temperature kept at 30 °C for 5 min, and then reached to 50 °C at a rate of 3 °C/min and kept for 3 min. After that, the oven temperature was raised to 150 °C at a rate of 5 °C/min, following by raising to 250 °C at a rate of 20 °C/min (held for 5 min). The ionization source with the energy of 70 eV at 230 °C was used, and the mass scan range was 40–400 mass unit with an emission current of 150 μA. The retention index (RI) of n-alkanes (C7–C30, Sigma-Aldrich, St. Louis, MO, USA) under the same GC conditions as Bao's study was calculated [10], and the volatile compounds were identified by comparing the calculated RI and the mass spectra of fragments in the NIT14 library. The peak area of each compound was used to semi-quantify the concentration of volatile compounds in the sample. Then, the odor activity value (OAV) was calculated based on OAV = C/OT, where C was the concentration of volatile compounds, and OT was the odor threshold, which was obtained from the literature [11]. All samples were conducted in triplicate.

2.5. Statistical Analysis

Statistical analysis was analyzed by one-way ANOVA and Spearman correlation using SPSS (version 22.0, IBM, Armonk, NY, USA) software. When $p < 0.05$, the data was considered to have a significant difference. The histograms were conducted using Origin 8.5 (Origin Lab Corp., Northampton, MA, USA). The heat maps were carried out by using R Studio (version 3.4.4). The heat maps of Pearson correlation coefficient were performed by TBtools (version 0.6652).

3. Results and Discussion

3.1. Changes of pH and Lactic Acid Contents during Fermentation

The changes in pH and lactic acid content were shown in Figure 1. The pH dropped rapidly in the first 7 d and slowed to 4.4 at the end of the fermentation. According to Zeng's study, *Suanzhayu* with a pH lower than 4.6 is generally considered as mature [12,13]; thus, both of the groups almost went on the same fermentation process. Additionally, the lactic acid content gradually increased from 2.7 to 7.6 and 6.8 mg/kg in U0 and U100 groups, respectively. The culture pH decreases with the increase in the concentration of lactic acid. This accumulation of lactic acid during the fermentation process can inhibit the growth of spoilage and pathogenic bacteria, thereby extending the shelf life of the product [14]. However, the lactic acid content of the U100 group at the end of fermentation was lower than that of the U0 group. Sun [15] and Knez [16] reported that lipase can catalyze the esterification of lactic acid and other fatty acids with alcohol. Therefore, the lower lactic acid content in U100 group might be due to the consumption of lactic acid by lipase-catalyzed esterification. This may also provide the possibility of flavor enhancement in the enzyme addition group. In all, the results indicate that the addition of lipase had no effect on the pH of *Suanzhayu*.

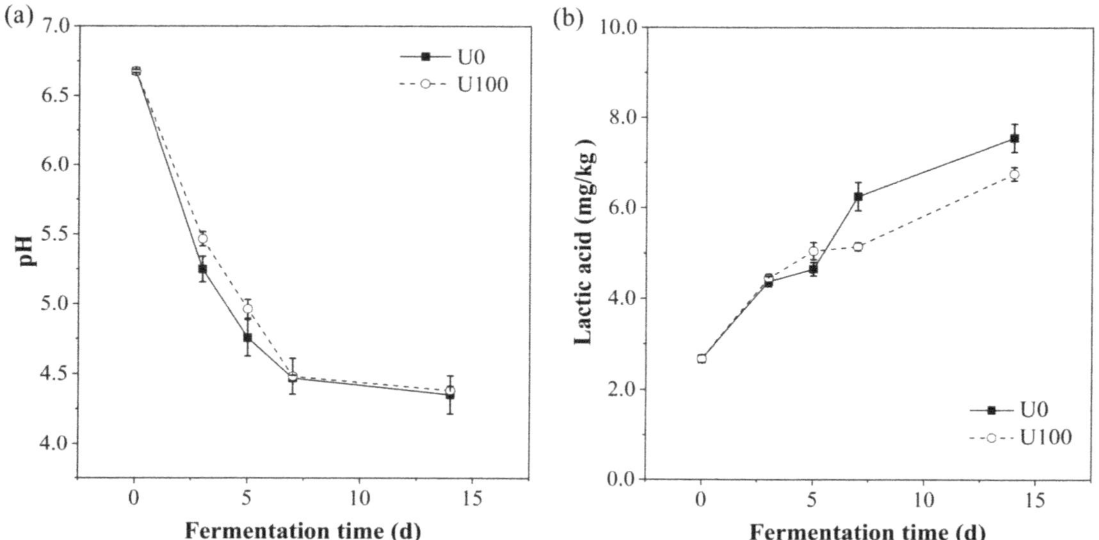

Figure 1. Changes of pH (**a**) and lactic acid content (**b**) in the samples during the fermentation of *Suanzhayu* without (U0) and with (U100) lipase. Samples were collected at 0, 3, 5, 7, and 14 d, respectively.

3.2. Microbial Succession during Fermentation

The sequencing coverage rate was above 99.8% (Supplementary Table S1), suggesting that the sequencing depth was sufficient to reflect the composition of the bacterial ecosystem in *Suanzhayu* samples. The sequences with a similarity of above 97% were clustered, resulting in 383 OTUs.

As for the phylum level (Figure 2a), at day 0, Firmicutes, Proteobacteria, and Bacteroidetes were the dominant phyla with a relative abundance of 28.8%, 26.3%, and 19.4%, respectively. With the extension of fermentation time, Firmicutes and Proteobacteria rapidly grew, and their relative abundance increased to more than 98.0%, which inhibited the growth of other phyla. Firmicutes in the U0 group became the only dominant phylum in the final product (71.42%). In contrast, the dominant bacteria in the final product of the U100 group were Firmicutes (58.91%) and Proteobacteria (40.39%). The results suggest that the addition of lipase could influence the distribution of bacterial communities in *Suanzhayu* at phylum level.

At the genus level (Figure 2b), *Escherichia-Shigella* was the dominant bacterium on day 0 with a relative abundance of 17.1%. *Escherichia-coli* is a spoilage bacterium found in most fermented foods. It can lead to the formation of harmful biogenic amines in fermented meat and fish products [17]. The relative abundance of *Lactobacillus* increased rapidly from 3 days of fermentation, from 2.08% to 25.35% (U0) and 31.64% (U100), respectively. As the fermentation proceeded, the abundance of *Lactobacillus* remained basically stable, and became the dominant genus. Compared with the U0, the abundance of *Proteus* in U100 increased significantly ($p < 0.01$), especially after 7 days of fermentation, reaching 17.69%.

It was inferred that the addition of lipase could promote the growth of *Proteus*. *Proteus*' growth was reported to facilitate flavor formation. In a cheese model, *Proteus vulgaris* was found to be closely related with 3-methyl-1-butanol and more volatile aroma substances, such as aldehydes and esters, were detected [18].

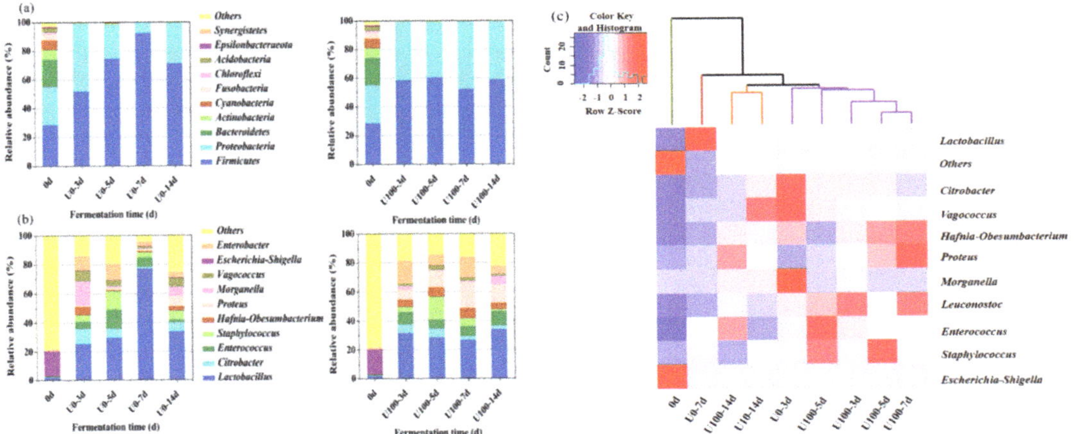

Figure 2. The relative abundances of bacteria at the phylum level (**a**) and the genus level (**b**) during the fermentation of *Suanzhayu* without (U0) and with (U100) lipase. (**c**) Heat map of the bacterial community compositions during the *Suanzhayu* fermentation without (U0) and with (U100) lipase. The colors indicate low (blue) to high (red).

The cluster analysis of the two groups was shown in Figure 2c. According to the results, *Suanzhayu* samples could be divided into three categories: (1) samples 0 d and U0-7 d; (2) samples U03-5 d and U100-3-7 d; and (3) samples U0-14 d and U100-14 d. The above results show that the fermentation process of the *Suanzhayu* sample mainly goes through three stages, early fermentation, middle fermentation and late fermentation. In addition, from the clustering results of U0 group and U100 group, it can be seen that the three stages of the two groups are relatively synchronized.

3.3. Volatile Components Generated during Fermentation

A total of 21 volatile compounds were detected in *Suanzhayu* samples (Table S2). Only 11 volatile compounds were found in the day 0 samples, with a total of 9327.4 μg/kg of volatile flavor compounds, indicating that the rest of the compounds were produced during fermentation. The content of volatile flavor compounds continued to accumulate as fermentation proceeded. At the end of fermentation, the content of volatile compounds in the U100 group (18,196.9 μg/kg) was significantly higher than that in the U0 group (15,428.5 μg/kg). Moreover, the contents of aldehydes and esters were significantly increased in the U100 group, which were 1.4 and 11.1 times higher than those in the U0 group, respectively. The above results indicate that the addition of lipase for fermentation could promote the formation of volatile substances (especially aldehydes and esters) in *Suanzhayu*.

Alcohols were one of the important components of volatile compounds contributing to the aroma of *Suanzhayu*. Only two alcohols, 1-hexanol and 1-octene-3-alcohols, were detected in unfermented samples. This suggests that most of the volatile alcohols detected in *Suanzhayu* were produced by fermentation. During the fermentation process, both of the content (Table S2) and percentage (Figure S1) of alcohols in the U0 group continued to increase, while the U100 group showed a trend of first increase and then decrease. This may be the result of the conversion of alcohols and acids (e.g., fatty acids, etc.) into esterification by microorganisms in the system [19]. This also coincided with the increase in the content of ethyl ester hexanoic acid and ethyl ester octanoic acid in the U100 group (Table S2). Some specific alcohols, for example, 3-methyl-1-butanol, was found to be 3.5 times higher in the U100 group than in the U0 group. It has been reported that 3-methyl-1-butanol is produced by the auto-oxidation of unsaturated fatty acids, which plays an important role in the faint aroma (e.g., mushroom or metallic taste) of fresh seafood [20]. Therefore, the addition of lipase may have increased the fatty acids, precursors of 3-methyl-1-butanol, and thus played a positive role in the formation of product aroma. At the end of fermentation, there was no significant difference ($p < 0.05$) in the alcohol content between U0 and U100 samples, which were 6857.8 μg/kg and 6870.9 μg/kg, respectively, indicating that the addition of lipase had little effect on the total alcohol content of *Suanzhayu*. Therefore, the addition of lipase had little effect on the total amount of alcoholic substances in *Suanzhayu* but promoted its characteristic flavor substances, such as 3-methyl-1-butanol.

Aldehydes were also one of the main volatile components of *Suanzhayu*. At the end of fermentation, the total content of aldehydes in the U100 group (9570 μg/kg) was higher than that in the U0 group (7002 μg/kg), indicating that the accumulation of aldehydes was promoted by the addition of lipase. Benzaldehyde, pentanal, hexanal, heptanal, and octanal were significantly increased 1.1, 1.5, 1.4, 1.1 and 2.2 times, respectively, in the U100 group compared to the U0 group. These aldehydes were reported as the main compounds of the typical fresh fish flavor [21]. Benzaldehyde offers almond flavor and typical mushroom flavor [22]. Pentanal, hexanal and octanal have an odor similar to the aroma of fat and green [23,24]. Heptanal and octanal were considered to as important flavor compounds in fermented sausages [25]. Pentanal, hexanal, heptanal, and octanal were fatty aldehydes, which were oxidation products of unsaturated fatty acids. The addition of exogenous lipase could promote the release of free fatty acids (FFAs). The degradation and oxidation of lipids promotes the production of aldehydes, creating the characteristic fatty and grassy taste of *Suanzhayu*.

Esters have a low threshold and produce fruit and flower odors [26]. They were generally result from the esterification of short-chain acids and alcohols. Only two esters, ethyl ester hexanoic acid and ethyl ester octanoic acid, were detected in the samples. At the beginning of the fermentation (day 0), no esters were observed. As fermentation proceeded, esters started to appear at day 5 and gradually accumulated. At the end of fermentation, esters in the U100 group (276.1 μg/kg) were 11 times higher than those in the U0 group (24.7 μg/kg). Lipase catalyzes the hydrolysis of triglycerides et al. [27], producing large amounts of FFAs. With microbial esterase, FFA esterifies with alcohols to produce esters, such as ethyl ester hexanoic acid and ethyl ester octanoic acid in this study, providing a special fruity and ester flavors for the product [28]. These results suggested that the addition of lipase could promote the production of esters and promote the formation of fruity and ester odors in *Suanzhayu*.

The odor activity value was calculated to assess the contribution of volatile compounds to *Suanzhayu*. Thirteen volatiles with OAV > 1 were considered to contribute significantly to the characteristic flavor of the product (Table 1). Octanal had the largest OAV value. Octanal was reported to provide a grassy flavor to the product [24]. In addition, pentanal, hexanal, nonanal and 1-hexanol also have relatively large OAV values (OAV > 1500), and these substances may contribute to the formation of fat, grass, mushroom, and flower aromas in *Suanzhayu* [24,29]. The main odor active compounds were similar in both groups, indicating that the addition of lipase did not change the main odor profile of the product. In addition, the OAV values of most substances in the U100 group were higher than those in the U0 group, indicating that the addition of lipase elevated the concentration of volatile compounds (OAV > 1) that could be perceived, i.e., enhanced the flavor of *Suanzhayu*.

Table 1. Concentration and OAV values of odor-active compounds in *Suanzhayu* (fermentation 14 day) with (U100) and without (U0) lipase.

Compound	Threshold Value (μg/Kg)	Content (μg/kg)		OAV*10		Significance
		U0	U100	U0	U100	
pentanal	0.019	437.2	671.2	2300.8	3532.5	*
hexanal	3.3	4992.6	6931.7	151.3	210.1	*
heptanal	0.011	319.3	359.4	2902.6	3267.5	
benzaldehyde	0.85	230.8	253.5	27.2	29.8	
octanal	0.0005	387.0	839.3	77,397.7	167,855.4	**
nonanal	0.02	391.7	300.1	1958.5	1500.7	*
3-methyl-1-butanol	28	546.2	1903.9	2.0	6.8	
1-pentanol	3.6	303.6	259.2	8.4	7.2	
1-Hexanol	0.25	3938.3	2795.9	1575.3	1118.4	*
1-octene-3-ol	0.03	1829.7	1794.1	6776.7	6644.9	
ethyl ester hexanoic acid	1	24.7	199.9	2.5	20	**
ethyl ester octanoic acid	40	0.0	76.2	0.0	0.2	**
2-pentyl furan	6	447.9	438.3	7.5	7.3	

Note: * stands for significant difference ($p < 0.05$), and ** stands for extremely significant difference ($p < 0.01$).

Hierarchical clustering analysis was performed with volatile compounds among different groups during the fermentation process and shown in Figure 3. The samples were divided into three categories: (1) 0d, U0–3d and U0–5d, the first stage of fermentation; (2) U0–7d and U100 group 3–5d, the second stage of fermentation; (3) U0–14d and U100 group 7–14d, the final stage of fermentation. The sample fermented for 7 days in the lipase addition group reached the flavor profile of the 14 d sample in the natural fermentation group, which could be considered as the maturation stage. Therefore, the addition of lipase might contribute to the formation of volatile substances. Similar reports had been reported in *Chouguiyu*, in which the volatile profile LW5 (inoculated with *Lactococcus lactis* M10 and *Weissella cibaria* M3 and fermented from 5 days 5) was the most similar to C7 (naturally fermented for 7 days) [10]. Therefore, the starter culture inoculation could promote the flavor formation in *Chouguiyu* products with time shortened. Ansorenaetal et al. also reported that the addition of lipase and protease reduced the fermentation time of dry sausages from 35 to 21 d [2]. Thus, lipase addition could be used as a method to promote the formation of volatile substances in *Suanzhayu*.

3.4. Correlation between Bacterial Community and Volatile Compounds, pH, and Lactic Acid Content

The correlations of the bacterial community with volatile compounds (OAV > 1), pH and lactic acid content were calculated and displayed in Figure 4. *Lactobacillus* showed a negative correlation with pH ($R = -0.645$) ($p < 0.05$) and a positive correlation with lactic acid content ($R = 0.611$) ($p < 0.05$), indicating that the accumulation of lactic acid and the decrease in pH were mainly caused by *Lactobacillus*. In contrast, *Escherichia-Shigella* had a positive correlation with pH ($R = -0.849$) ($p < 0.01$) and a negative correlation with lactic acid content ($R = -0.653$) ($p < 0.05$). *Escherichia-Shigella* was also found to be negatively correlated with *Enterococcus* ($R = -0.619$) ($p < 0.05$) and *Proteus* ($R = -0.565$) ($p < 0.05$), indicating that *Lactobacillus*, *Enterococcus* and *Proteus* had an inhibitory effect on the growth of *Escherichia-Shigella*, a typical spoilage bacterium [30]. *Lactobacillus*, *Enterococcus* and *Proteus* were important microorganisms for the safety of *Suanzhayu*, probably by producing acid and lowering the pH value, thus inhibiting of *Escherichia-Shigella*.

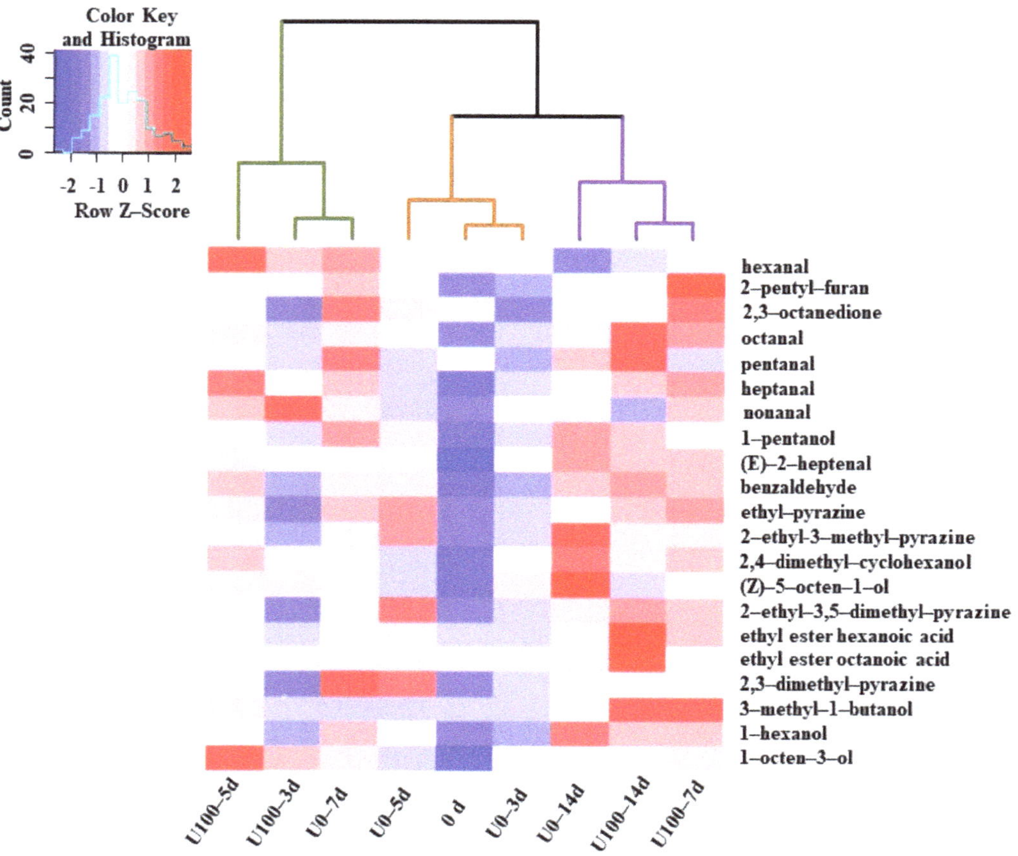

Figure 3. The heat map of volatile compositions during the fermentation *Suanzhayu* samples without (U0) and with (U100) lipase. The colors indicate low (**blue**) to high (**red**).

For flavor, *Lactobacillus*, *Enterococcus* and *Proteus* showed significant positive correlations with the 13 compounds (OAV > 1, Table 1), suggesting that these strains played an important role in the production of the characteristic compounds. These results were similar to the report in sourdough fermentation, that the group added with *Lactobacillus plantarum* had the highest content of heptanal [31]. In another study, 1-pentanol was produced during growth of all LAB species in sourdough [32]. In the fermented fish sausages, *Enterococcus* played a significant role in the formation of flavor characteristics [33], such as hexanal, heptanal, octanal, benzaldehyde, etc. *Proteus* species were reported to play important roles during cheese ripening [18], especially in the formation of flavor substances, such as 3-methyl-1-butanol, ethyl ester hexanoic acid, etc. Deetae et al. also found in the cheese model that *Proteus* was closely related to aldehyde content and played an important role in the formation of 3-methyl-1-butanol [18]. The above correlation analysis results showed that *Lactobacillus*, *Enterococcus* and *Proteus* played significant roles not only in product safety, but also in the formation of product flavor.

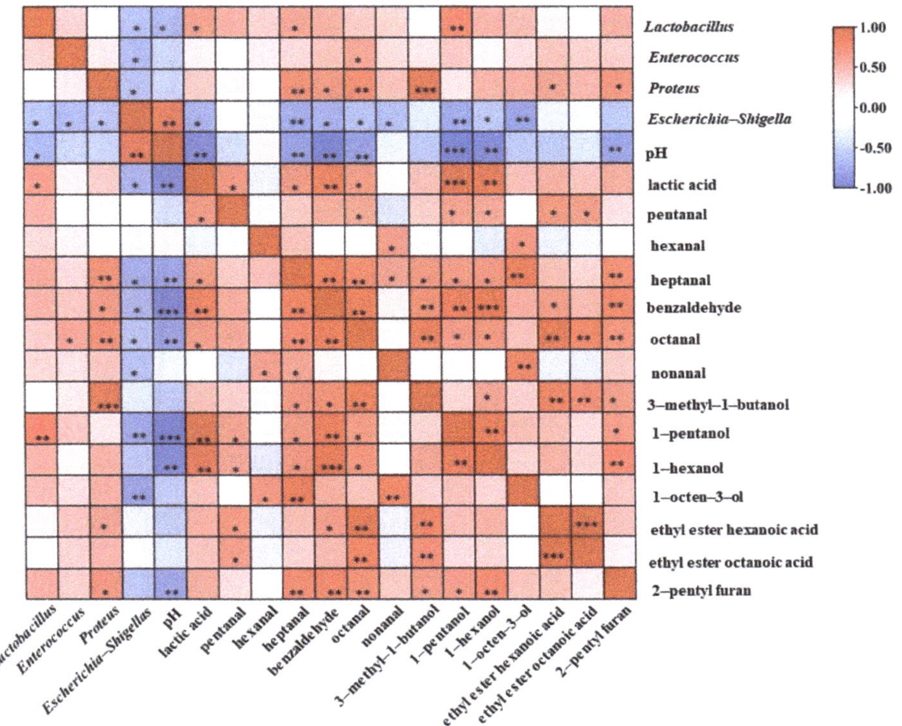

Figure 4. Correlations between the bacterial communities with volatile compounds (OAV > 1), pH and lactic acid in *Suanzhayu* samples. Red indicates positive correlations, while blue indicates negative correlations. Significant at: * ($p < 0.05$); ** ($p < 0.01$); *** ($p < 0.001$).

4. Conclusions

The addition of lipase had no significant effect on the pH of the *Suanzhayu*. The addition of lipase promoted the growth of *Proteus* and the formation of volatile substances (especially aldehydes and esters). The correlation analysis showed that *Lactobacillus*, *Enterococcus* and *Proteus* played an important role not only in the safety of the product but also in the formation of flavor. Addition of lipase could be used as a novel means to enhance the quality of *Suanzhayu*.

Supplementary Materials: The following are available online at https://www.mdpi.com/article/10.3390/foods10112529/s1, Figure S1: The percentage of volatile compounds during the fermentation *Suanzhayu* samples without (U0) and with (U100) lipase, Table S1: Alpha diversity indices of bacteria in *Suanzhayu* with and without lipase, Table S2: Concentration of volatile compounds in *Suanzhayu* with or without lipase.

Author Contributions: Methodology, C.J. (Cuicui Jiang); investigation, C.J. (Cuicui Jiang); writing—original draft preparation, C.J. (Cuicui Jiang); writing—review and editing, M.L. and X.Y.; investigation, R.B. and A.L.; data curation, W.W.; visualization, Z.Z. and H.L.; Validation, C.J. (Chaofan Ji); formal analysis, S.Z.; funding acquisition, X.L. All authors have read and agreed to the published version of the manuscript.

Funding: This research was funded by the National Natural Science Foundation of China, grant number 31972204, 32072289 and the Natural Science Foundation of Liaoning Province, grant number 2020-MS-277.

Data Availability Statement: The data that support the findings of this study are available from the corresponding author upon reasonable request.

Conflicts of Interest: The authors declare that there is no conflict of interest.

References

1. Xu, Y.; Li, L.; Mac Regenstein, J.; Gao, P.; Zang, J.; Xia, W.; Jiang, Q. The contribution of autochthonous microflora on free fatty acids release and flavor development in low-salt fermented fish. *Food Chem.* **2018**, *256*, 259–267. [CrossRef]
2. Ansorena, D.; Astiasarán, I.; Bello, J. Influence of the Simultaneous Addition of the Protease Flavourzyme and the Lipase Novozym 677BG on Dry Fermented Sausage Compounds Extracted by SDE and Analyzed by GC-MS. *J. Agric. Food Chem.* **2000**, *48*, 2395–2400. [CrossRef]
3. Yang, J.; Jiang, C.; Bao, R.; Liu, M.; Lin, X. Effects of flavourzyme addition on physicochemical properties, volatile com-pound components and microbial community succession of Suanzhayu. *Int. J. Food Microbiol.* **2020**, *334*, 108839. [CrossRef]
4. Suzuki, T.; Kim, S.-J.; Mukasa, Y.; Morishita, T.; Noda, T.; Takigawa, S.; Hashimoto, N.; Yamauchi, H.; Matsuura-Endo, C. Effects of lipase, lipoxygenase, peroxidase and free fatty acids on volatile compound found in boiled buckwheat noodles. *J. Sci. Food Agric.* **2010**, *90*, 1232–1237. [CrossRef]
5. Kendirci, P.; Salum, P.; Bas, D.; Erbay, Z. Production of enzyme-modified cheese (EMC) with ripened white cheese flavour: II-effects of lipases. *Food Bioprod. Process.* **2020**, *122*, 230–244. [CrossRef]
6. Wang, B.; Xu, S. Effects of different commercial lipases on the volatile profile of lipolysed milk fat. *Flavour Fragr. J.* **2009**, *24*, 335–340. [CrossRef]
7. Zhou, H.; Tao, N.; Jia, L. Antifungal activity of citral, octanal and α-terpineol against Geotrichum citri-aurantii. *Food Control.* **2014**, *37*, 277–283. [CrossRef]
8. Rani, S.; Jagtap, S. Acceleration of Swiss cheese ripening by microbial lipase without affecting its quality characteristics. *J. Food Sci. Technol.* **2019**, *56*, 497–506. [CrossRef] [PubMed]
9. Lv, J.; Yang, Z.; Xu, W.; Li, S.; Liang, H.; Ji, C.; Yu, C.; Zhu, B.; Lin, X. Relationships between bacterial community and metabolites of sour meat at different temperature during the fermentation. *Int. J. Food Microbiol.* **2019**, *307*, 108286. [CrossRef]
10. Bao, R.; Liu, S.; Ji, C.; Liang, H.; Yang, S.; Yan, X.; Zhou, Y.; Lin, X.; Zhu, B. Shortening Fermentation Period and Quality Improvement of Fermented Fish, Chouguiyu, by Co-inoculation of Lactococcus lactis M10 and Weissella cibaria M3. *Front. Microbiol.* **2018**, *9*, 3003. [CrossRef] [PubMed]
11. Van Gemert, L.J. *Compilations of Odour Threshold Values in Air, Water and Other Media*, 2nd ed.; Oliemans Punter&Partners BV: Huizen, The Netherlands, 2011.
12. Zeng, X.; Xia, W.; Yang, F.; Jiang, Q. Changes of biogenic amines in Chinese low-salt fermented fish pieces (Suan yu) inoculated with mixed starter cultures. *Int. J. Food Sci. Technol.* **2013**, *48*, 685–692. [CrossRef]
13. Zeng, X.; Zhang, W.; Zhu, Q. Effect of starter cultures on the quality ofSuan yu, a Chinese traditional fermented freshwater fish. *Int. J. Food Sci. Technol.* **2016**, *51*, 1774–1786. [CrossRef]
14. Özyurt, G.; Gökdoğan, S.; Şimşek, A.; Yuvka, I.; Ergüven, M.; Boga, E.K. Fatty acid composition and biogenic amines in acidified and fermented fish silage: A comparison study. *Arch. Anim. Nutr.* **2015**, *70*, 72–86. [CrossRef]
15. Sun, J.; Jiang, Y.; Zhou, L.; Gao, J. Optimization and kinetic study of immobilized lipase-catalyzed synthesis of ethyl lactate. *Biocatal. Biotransformation* **2010**, *28*, 279–287. [CrossRef]
16. Knez, Ž.; Kavčič, S.; Gubicza, L.; Bélafi-Bakó, K.; Németh, G.; Primožič, M.; Habulin, M. Lipase-catalyzed esterification of lactic acid in supercritical carbon dioxide. *J. Supercrit. Fluids* **2012**, *66*, 192–197. [CrossRef]
17. Anastasio, A.; Draisci, R.; Pepe, T.; Mercogliano, R.; Quadri, F.D.; Luppi, G.; Cortesi, M.L. Development of Biogenic Amines during the Ripening of Italian Dry Sausages. *J. Food Prot.* **2010**, *73*, 114–118. [CrossRef]
18. Deetae, P.; Bonnarme, P.; Spinnler, H.E.; Helinck, S. Production of volatile aroma compounds by bacterial strains isolated from different surface-ripened French cheeses. *Appl. Microbiol. Biotechnol.* **2007**, *76*, 1161–1171. [CrossRef]
19. Ziadi, M.; Bergot, G.; Courtin, P.; Chambellon, E.; Hamdi, M.; Yvon, M. Amino acid catabolism by Lactococcus lactis during milk fermentation. *Int. Dairy J.* **2010**, *20*, 25–31. [CrossRef]
20. Zhang, Z.; Li, G.; Luo, L.; Chen, G. Study on seafood volatile profile characteristics during storage and its potential use for freshness evaluation by headspace solid phase microextraction coupled with gas chromatography–mass spectrometry. *Anal. Chim. Acta* **2010**, *659*, 151–158. [CrossRef]
21. Lindsay, R.C. *Flavour of Fish. Seafoods: Chemistry, Processing Technology and Quality*; Springer: Berlin/Heidelberg, Germany, 1994.
22. Xiao, Y.; Liu, Y.; Chen, C.; Xie, T.; Li, P. Effect of Lactobacillus plantarum and Staphylococcus xylosus on flavour development and bacterial communities in Chinese dry fermented sausages. *Food Res. Int.* **2020**, *135*, 109247. [CrossRef]
23. Mallia, S.; Escher, F.; Dubois, S.; Schieberle, P.; Schlichtherle-Cerny, H. Characterization and Quantification of Odor-Active Compounds in Unsaturated Fatty Acid/Conjugated Linoleic Acid (UFA/CLA)-Enriched Butter and in Conventional Butter during Storage and Induced Oxidation. *J. Agric. Food Chem.* **2009**, *57*, 7464–7472. [CrossRef]
24. Zhang, Y.; Fraatz, M.A.; Müller, J.; Schmitz, H.-J.; Birk, F.; Schrenk, D.; Zorn, H. Aroma Characterization and Safety Assessment of a Beverage Fermented by Trametes versicolor. *J. Agric. Food Chem.* **2015**, *63*, 6915–6921. [CrossRef]

25. Wang, Y.; Liu, R.; Ge, Q.; Wu, M.; Xu, B.; Xi, J.; Yu, H. Effects of branched-chain amino acids and Lactobacillus plantarum CGMCC18217 on volatiles formation and textural properties of dry-cured fermented sausage. *Int. J. Food Sci. Technol.* **2020**, *56*, 374–383. [CrossRef]
26. Flores, M.; Durá, M.A.; Marco, A.; Toldrá, F. Effect of debaryomyces spp. on aroma formation and sensory quality of dry-fermented sausages. *Meat Sci.* **2004**, *68*, 439–446. [CrossRef]
27. Maddikeri, G.L.; Pandit, A.B.; Gogate, P.R. Intensification Approaches for Biodiesel Synthesis from Waste Cooking Oil: A Review. *Ind. Eng. Chem. Res.* **2012**, *51*, 14610–14628. [CrossRef]
28. Marušić, N.; Vidaček, S.; Janči, T.; Petrak, T.; Medić, H. Determination of volatile compounds and quality parameters of traditional Istrian dry-cured ham. *Meat Sci.* **2014**, *96*, 1409–1416. [CrossRef]
29. De Schutter, D.P.; Saison, D.; Delvaux, F.; Derdelinckx, G.; Rock, J.-M.; Neven, H.; Delvaux, F.R. Characterization of Volatiles in Unhopped Wort. *J. Agric. Food Chem.* **2008**, *56*, 246–254. [CrossRef]
30. Zhao, D.; Lu, F.; Qiu, M.; Ding, Y.; Zhou, X. Dynamics and Diversity of Microbial Community Succession of Surimi During Fermentation with Next-Generation Sequencing. *J. Food Saf.* **2015**, *36*, 308–316. [CrossRef]
31. Liu, T.; Li, Y.; Yang, Y.; Yi, H.; Zhang, L.; He, G. The influence of different lactic acid bacteria on sourdough flavor and a deep insight into sourdough fermentation through RNA sequencing. *Food Chem.* **2020**, *307*, 125529. [CrossRef]
32. Kaseleht, K.; Paalme, T.; Mihhalevski, A.; Sarand, I. Analysis of volatile compounds produced by different species of lactobacilli in rye sourdough using multiple headspace extraction. *Int. J. Food Sci. Technol.* **2011**, *46*, 1940–1946. [CrossRef]
33. Yue, Z.A.; Yw, A.; Cla, B.; Lla, B.; Xya, B.; Ywa, B.; Sca, B.; Yza, B. Novel insight into physicochemical and flavor formation in naturally fermented tilapia sausage based on microbial metabolic network. *Food Res. Int.* **2021**, *141*, 110122.

Article

Solid-State Fermentation of *Arthrospira platensis* to Implement New Food Products: Evaluation of Stabilization Treatments and Bacterial Growth on the Volatile Fraction

Francesco Martelli, Martina Cirlini *, Camilla Lazzi, Erasmo Neviani and Valentina Bernini

Department of Food and Drug, University of Parma, Parco Area delle Scienze 49/A, 43124 Parma, Italy; francesco.martelli@studenti.unipr.it (F.M.); camilla.lazzi@unipr.it (C.L.); erasmo.neviani@unipr.it (E.N.); valentina.bernini@unipr.it (V.B.)
* Correspondence: martina.cirlini@unipr.it

Abstract: *Arthrospira platensis* is a cyanobacterium widely used in food formulation and mainly consumed as a food supplement because of its high amount of proteins, vitamins and minerals. Different probiotic food supplements are present in the market, and a lactic acid fermented food product like dried spirulina could be useful not only to introduce lactic acid bacteria (LAB) with beneficial effects to the diet of consumers, but also to improve or change the aromatic profile of the substrate. Therefore, the aim of this study was the evaluation of lactic acid fermentation of *A. platensis* biomass, focusing on the consequent changes in the aromatic profile. For this purpose, two different stabilization treatments (UV light treatment and sterilization) were applied prior to fermentation with two LAB strains, *Lacticaseibacillus casei* 2240 and *Lacticaseibacillus rhamnosus* GG. The biomass proved to be a suitable matrix for solid-state fermentation, showing a LAB growth of more than 2 log CFU/g in 48 h. The fermentation process was also useful for off-flavor reduction. In particular, the fermentation process significantly influenced the concentration of those compounds responsible for aldehydic/ethereal, buttery/waxy (acetoin and diacetyl), alkane and fermented aromatic notes (isoamyl alcohol). The heat treatment of the matrix, in addition to guaranteed safety for consumers, led to an improved aroma after fermentation. In conclusion, a fermented spirulina powder with a different aromatic profile was obtained with the applied heat treatment. Fermentation with lactic acid bacteria can be an interesting tool to obtain cyanobacterial biomasses with more pleasant sensory properties for potential use in food formulations.

Keywords: *Arthrospira platensis*; fermentation; lactic acid bacteria; food supplement; aromatic profile

Citation: Martelli, F.; Cirlini, M.; Lazzi, C.; Neviani, E.; Bernini, V. Solid-State Fermentation of *Arthrospira platensis* to Implement New Food Products: Evaluation of Stabilization Treatments and Bacterial Growth on the Volatile Fraction. *Foods* **2021**, *10*, 67. https://doi.org/10.3390/foods10010067

Received: 14 December 2020
Accepted: 27 December 2020
Published: 30 December 2020

Publisher's Note: MDPI stays neutral with regard to jurisdictional claims in published maps and institutional affiliations.

1. Introduction

Arthrospira platensis, commercially known as Spirulina, is a cyanobacterium commonly consumed as a food supplement because of its high nutritional value [1,2]. It is characterized by a high percentage of proteins (60%), providing all of the essential amino acids, followed in abundance by carbohydrates and polyunsaturated fatty acids (ω-3 and ω-6). Moreover, discrete quantities and varieties of minerals, vitamins and pigments (C-phycocyanin, chlorophyll and carotenoids) [3] are present. To date, microalgae and spirulina, defined as the novel food of the future [4], are used for various purposes in nutraceutical, cosmetic [5], feed [6] and pharma [7] sectors, and they continue to increasingly attract the interest of consumers and companies because of several bioactivities that this cyanobacterium is being proved to possess in many studies [8,9]. Applications with the aims of conferring macro- and micro-nutrients and improving color are increasing, and many food products, such as bread, cookies and pasta [10–12], cheeses [13], yogurt [14,15] and beverages [16], supplemented with *A. platensis* have been developed. To date, most of the algal biomass produced is consumed in the form of powder or tablets as protein- and micronutrient-rich supplements. Positive results have also been obtained using this cyanobacterium as a prebiotic for lactic acid bacteria (LAB). Indeed, it was able not only to preserve LAB viability in food matrixes [17,18] but also to stimulate their growth in broth [19–21]. The effect of *A. platensis* biomass on LAB could be exploited for the production of probiotic food supplements or ingredients in particular. Fortification of foods with probiotic strains shows beneficial effects such as anti-inflammatory, antioxidant and immunomodulatory effects; protection against colitis; and damage to epithelial cells [22]. Moreover, probiotics, once ingested, are believed to play an important role in the intestinal tract against foodborne pathogens [23] and can reduce symptoms due to antibiotic therapies, thereby relieving food allergies and reducing atypical dermatitis [24]. A lactic acid fermentation process applied to spirulina dried powder could be useful not only to supply consumers of LAB with beneficial effects to their diet, but also to improve or change the aromatic profile as seen in the case of vegetable matrices [25]. Indeed, a biological process to reduce unpleasant smells may represent an important tool for applications of algal biomass in complex foods for the avoidance of off-flavors. During fermentation, the metabolic activity of LAB leads not only to rapid acidification of the substrate and fast consumption of easily fermentable sugars, with a competitive advantage in the use of LAB in nutrient-rich environments, but also to the production of volatile compounds belonging to different chemical classes, such as alcohols, aldehydes, ketones, acids, esters and sulfur compounds. These compounds mainly derive from the catabolism of citrate and from the degradation of proteins and lipids. The formation of aromatic compounds is a complex process in which precursors are initially generated and subsequently converted into aromatic compounds [26].

On the other hand, complex issues arise in the production of spirulina-based products that can be considered safe for consumer health. Because of cultivation conditions and manipulation during downstream processing, *A. platensis* can be contaminated by alterative and pathogenic bacteria [27] that may duplicate during fermentation, thus compromising quality and safety.

The aromatic fraction of *A. platensis* is characterized by several volatiles present in different concentrations and with different odor thresholds formed in the matrix during growth and maturation [28]. The most abundant volatiles identified as aromatic components of *A. platensis* are hydrocarbons, especially heptadecane, followed by furanic compounds, pyrazines, sulfur compounds, aldehydes, ketones and alcohols [28–30]. All of these substances, naturally present in the matrix and ascribed to amino acid and fatty acid microorganism metabolism, contribute to the typical fishy odor that characterizes algae and algae-derived products [31,32]. This unpleasant flavor could be reduced by applying a fermentation step to the product, as described in recent works in which yeasts and bacteria were used to ferment different algae matrixes [30,33].

On the basis of these considerations, the aim of the present study was to evaluate solid-state fermentation of *A. platensis* biomass considering two LAB species, *Lacticaseibacillus rhamnosus* and *Lacticaseibacillus casei*, belonging to the *Lacticaseibacillus casei* group, for the production of a lactic acid-fermented food product. In order to reach this goal, to stabilize the biomass, two different treatments based on UV light irradiation and thermal sterilization were applied prior to fermentation. The effects on LAB growth were evaluated and the aromatic fraction of *A. platensis* was characterized in order to assess the improvements in the volatile profile given by LAB fermentation.

2. Materials and Methods

2.1. *Arthrospira Platensis* Stabilization Treatments

The dehydrated *A. platensis*, kindly provided by Bertolini Farm (Fidenza, PR, Italy), was used as a substrate for fermentation. It is a commercial product cultivated in raceway pond and marketed as "Organic Spirulina" according to EU Organic Aquaculture Regulation (EC No. 834/07). In order to reduce the microbial contamination present on the biomass, two different treatments were applied separately. The first consisted of UV light irradiation for 15 min to reduce the microbial total charge applied under a Faster BH-EN

2004 Class II Microbiological Safety Cabinet (S/N 1113) (Richmond Scientific, Chorley, UK) with a lamp emitting light at an intensity of 253.7 nm in the spectrum of UV-C (UV), while the second was based on a thermal treatment of 121 °C for 20 min applied in an autoclave (3870MLV, Tuttnauer, NY, USA) to sterilize the algal biomass (ST). Specifically, the time for the complete sterilization cycle was of approximately 50 min. To reach the temperature of 121 °C, 15 min were necessary, while 20 min were required to perform the sterilization process and a further 15 min to cool the autoclave after treatment. To evaluate the efficacy of the stabilization treatments, microbial plate counts were performed following treatments and after 48 h of incubation at 37 °C in order to determine the residual microbial load. To this purpose, samples were ten-fold serially diluted in Ringer solution (Oxoid, Basingstoke, UK) and plated on plate count agar (PCA) (Oxoid, Basingstoke, UK) incubated at 37 °C for 48 h. Analyses were performed in duplicate and average values ± standard deviations were reported as log CFU/g.

2.2. Lactic Acid Bacteria Strains

Two LAB strains were used to ferment *A. platensis* biomass: *Lacticaseibacillus casei* 2240, isolated from Parmigiano Reggiano cheese, belonging to the collection of the Department of Food and Drug (University of Parma), and *Lacticaseibacillus rhamnosus* GG, a probiotic commercial strain. They were maintained at −80 °C in de Man, Rogosa and Sharpe (MRS) cultivation medium (Oxoid, Basingstoke, UK), with 12.5% glycerol (*v/v*) added before use.

2.3. Arthrospira Platensis Biomass Fermentation

LAB strains were revitalized twice in MRS broth (Oxoid) (inoculation of 3% *v/v*) incubated for 16 h at 37 °C under aerobic conditions. They were then inoculated in fresh MRS broth (3% *v/v*) and incubated for 15 h at 37 °C to obtain a bacterial concentration of 9 log CFU/mL. After centrifugation (Eppendorf centrifuge 5810 R, Eppendorf, Hamburg, Germany) (12,857× *g* for 10 min at 4 °C), cells were collected, washed twice in Ringer solution (Oxoid, Milan, Italy) and suspended in sterile bidistilled water.

A. platensis biomass was rehydrated with 70% *w/w* sterile water, and then inoculated individually with each bacterial suspension in order to obtain an estimated LAB concentration of 7 log CFU/mL in each sample. LAB concentration was evaluated following the inoculation (T0), and after 24 h (T1) and 48 h of fermentation (T2). Serial dilutions of the samples in Ringer (Oxoid) were plated on MRS agar (Oxoid) and incubated for 48 h at 37 °C in aerobic conditions. pH of all the samples, before and after the fermentation step, was also measured (Mettler Toledo, Greifensee, Switzerland). Fermentations were carried out in duplicate, and for each sampling time analyses were performed in duplicate. Colonies were counted manually, and to calculate the CFU/g concentration, the following equation was applied:

$$\frac{\Sigma c}{(1 \times na + 0.1 \times nb + 0.01 \times nc)d} \quad (1)$$

where $\sum C$ is the summation of all the counted colonies, *na* is the number of plates of the first countable serial dilution, *nb* is the number of plates of the second countable serial dilution, *nc* is the number of plates of the third countable serial dilution, and *d* is the serial dilution factor of the first countable serial dilution.

Average values ± standard deviation were reported as log CFU/g. Treated but not inoculated samples were also incubated at 37 °C and analyzed at the same sampling times as control samples. The 48 h fermented and stabilized but not fermented samples were then lyophilized by a Freeze dryer Lio-5P (5Pascal, Milan, Italy) for 48 h, and then the LAB concentration was determined by plate counting on MRS.

2.4. HS-SPME/GC-MS Analysis

The characterization of the volatile fraction was conducted on all the fermented and stabilized but not fermented *A. platensis* samples. To this purpose, the protocol reported by Ricci et al. 2019 [25] was applied with some modifications. Briefly, 2 g of biomass and

5 μL of an aqueous Toluene standard solution (100 μg/mL in 10 mL) were used for the analyses. The headspace of the samples was extracted by a divinylbenzene–Carboxen–polydimethylsiloxane (DVB/Carboxen/PDMS) SPME fiber (Supelco, Bellefonte, PA, USA) for 30 min at 40 °C after an equilibration time of 15 min at the same temperature. The separation of the volatile compounds was achieved using a SUPELCOWAX 10 capillary column (Supelco, Bellefonte, PA, USA; 30 m × 0.25 mm × 0.25 μm) placed on a Thermo Scientific Trace 1300 gas chromatograph coupled with a Thermo Scientific ISQ single quadrupole mass spectrometer equipped with an electronic impact (EI) source. All the parameters applied for the analyte separation and detection, such as the injector, transfer line and column compartment temperatures, injection mode and gas carrier flow, were the same as those reported by Ricci et al. 2019 [25]. The detected volatile compounds were then identified on the basis of their linear retention indexes, calculated using as reference a C8–C20 alkane solution analyzed under the same chromatographic conditions applied for sample analysis, and by the comparison of the registered mass spectra with those reported in the instrument libraries (NIST 14). The semiquantification of all the identified volatiles was achieved on the basis of the use of a reference compound (Toluene). Data were reported as μg/g of wet weight.

2.5. Statistical Analysis

In order to determine the actual growth of tested LAB strains in *A. platensis* samples, results obtained from microbial counts were statistically treated applying one-way ANOVA test, comparing different growing times and lyophilized samples.

One-way ANOVA was carried out also considering data obtained from semiquantification of all the detected volatiles to underline analogies and differences among the considered samples in terms of production/diminution and/or release of volatiles. Moreover, two-way ANOVA was applied in order to evaluate the influence of two different factors, fermentation and stabilization treatment, on the volatile profile. All the analyses were performed using IBM SPSS Statistics 23.0 software (SPSS Inc., Chicago, IL, USA) applying Tukey's test as a post hoc test ($p \leq 0.05$).

3. Results and Discussion

3.1. *Arthrospira Platensis* Fermentation

One of the aims of the study was the evaluation of solid-state fermentation of *A. platensis* with *L. casei* bacteria that could lead not only to the implementation of new lactic acid-fermented food products but also to modifications of algae biomass aromatic fraction. Fermenting a matrix having a high bacterial charge could negatively affect the finished product considering that spoilage and/or pathogenic bacteria may also grow during the process. In this specific case, the biomass utilized for the experiment was cultivated in a raceway pond. This type of cultivation is associated with undesired microflora, such as *Bacillus* spp., *Alteromonas* spp., *Flavobacterium* spp. and *Pseudomonas* spp. [27], and previous studies have reported a variable concentration of bacteria, ranging from 2 to 7 log CFU/g, in this type of matrix [3]. For this reason, prior to fermentation, carried out for 48 h in sterilized glass cans, samples were subjected to UV radiation treatment (which does not seem to affect *A. platensis* composition [34]) or sterilization treatment in an autoclave in order to reduce the presence of microbiological contaminants. The UV treatment was ineffective for the reduction of microbial contamination, which was maintained at the same level of untreated samples (5.10 ± 0.2 log CFU/g). In order to predict the UV disinfection rates on food surfaces, it is necessary to consider the interactions between microorganisms and surface materials while trying to avoid the shielding effects from incident UV and predict the dependency on the surface structure or topography [35]. For this reason, UV treatment did not reduce the initial microbial contamination because of an uneven distribution of rays in the dehydrated samples. Furthermore, the presence of a high amount of pigments, phenolic compounds and other components with a protective effect against UV rays may contribute to microbial survival [36]. After UV treatment, a fermentation process of 48 h was conducted considering two *Lacticaseibacillus casei* bacteria. *L. casei* 2240 and *L. rhamnosus* GG concentrations were determined following inoculation, and growth was determined after 24 and 48 h of incubation at 37 °C. The endogenous contamination did not affect the counts, giving a much more higher LAB inoculation of 7 log CFU/g.

As shown in (Figure 1A), the two strains showed a similar growth trend in the UV-treated matrix. After 24 h of fermentation, both species showed good replication capacity, reaching a concentration higher than 9 log CFU/g. In particular, *L. rhamnosus* GG increased by 2.23 log CFU/g (from 7.13 ± 0.09 to 9.36 ± 0.03 log CFU/g; $p < 0.05$), and to a higher extent than *L. casei* 2240 with a growth of 1.78 log CFU/g (from 7.52 ± 0.06 to 9.30 ± 0.21 log CFU/g).

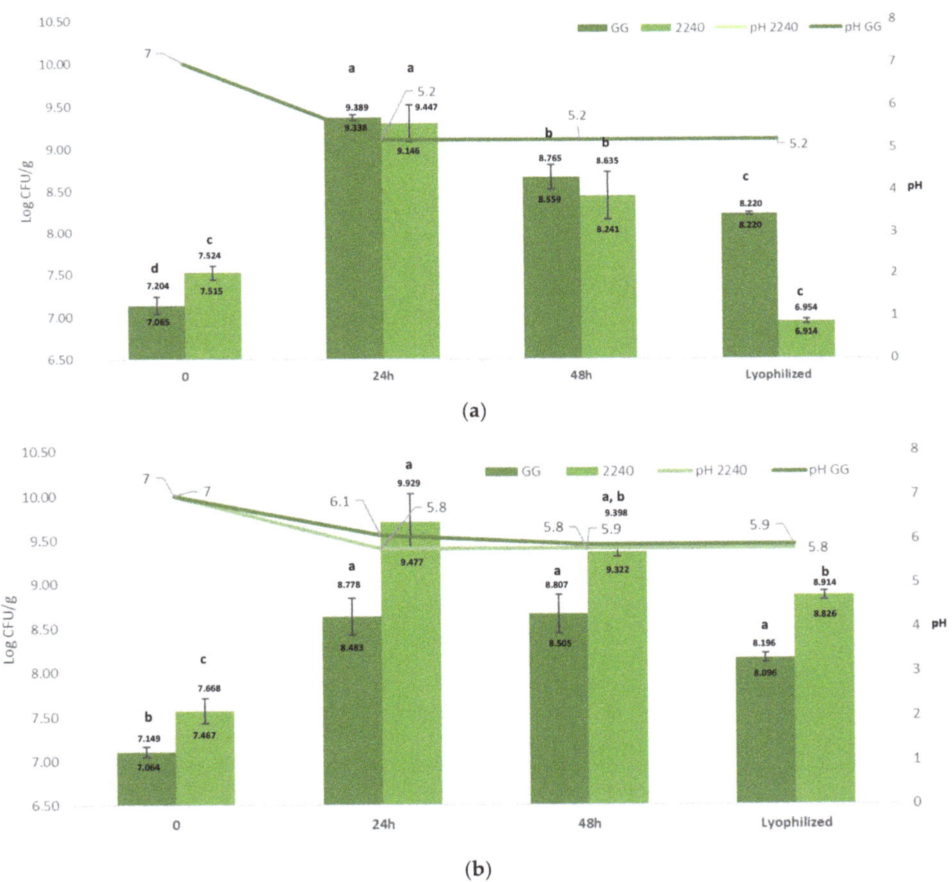

Figure 1. Lactic acid bacteria (LAB) growth (log CFU/g) and pH values in *Arthrospira platensis* biomass after 24 and 48 h of fermentation at 37 °C for *Lacticaseibacillus rhamnosus* GG (GG) and *Lacticaseibacillus casei* 2240 (2240) bacteria: (**a**) UV-treated samples; (**b**) heat-treated samples. Data are represented as average ± SD (bars); two replicates for each sample were measured. Letters indicate significant differences ($p < 0.05$).

After 48 h of fermentation, a decrease in the concentration of both strains was observed. *L. casei* 2240 decreased by 0.86 log CFU/g (from 9.30 ± 0.21 log CFU/g to 8.44 ± 0.28 log CFU/g), while *L. rhamnosus* GG decreased by 0.70 log CFU/g (from 9.36 ± 0.03 log CFU/g to 8.66 ± 0.14 log CFU/g) (Figure 1A) ($p < 0.05$). However, the values remained higher than the inoculation. The *L. casei* 2240 strain has previously been used to ferment other matrixes, such as *Himanthalia elongata*, but the growth ability on this brown seaweed was lower than that observed in this study [37]. The best capacity of this strain to grow on *A. platensis* can be linked to the high amount of proteins and small peptides [2]. *L. rhamnosus* GG is a probiotic commonly supplemented to fermented milk [38], but applications to other matrices have also been evaluated. For instance, a sausage supplemented with *L. rhamnosus* GG has been produced [39] though this strain grew to a lesser extent than in *A. platensis*.

During fermentation, the initial pH of the hydrated *A. platensis* biomass decreased from 7.0 ± 0.1 to 5.2 ± 0.1 in the first 24 h for both samples and remained stable until the end of the fermentation process. This decrease in pH during fermentation reflects the growth of the LAB strains with consequent production of organic acids, mainly lactic acid.

Considering the possibility to use lactic acid-fermented *A. platensis* in a food supplement formulation or as an ingredient, the survival of LAB was evaluated after a lyophilization process.

L. rhamnosus GG was reduced at 8.22 ± 0.02 log CFU/g ($p < 0.05$), and *L. casei* 2240 was reduced by more than 1 log CFU/g (from 8.44 ± 0.28 to 6.9 ± 0.03 log CFU/g) ($p < 0.05$) (Figure 1A).

The second fermentation process was carried out on *A. platensis* samples treated at 121 °C for 20 min in order to eliminate the presence of any microbial contamination. The absence of residual contamination after treatment was confirmed by plate count below the detection limit (1 log CFU/g). The sterilization treatment involves the use of a particular apparatus, the autoclave, and high energy to reach the temperatures and pressures required by the process. This treatment is therefore longer and more expensive than that based on UV rays, but the results show that it is certainly more effective in reducing the initial microbial concentration of the product. Following the heat treatment, the characteristic green-blue color of *A. platensis* changed to a dark green, brownish color. This may be due to the degradation of pigments such as carotenoids, xanthophylls, phycocyanins and chlorophyll caused by the high temperatures reached [40–42]. The growth trend of LAB strains is presented in (Figure 1B). In particular, after 24 h, *L. casei* 2240 increased by 2.13 log CFU/g (from 7.57 ± 0.142 to 9.70 ± 0.320 log CFU/g) ($p < 0.05$) and *L. rhamnosus* GG by 1.5 log (from 7.10 ± 0.06 to 8.63 ± 0.21 log CFU/g) ($p < 0.05$) and then remained constant (with no significant differences) up to 48 h.

The ability of these *Lacticaseibacillus* species to grow on different matrixes has been challenged over the years, with different results [25,39,43,44]. In particular *L. casei* and *L. rhamnosus* strains grown on sterilized vegetables' by-products increased by more than that observed on *A. platensis* [45]. However, the ability of this cyanobacterium to boost the growth of LAB is known in the literature [20], and because of its richness in small peptides and proteins, *A. platensis* can be considered a good matrix to allow for the growth of LAB species such as *L. casei* and *L. rhamnosus*.

After fermentation of the sterilized biomass, a lower acidification rate than that of the UV-treated fermented samples was observed. The pH value of the sterilized fermented biomasses after 24 h was 5.8 ± 0.1 and 6.1 ± 0.1, respectively, for *L. casei* 2240 and *L. rhamnosus* GG and also remained stable at 48 h of fermentation (Figure 1B).

Analogously to the UV-treated samples, a lyophilization process was applied to the sterilized products after fermentation. A decrease in the microbial concentration of 0.5 log CFU/g was observed for both strains; in particular, *L. rhamnosus* GG was reduced to 8.15 ± 0.05 log CFU/g and *L. casei* 2240 to 8.86 ± 0.05 log CFU/g ($p < 0.05$) (Figure 1B).

L. rhamnosus GG better survived the lyophilization process than *L. casei* 2240, whose viability showed a decrease of more than 1 log CFU/g. The survivability of freeze-dried strains is of particular importance for the production of foods containing live cells, and, for this reason, many studies focus on enhancing the viability of LAB after this process [46,47]. The better survivability of *L. casei* 2240 to lyophilization in sterilized biomass than UV-treated samples was observed on the basis of the applied statistical model (Figure 1). Several authors [17,48,49] tested the addition of *A. platensis* to dairy products, such as yogurt, cheese and fermented milk, with positive results, including an increase in the number of LAB and an improvement in the nutritional quality of the fermented product during storage.

The fermentation process leads to increased production of phenolic compounds and phycocyanobilins and consequently increased radical scavenger properties of the cyanobacterium [50]. Lactic acid fermentation is an appropriate method to enhance the functional properties of spirulina and also to integrate beneficial bacteria into consumers' diets, thereby providing further advantages to the final product. However, to meet the definition of probiotic products, microorganisms must be viable for the entire shelf life of the product, and in such quantities to be able to multiply and integrate the intestinal flora. The activity of *A. platensis* to enhance LAB vitality, such as that of *L. casei*, *Streptococcus thermophilus*, *Lacticaseibacillus acidophilus*, and *Bifidobacteria*, has been documented [3,19–21]. There is no consensus regarding the minimum quantity of probiotic microorganisms to be ingested to guarantee their functionality in the human intestine. Usually, to observe a positive effect on health, 6 to 7 log CFU/g of live probiotic microorganisms, able to colonize the intestine, should be consumed daily [51]. All of the fermented biomasses obtained in this study presented a sufficient amount of bacteria, allowing the production of fermented foods with a high functional value. An image of the fermented biomasses produced after lyophilization is presented in Figure 2.

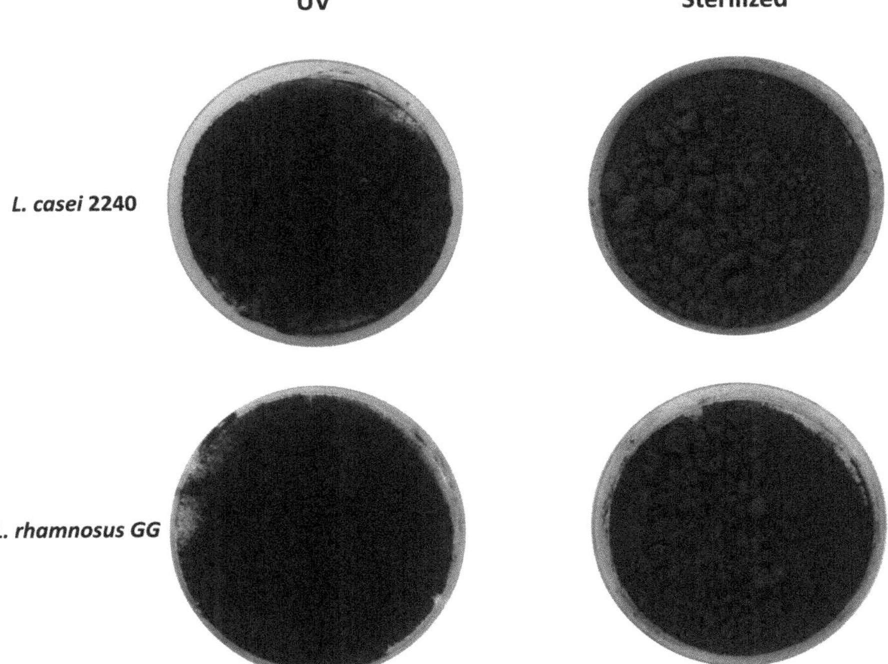

Figure 2. Representative image of fermented and lyophilized *Arthrospira platensis* biomasses subjected to UV and sterilization treatments.

3.2. *Volatile Profile Characterization of A. platensis and Changes in Volatile Components after Fermentation*

Recently, several studies have been conducted on the characterization of the aromatic fraction of *A. platensis*, and, in particular, a number of studies were carried out with the aim of identifying how the characteristic fishy odor of this product can be reduced in order to use it as additive or ingredient in food [29,30,33]. The reduction of unpleasant aromatic notes can be achieved by solvent extraction [29], such as by fermentation using fungi or bacteria [30,33]. The deodorization of *A. platensis* can be performed by applying an extraction procedure to the biomass; Cuellar-Bermúdez et al. stated that the fishy odor can be reduced by treating spirulina with a solvent (e.g., ethanol, acetone or hexane) capable of dissolve aromatic compounds, thereby preserving the nutritional characteristic of the product [29]. On the other hand, fermenting *A. platensis* with lactic acid bacteria, such as *Lacticaseibacillus plantarum* or *Bacillus subtilis*, may remove the volatile compounds initially present in the matrix and lead to the formation of new components, such as acetoin, that provide fermented and creamy aromatic notes [30]. In the present study, lactic acid bacteria fermentation was applied to stabilized *A. platensis* materials, and changes in the volatile fraction were determined.

A total of 61 different volatile compounds were identified in the volatile fraction of treated but not fermented and fermented *A. platensis* samples (Table 1). In particular, 7 aldehydes, 9 ketones, 4 esters, 9 terpenes/norisoprenoids, 7 alcohols, 4 furans, 11 hydrocarbons, 7 pyrazines and 3 sulfur compounds were detected. These results are consistent with data reported by Bao et al. (2018), who detected the same classes of volatile compounds in spirulina samples fermented with different strains of *L. plantarum*, *L. acidophilus* and *Bacillus subtilis* [30].

The class that quantitatively mainly represents the aromatic fraction of *A. platensis*, both before and after the fermentation process, is that of hydrocarbons (Supplementary Table S1), as also demonstrated in previous studies [30]. The concentration of these compounds was higher in respect to the amount of all the other components. Hydrocarbon release was significantly different when comparing fermentations ($p = 0.007$) and opposing technological treatments ($p < 0.001$).

Statistical differences were observed in hydrocarbon quantity among the UV- and heat-treated samples ($p < 0.001$), among samples subjected to the same process, and between fermented and not fermented samples ($p = 0.007$), but no interaction between factors (type of treatment and fermentation) was observed (Figure 3, Supplementary Table S1).

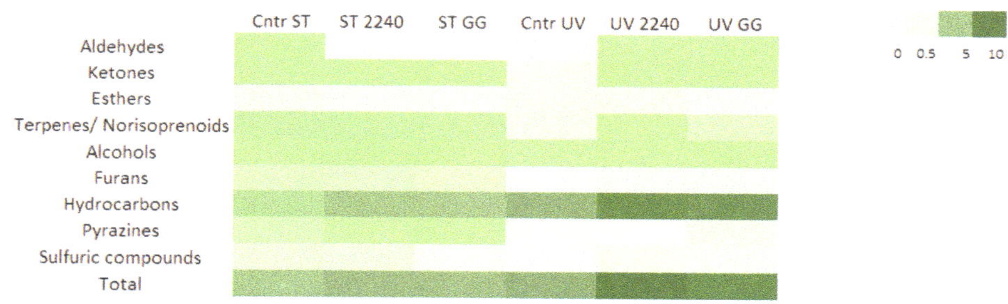

Figure 3. Heat map performed on volatile chemical classes' concentrations detected in sterilized (ST) or UV-treated (UV) *A. platensis* biomasses fermented with *L. casei* 2240 (2240) and *L. rhamnosus* GG (GG) as well as treated but unfermented biomasses (Cntr). A scale ranging from a maximum of 10 μg/g (dark green) to a minimum of 0 μg/g (light green) was used.

Table 1. Volatile compounds found on *Arthrospira platensis* treated but not fermented and fermented samples. For each volatile compound's aromatic note, calculated and tabulated linear retention indices (LRIs), references and effect given by treatments, stabilization and fermentation (statistical difference = positive = p; no statistical difference = negative = n; not determinable = nd) are reported.

Chemical Class, Compound Name	Odor Type	Calculated LRI	Reference LRI	Identification Method	Reference	Effect of Stabilization	Effect of Fermentation
Aldehydes							
Isobutyraldehyde	aldehydic	805	814	MS + LRI	[52]	n	n
2-Methylbutanal	chocolate	904	903	MS + LRI	[53]	nd	nd
Isovaleraldehyde	aldehydic	907	888	MS + LRI	[54]	p	p
Hexanal	green	1075	1086	MS + LRI	[55]	n	n
Methional	vegetable	1452	1468	MS + LRI	[56]	p	p
Benzaldehyde	fruity	1523	1537	MS + LRI	[55]	n	n
2,5-Dimethyl benzaldehyde		1733	1705	MS + LRI	[53]	p	p
Ketones							
Acetone	solvent	810	901	MS + LRI	[52]	p	p
2-Butanone	ethereal	894	901	MS + LRI	[52]	p	p
diacetyl	buttery	971	973	MS + LRI	[54]	n	p
6-Methyl-2-heptanone	camphoreous	1229	1236	MS + LRI	[57]	p	p
3-Octanone	herbal	1245	1261	MS + LRI	[58]	n	p
2-Octanone	earthy	1277	1287	MS + LRI	[53]	p	n
Acetoin	buttery	1282	1300	MS + LRI	[54]	n	p
2,2,6-Trimethylcyclohexanone	thujonic	1306	1308	MS + LRI	[59]	p	n
Sulcatone	citrus	1329	1335	MS + LRI	[60]	p	n
Esters							
Ethyl acetate	ethereal	872	869	MS + LRI	[61]	n	p
Ethyl caprylate	waxy	1430	1438	MS + LRI	[61]	nd	nd
Ethyl decanoate	waxy	1628	1645	MS + LRI	[61]	p	n
Phenethyl acetate	floral	1804	1803	MS + LRI	[62]	nd	nd
Terpenes, norisoprenoids and similar							
p-Xylene		1133	1149	MS + LRI	[63]	p	n
Myrcene	spicy	1155	1143	MS + LRI	[54]	n	n
α-Cyclocitral	citrus	1427	1420	MS + LRI	[53]	p	p
β-Cyclocitral	tropical	1609	1612	MS + LRI	[59]	p	n
Safranal	herbal	1635	1637	MS + LRI	[59]	p	p
α-Ionene	fruity	1675		MS		n	n
α-Ionone	floral	1841	1848	MS + LRI	[64]	p	p
β-Ionone	floral	1918	1935	MS + LRI	[64]	n	n
β-Ionone-5,6-epoxide	fruity	1950	1989	MS + LRI	[53]	p	n
Alcohols							
Ethanol	alcoholic	923	903	MS + LRI	[54]	p	p
Isobutyl alcohol	ethereal	1080	1100	MS + LRI	[60]	n	n
Isoamyl alcohol	fermented	1195	1210	MS + LRI	[60]	p	p

Table 1. *Cont.*

Chemical Class, Compound Name	Odor Type	Calculated LRI	Reference LRI	Identification Method	Reference	Effect of Stabilization	Effect of Fermentation
Alcohols							
1-Pentanol	fermented	1239	1260	MS + LRI	[55]	p	n
1-Hexanol	herbal	1341	1357	MS + LRI	[61]	p	n
1-Octen-3-ol	earthy	1437	1455	MS + LRI	[55]	p	n
Benzyl alcohol	floral	1882	1896	MS + LRI	[61]	n	n
Furans							
2-Methylfuran	chocolate	853	876	MS + LRI	[52]	p	p
3-Methylfuran		881	877	MS + LRI	[53]	n	n
2-Butylfuran	spicy	1123	1140	MS + LRI	[65]	p	p
2-Pentylfuran	fruity	1220	1232	MS + LRI	[64]	p	n
Hydrocarbons							
1,2,4,4-Tetramethylcyclopentene		920		MS		n	n
2,2,4,6,6-Pentamethylheptane		944		MS		n	n
Ethyl benzene		1119	1127	MS + LRI	[66]	n	n
Tridecane	alkane	1300	1300	MS + LRI	[56]	p	p
Tetradecane	waxy/alkane	1396	1400	MS + LRI	[56]	p	p
2,6,10-Trimethyltridecane		1434	1442	MS + LRI	[53]	n	n
Pentadecane	waxy	1492	1500	MS + LRI	[56]	p	p
Hexadecane	alkane	1590	1600	MS + LRI	[56]	p	p
N-acetyl-4(H)-Pyridine		1644		MS		p	p
Heptadecane	alkane	1687	1700	MS + LRI	[56]	p	p
6,9-Heptadecadiene		1743				p	p
Pyrazines							
2-Methylpyrazine	nutty	1267	1267	MS + LRI	[52]	n	n
2,5-Dimethylpyrazine	chocolate	1318	1321	MS + LRI	[52]	p	n
2-Methyl-5-ethylpyrazine	coffee	1368	1406	MS + LRI	[52]	n	n
2-Ethyl-6-methylpyrazine	potato	1383	1402	MS + LRI	[52]	p	p
Trimethyl pyrazine	nutty	1398	1401	MS + LRI	[52]	p	p
2,3-Dimethyl-5-ethylpyrazine	burnt	1452	1460	MS + LRI	[53]	p	n
Tetramethyl pyrazine	nutty	1468	1474	MS + LRI	[52]	p	n
Sulfur compounds							
Dimethyl disulfide	sulfurous	1063	1073	MS + LRI	[54]	p	n
2-Ethyl-4-methylthiazole	nutty	1336	1322	MS + LRI	[52]	n	n
Dimethyl trisulfide	alliaceous	1369	1375	MS + LRI	[54]	n	n

Pentadecane, hexadecane and heptadecane are hydrocarbons derived from the decarboxylation of palmitic and stearic acids, respectively [29,67], and they were the most representative among all of the hydrocarbons identified in this work. In particular, heptadecane was the volatile found in the highest amount in all of the considered samples as observed in a recent study focused on the fermentation of spirulina using *Lacticaseibacillus plantarum* and *Bacillus subtilis* [30]. The presence of this compound could contribute to the off-flavor of algae associated with crude fish notes, despite the fact that it presents a high odor threshold [29]. Heptadecane was significantly lower in the sterilized samples when compared to the UV-treated ones, despite the fact that the fermentation process seemed to increase its concentration in both cases; indeed, both considered factors, stabilization treatment ($p < 0.001$) and fermentation ($p = 0.006$), appeared to influence its release/formation. In other studies considering *L. plantarum* and *B. subtilis,* the fermentation caused a 27% reduction in the relative content of heptadecane, but no information regarding the actual quantity found in the samples was reported [30]. The LAB species considered in the present work, *L. casei* and *L. rhamnosus*, provided significantly different results in the production of heptadecane in the differently treated substrates, which is indeed linkable to LAB fermentation [68]. It can be speculated that the production and/or release of this compound may depend on the applied LAB species. The same phenomenon can be observed also for pentadecane. A higher amount was detected in the UV-treated samples when compared to the sterilized ones. It can be observed that the quantity of hydrocarbon grew following fermentation when both the strains were used (Supplementary Table S1). The significant difference observed in the fermented biomasses (i.e., heptadecane: UV 2240 = 7.65 ± 1.95 μg/g; ST 2240 = 2.53 ± 0.66 μg/g) can be ascribed to the different stabilization processes applied; the higher concentration of these volatiles in the UV-treated biomass could be ascribed to the metabolism of the endogenous microflora typical of the spirulina biomass. Since the UV treatment, unlike the heat treatment, did not completely eliminate contaminating microflora, this behavior can be ascribed to the residual microflora deriving from cultivation in open ponds.

Terpenes, the second most important class in terms of concentration, were positively influenced by sterilization treatment, and an interaction between the two factors was observed ($p = 0.013$). Terpenes and norisoprenoids are fundamental compounds in food aroma. β-cyclocitral and β-ionone, with pleasant fruity and floral notes, are two volatiles typical of cyanobacteria. The formation of these compounds is due to the oxidation of carotenoids following carotene oxygenases. Norcarotenoids are an important group of compounds formed by several species of cyanobacteria, generated by enzymatic degradation of carotenes and carotenoids. For example, the oxygenase reaction of carotenoids was first described in Microcystis. β-cyclocitral is formed by the cleavage reaction of β-carotene catalyzed by the enzyme [59]. β-ionone and norcarotenoids of the β-ionone-type are an important group of compounds that were found in axenic cyanobacterial cultures and a monoxenic culture of *Phormidium* sp. [69]. Significant differences in the presence of β-cyclocitral, the compound that mainly characterizes the class of terpenes and norisoprenoids, were found because of the biomass stabilization treatment ($p < 0.001$), while no significant differences were found in β-ionone production.

Regarding the production of aldehydes and ketones and/or release, both fermentations and treatments produced significant differences in the samples ($p < 0.02$); in addition, a strong interaction was noticed between the two factors for these two volatile classes ($p < 0.005$). While differenced were noted among the two stabilization treatments ($p = 0.012$), ketone concentration was mainly affected by fermentation, (Figure 3, Supplementary Table S1). In particular for aldehydes, a reduction following the fermentation of the sterilized biomass using both the tested strains was observed. Conversely, by fermentation of the UV-treated biomass, the amount of aldehydes increased. This behavior could be due to the presence of the epiphytic bacteria typical of the spirulina biomass. Considering these data, it can be suggested that fermentative LAB can reduce the amount of these compounds, but the typical microflora of *A. platensis* can produce aldehydes, thereby giving the fermented biomass a different aroma profile. Hexanal was found in several microalgae [70–73]. Fermentation of sterilized biomass by both of the strains led to a significant diminution of hexanal content ($p < 0.01$), while only fermentation of the *L. rhamnosus* GG strain seemed to produce this compound in UV-treated samples. Since the presence of C6 aldehydes may be associated with fish odor [32], this can be considered an undesired aromatic notes, especially in certain food preparations, such as those of dairy products; thus, fermentation could provide a valuable improvement of in aromatic profile of the sample. Methional was found in both of the UV-treated fermented samples, suggesting, as already suggested by other studies [74,75], that this aldehyde is produced by LAB. This compound was also previously isolated from *Rhodomonas*, another microalgae [71]. Both fermentation and treatments applied prior to it, as well as their combination, produced significant differences in methional production ($p < 0.001$). Benzaldehyde is a typical aroma formed in several species of microalgae by enzymatic and chemical degradation of phenylalanine (generating benzaldehyde) [71]. This compound is often associated with pleasant notes of almond. It was found in the sterilized *A. platensis* biomass, and it was produced by the fermentation process applied to the UV-stabilized samples without differences among the two strains used. A similar trend was noticed for isovaleraldehyde (Supplementary Table S1), an aldehyde produced by LAB via the metabolization of the amino acid leucine [76].

Regarding ketones, a growing amount of these compounds was noticed following the fermentation of both the tested biomasses. In general, the number of ketones in sterilized biomass was higher than in the UV-treated biomass, proving the role of thermal treatment in the formation of these compounds (Supplementary Table S1). Diacetyl and acetoin are two typical aromatic compounds produced during fermentation with the characteristic buttery aroma [77]. Fermentation produced significant differences in the presence of these molecules. In particular, the fermentation with *L. casei* 2240 produced a higher amount of diacetyl compared to *L. rhamnosus* GG ($p < 0.05$) in both the UV-treated and sterilized materials. Production of these ketones derives from metabolization of pyruvate and citrate and is dependent on the strains used for fermentation [78]. Sulcatone is a citrus-like aroma. This compound was found after sterilization of the biomass with a significant difference when compared to the UV-treated samples, and its concentration was maintained after the fermentation step with both the strains. Interestingly, the presence of this aroma was never linked to *A. platensis* biomass or microalgae in general.

Fermentation was the only factor affecting the total content of alcohols and esters. Alcohols were found in all of the analyzed samples. The most representative was 1-octen-3-ol with the typical mushroom aroma, previously isolated from the microalgae aromatic fraction [72]. The statistical model underlined that this compound was significantly improved after sterilization when compared to UV-treated but not fermented samples (0.13 ± 0.03 and 0.04 ± 0.00 µg/g, respectively), but fermentation did not affect its amount. Significant differences were found for ethanol. A significantly higher amount of ethanol was found in the UV-treated biomasses compared to the heat-treated samples ($p = 0.001$). The higher amount of this alcohol in the samples could be linked to the sterilization treatment that affected the composition of the starting biomass and therefore the presence of precursors for the production of alcohols, or to the presence of epiphytic bacteria and yeasts that survived the UV treatment. Significant differences were also found in the production of ethanol during fermentation ($p = 0.03$). Both *L. casei* and *L. rhamnosus* are heterofermentative species and produced slight amounts of ethanol in the biomass during fermentation. Isoamyl alcohol, a volatile compound typical of beverages and fermented foods, is formed from leucine during fermentation [72]. This alcohol was found in sterilized samples following lactic acid fermentation. Although isoamyl alcohol has also been reported in phototrophic cultures of *Chlorella vulgaris* [71], in this study, it was not found in treated biomass, but rather in unfermented *A. platensis* biomass.

The UV treatment and the sterilization process produced significant differences in the total amount of furans, pyrazines and sulfur compounds.

Four furans were found as a component of the aromatic profile of fermented *A. platensis* biomass. 2-Pentylfuran is an important product of lipid degradation and is responsible for beany and licorice-like sensory qualities in various food products [79]; it has been associated with the typical beany, green and metallic odor of spirulina, and it was detected in considerable percentages of about 10% in fermented spirulina samples in a previous study [30]. In our case, this furan was formed following the sterilization of the biomass (0.10 ± 0.04 μg/g), but it decreased following fermentation, while in UV-treated samples, it was not detected in the not fermented substrate, but it increased after fermentation.

Pyrazines are molecules typical of roasted and thermally treated foods. Seven pyrazines were found in the analyzed samples, and most of them were connectable to the sterilization process. For example, 2,5-dimethylpyrazine, a pyrazine with a typical chocolate aroma, was found in the sterilized biomass and not in the UV-treated samples. Interestingly, some of these compounds were also produced following lactic acid fermentation. Small amounts of 2-methylpyrazine were produced by *L. rhamnosus* GG (0.02 ± 0.00 μg/g) in the sterilized biomass. This is not the first time that the production of pyrazine by *L. casei* group bacteria was underlined [80]. Among all of the pyrazines detected in the tested samples, 2,5-dimethylpyrazine and 2-methylpyrazine are of particular interest, because these compounds have been associated with the off-flavor of *A. platensis* [30].

Similar behavior was observed for sulfur compounds, prevalently found in the sterilized samples. Dimethyl disulfide is a sulfuric compound formed following thermal oxidation of other volatile sulfur compounds such as methanethiol [72]. The number of sulfur compounds seemed to be related especially to the stabilization treatment, as the concentration of these compounds was higher in the sterilized samples when compared to those treated with UV ($p = 0.019$). Generally, sulfur compounds may contribute to the unpleasant odor of algae products because of their low threshold value. Seo et al. (2017) stated that sulfides were responsible for about 26% of the total odor profile of sea tangle extract, and fungal fermentation lead to a total reduction of dimethyl disulfide after two days from the inoculation [33]. In our case, the fermentation step did not reduce the amount of this compound detected only in UV-treated materials (Supplementary Table S1).

Esters, with floral and waxy notes, are the category that presented the lower quantities in the analyzed samples. Four compounds were identified, and this is in line with what has been reported in other studies [30,33]. The stabilization treatment did not seem to affect their content, while fermentation led to a decrease in the initial quantity ($p = 0.021$) (Supplementary Table S1).

3.3. Sensory Properties of the Detected Volatile Components

In order to evaluate the modifications in the sensory properties of fermented *A. platensis* in respect to the unfermented materials and to determine the changes in aromatic notes due to the different stabilization processes applied, all of the volatile compounds and their relative concentrations were grouped based on the odor type that they were able to provide. The following ten main different aroma attributes were identified: aldehydic/ethereal, sulfurous, green/herbal, buttery/waxy, spicy, fruity, floral, nutty/roasted, alkane and fermented. On the basis of this classification, analogies and differences in odor type production among different samples were observed by applying a two way ANOVA test considering fermentation and stabilization treatments as factors (Figure 4, Supplementary Table S2). Lactic acid fermentation of spirulina biomass caused an enhancement in the aromatic profile.

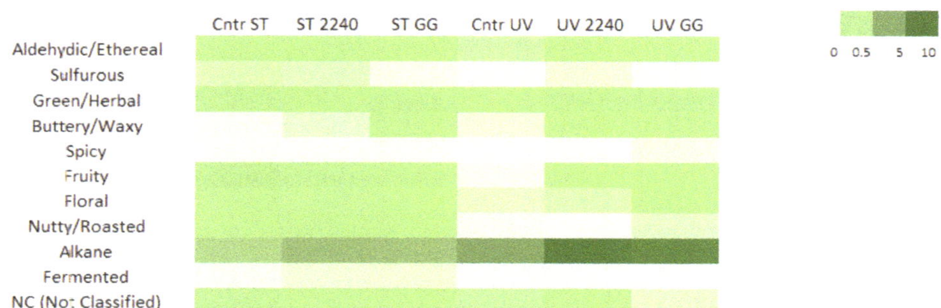

Figure 4. Heat map performed on the volatile concentration of volatiles grouped on the basis of their odor type detected in sterilized (ST) or UV-treated (UV) *A. platensis* biomasses fermented with *L. casei* 2240 (2240) and *L. rhamnosus* GG (GG) as well as treated but unfermented biomasses (Cntr). A scale ranging from a maximum of 10 μg/g (dark green) to a minimum of 0 μg/g (light green) was used.

In particular, the fermentation process significantly influenced the concentration of those compounds responsible for aldehydic/ethereal, buttery/waxy, alkane and of course fermented aromatic notes ($p < 0.01$). As expected, buttery/waxy and fermented notes increased with the fermentation process (Figure 4). These aromatic properties can be mainly associated with the presence of diacetyl, acetoin (buttery/waxy) and isoamyl alcohol (fermented). The concentration of all these volatiles showed an improvement in both of the stabilized materials after the fermentation process, but while in sterilized spirulina, *L. casei* 2240 produced a higher amount of diacetyl and acetoin, in UV-treated material, *L. rhamnosus* GG affected the content of these compounds more prominently (Supplementary Table S1). In addition, buttery/waxy and fermented aromatic notes were also influenced by stabilization treatment; in particular, a statistically significant interaction was observed for fermented notes of the two considered factors ($p < 0.001$) (Supplementary Table S2). Moreover, aldehydic/ethereal and alkane attributes, provided mainly by the presence of aldehydes and hydrocarbons, were strongly influenced by the fermentation step, especially in UV-treated samples (Figure 4). It must be emphasized that hydrocarbons, and in particular heptadecane, could be associate with the off-flavor of algae [41]. Alkane attribute concentrations were also influenced by the applied stabilization method ($p < 0.01$), as clearly represented in Figure 4. Thus, to limit the formation of these sensory attributes, sterilization must be chosen as the stabilization treatment.

On the other hand, significant differences were found in the production of sulfurous, green/herbal, fruity, and nutty/roasted notes ($p < 0.02$) among the stabilization treatments (Supplementary Table S2). The concentrations of all of these aromatic classes were statistically higher in the sterilized samples ($p < 0.02$). High temperature can cause the formation of pyrazine and sulfur compounds associated with nutty/roasted and sulfurous notes, as well as the degradation of carotenoids and fatty acids leading to an increase in hexanal, methional, 1-hexanol and 1-octen-3-ol (green/herbal and earthy notes), and benzaldehyde, β-cyclocitral, β-ionone 5,6-epoxide and other compounds associated with fruity aromatic notes. In addition, a statistically significant interaction between these two factors was noted for green/herbal and fruity notes ($p < 0.001$). Since fruity attributes are generally considered as pleasant aromatic properties, in this case, it is also possible to speculate that sterilization may represent the better stabilization treatment because it leads to an augment of this odor class [59,69,77,78].

In conclusion, despite the fact that sterilization may lead to some modifications of the aromatic characteristic of the starting material, *A. platensis*, it seemed to be the best choice to reduce the initial microbial count naturally occurring in the tested samples. Fermentation by *L. casei* bacteria can generate or enhance some volatile compounds responsible for pleasant aromatic notes, such as fruity and creamy (buttery) notes, associated with volatiles produced by LAB metabolism.

4. Conclusions

In this study, *A. platensis* biomass, treated by UV or sterilization at 121 °C, was fermented by LAB (*L. casei 2240* and *L. rhamnosus GG*) in order to evaluate the fermentative and aromatic potential. The LAB growth was not affected by the applied treatment, confirming *A. platensis* as a fermentable substrate that may be used for the development of new fermented foods and supplements with high functional values. Furthermore, the survivability of LAB to freeze-drying may allow the production of food products which, in addition to integrating proteins and vitamins typical of *A. platensis*, also include LAB or probiotic bacteria in the diet.

Considering the importance of aroma and flavor for consumers' acceptability, the main volatile compounds involved in *A. platensis* biomass fermentation were screened. An overall improvement of smell was obtained.

In particular, a greater presence of fermentation aroma was found during the fermentation of the sterilized sample, highlighting that in addition to guaranteeing safety for the consumer, this process may enhance applications, thereby avoiding side effects due to off-flavor. In conclusion, a fermented lyophilized spirulina powder with an interesting aromatic profile and high LAB concentration was obtained with the applied heat treatment, opening up perspectives for new applications in food productions.

Supplementary Materials: The following are available online at https://www.mdpi.com/2304-8158/10/1/67/s1, Table S1: Concentration (μg/g) of compounds identified in UV-treated and sterilized *Arthrospira platensis* fermented with *L. casei* 2240 and *L. rhamnosus* GG and in controls (not fermented UV-treated and sterilized *Arthrospira platensis*) after 48 h. Table S2: Concentration (μg/g) of odor type identified in UV-treated and sterilized *Arthrospira platensis* fermented with *L. casei* 2240 and *L. rhamnosus* GG, and in controls (not fermented UV-treated and sterilized *Arthrospira platensis*) after 48 h.

Author Contributions: Conceptualization, V.B. and C.L.; investigation, F.M. and M.C.; writing—original draft preparation, F.M. and M.C.; writing—review and editing, V.B. and C.L.; supervision, E.N. All authors have read and agreed to the published version of the manuscript.

Funding: This research received no external funding.

Acknowledgments: The authors are grateful to Bertolini Farm (Fidenza, Parma, Italy) for providing *Arthrospira platensis*.

Conflicts of Interest: The authors declare no conflict of interest.

References

1. Batista, A.P.; Niccolai, A.; Fradinho, P.; Fragoso, S.; Bursic, I.; Rodolfi, L.; Biondi, N.; Tredici, M.R.; Sousa, I.; Raymundo, A. Microalgae biomass as an alternative ingredient in cookies: Sensory, physical and chemical properties, antioxidant activity and in vitro digestibility. *Algal Res.* **2017**, *26*, 161–171. [CrossRef]
2. Andrade, L.M.; Andrade, C.J.; Dias, M.; Nascimento, C.A.; Mendes, M.A. *Chlorella* and *Spirulina* Microalgae as Sources of Functional Foods, Nutraceuticals, and Food Supplements; an Overview. *MOJ Food Process. Technol.* **2018**, *6*, 45–58. [CrossRef]
3. Niccolai, A.; Shannon, E.; Abu-Ghannam, N.; Biondi, N.; Rodolfi, L.; Tredici, M.R. Lactic acid fermentation of *Arthrospira platensis* (spirulina) biomass for probiotic-based products. *J. Appl. Phycol.* **2019**, *31*, 1077–1083. [CrossRef]
4. Zarbà, C.; Chinnici, G.; D'Amico, M. Novel Food: The Impact of Innovation on the Paths of the Traditional Food Chain. *Sustainability* **2020**, *12*, 555. [CrossRef]
5. Ariede, M.B.; Candido, T.M.; Jacome, A.L.M.; Velasco, M.V.R.; de Carvalho, J.C.M.; Baby, A.R. Cosmetic attributes of algae—A review. *Algal Res.* **2017**, *25*, 483–487. [CrossRef]
6. Holman, B.W.B.; Malau-Aduli, A.E.O. *Spirulina* as a livestock supplement and animal feed: *Spirulina* supplementation in livestock. *J. Anim. Physiol. Anim. Nutr.* **2013**, *97*, 615–623. [CrossRef]
7. Hamed, I.; Özogul, F.; Özogul, Y.; Regenstein, J.M. Marine Bioactive Compounds and Their Health Benefits: A Review. *Compr. Rev. Food Sci. Food Saf.* **2015**, *14*, 446–465. [CrossRef]

8. Martelli, F.; Cirlini, M.; Lazzi, C.; Neviani, E.; Bernini, V. Edible Seaweeds and Spirulina Extracts for Food Application: In Vitro and In Situ Evaluation of Antimicrobial Activity towards Foodborne Pathogenic Bacteria. *Foods* **2020**, *9*, 1442. [CrossRef] [PubMed]
9. Bancalari, E.; Martelli, F.; Bernini, V.; Neviani, E.; Gatti, M. Bacteriostatic or bactericidal? Impedometric measurements to test the antimicrobial activity of *Arthrospira platensis* extract. *Food Control* **2020**, 107380. [CrossRef]
10. Massoud, R.; Khosravi-Darani, K.; Nakhsaz, F.; Varga, L. Evaluation of physicochemical, microbiological and sensory properties of croissants fortified with *Arthrospira platensis* (Spirulina). *Czech J. Food Sci.* **2017**, *34*, 350–355. [CrossRef]
11. Ak, B.; Avşaroğlu, E.; Işık, O.; Özyurt, G.; Kafkas, E.; Uslu, L. Nutritional and Physicochemical Characteristics of Bread Enriched with Microalgae *Spirulina platensis*. *Int. J. Eng. Res. Appl.* **2016**, *6*, 9.
12. Zouari, N.; Abid, M.; Fakhfakh, N.; Ayadi, M.A.; Zorgui, L.; Ayadi, M.; Attia, H. Blue-green algae (*Arthrospira platensis*) as an ingredient in pasta: Free radical scavenging activity, sensory and cooking characteristics evaluation. *Int. J. Food Sci. Nutr.* **2011**, *62*, 811–813. [CrossRef] [PubMed]
13. Golmakani, M.-T.; Soleimanian-Zad, S.; Alavi, N.; Nazari, E.; Eskandari, M.H. Effect of Spirulina (*Arthrospira platensis*) powder on probiotic bacteriologically acidified feta-type cheese. *J. Appl. Phycol.* **2019**, *31*, 1085–1094. [CrossRef]
14. Mohammadi-Gouraji, E.; Soleimanian-Zad, S.; Ghiaci, M. Phycocyanin-enriched yogurt and its antibacterial and physicochemical properties during 21 days of storage. *LWT* **2019**, *102*, 230–236. [CrossRef]
15. Yamaguchi, S.K.F.; Moreira, J.B.; Costa, J.A.V.; de Souza, C.K.; Bertoli, S.L.; Carvalho, L.F.D. Evaluation of Adding *Spirulina* to Freeze-Dried Yogurts Before Fermentation and After Freeze-Drying. *Ind. Biotechnol.* **2019**, *15*, 89–94. [CrossRef]
16. Camacho, F.; Macedo, A.; Malcata, F. Potential Industrial Applications and Commercialization of Microalgae in the Functional Food and Feed Industries: A Short Review. *Mar. Drugs* **2019**, *17*, 312. [CrossRef]
17. Guldas, M.; Irkin, R. Influence of *Spirulina platensis* powder on the microflora of yoghurt and acidophilus milk. *Mljekarstvo* **2010**, *60*, 237–243.
18. Barkallah, M.; Dammak, M.; Louati, I.; Hentati, F.; Hadrich, B.; Mechichi, T.; Ayadi, M.A.; Fendri, I.; Attia, H.; Abdelkafi, S. Effect of *Spirulina platensis* fortification on physicochemical, textural, antioxidant and sensory properties of yogurt during fermentation and storage. *LWT* **2017**, *84*, 323–330. [CrossRef]
19. Bhowmik, D.; Dubey, J.; Mehra, S. Probiotic Efficiency of *Spirulina platensis*-Stimulating Growth of Lactic Acid Bacteria. *World J. Dairy Food Sci.* **2009**, *4*, 160–164.
20. Martelli, F.; Alinovi, M.; Bernini, V.; Gatti, M.; Bancalari, E. *Arthrospira platensis* as Natural Fermentation Booster for Milk and Soy Fermented Beverages. *Foods* **2020**, *9*, 350. [CrossRef]
21. Parada, J. Lactic acid bacteria growth promoters from *Spirulina platensis*. *Int. J. Food Microbiol.* **1998**, *45*, 225–228. [CrossRef]
22. Bron, P.A.; Kleerebezem, M.; Brummer, R.-J.; Cani, P.D.; Mercenier, A.; MacDonald, T.T.; Garcia-Ródenas, C.L.; Wells, J.M. Can probiotics modulate human disease by impacting intestinal barrier function? *Br. J. Nutr.* **2017**, *117*, 93–107. [CrossRef] [PubMed]
23. Gardiner, G.E.; Bouchier, P.; O'Sullivan, E.; Kelly, J.; Kevin Collins, J.; Fitzgerald, G.; Paul Ross, R.; Stanton, C. A spray-dried culture for probiotic Cheddar cheese manufacture. *Int. Dairy J.* **2002**, *12*, 749–756. [CrossRef]
24. Wong, S.-S.; Quan Toh, Z.; Dunne, E.M.; Mulholland, E.K.; Tang, M.L.; Robins-Browne, R.M.; Licciardi, P.V.; Satzke, C. Inhibition of *Streptococcus pneumoniae* adherence to human epithelial cells in vitro by the probiotic *Lacticaseibacillus rhamnosus* GG. *BMC Res. Notes* **2013**, *6*, 135. [CrossRef]
25. Ricci, A.; Cirlini, M.; Maoloni, A.; Del Rio, D.; Calani, L.; Bernini, V.; Galaverna, G.; Neviani, E.; Lazzi, C. Use of Dairy and Plant-Derived Lactobacilli as Starters for Cherry Juice Fermentation. *Nutrients* **2019**, *11*, 213. [CrossRef]
26. Smid, E.J.; Kleerebezem, M. Production of Aroma Compounds in Lactic Fermentations. *Annu. Rev. Food Sci. Technol.* **2014**, *5*, 313–326. [CrossRef]
27. Wang, H.; Zhang, W.; Chen, L.; Wang, J.; Liu, T. The contamination and control of biological pollutants in mass cultivation of microalgae. *Bioresour. Technol.* **2013**, *128*, 745–750. [CrossRef]
28. Aguero, J.; Lora, J.; Estrada, K.; Concepcion, F.; Nunez, A.; Rodriguez, A.; Pino, J.A. Volatile Components of a Commercial Sample of the Blue-Green Algae *Spirulina platensis*. *J. Essent. Oil Res.* **2003**, *15*, 114–117. [CrossRef]
29. Cuellar-Bermúdez, S.P.; Barba-Davila, B.; Serna-Saldivar, S.O.; Parra-Saldivar, R.; Rodriguez-Rodriguez, J.; Morales-Davila, S.; Goiris, K.; Muylaert, K.; Chuck-Hernández, C. Deodorization of *Arthrospira platensis* biomass for further scale-up food applications. *J. Sci. Food Agric.* **2017**, *97*, 5123–5130. [CrossRef]
30. Bao, J.; Zhang, X.; Zheng, J.-H.; Ren, D.-F.; Lu, J. Mixed fermentation of *Spirulina platensis* with *Lacticaseibacillus plantarum* and *Bacillus subtilis* by random-centroid optimization. *Food Chem.* **2018**, *264*, 64–72. [CrossRef]
31. Högnadóttir, Á. Flavor Perception and Volatile Compounds in Fish. 2000, p. 25. Available online: http://www.matis.is/media/matis/utgafa/Skyrsla_01-00_Flavor_Perception_and_Volatile_Compounds_in_Fish.pdf (accessed on 30 December 2020).
32. Fink, P. Ecological functions of volatile organic compounds in aquatic systems. *Mar. Freshw. Behav. Physiol.* **2007**, *40*, 155–168. [CrossRef]
33. Seo, Y.-S.; Bae, H.-N.; Eom, S.-H.; Lim, K.-S.; Yun, I.-H.; Chung, Y.-H.; Jeon, J.-M.; Kim, H.-W.; Lee, M.-S.; Lee, Y.-B.; et al. Removal of off-flavors from sea tangle (*Laminaria japonica*) extract by fermentation with *Aspergillus oryzae*. *Bioresour. Technol.* **2012**, *121*, 475–479. [CrossRef]
34. Pala, Ç.U.; Toklucu, A.K. Effect of UV-C light on anthocyanin content and other quality parameters of pomegranate juice. *J. Food Compos. Anal.* **2011**, *24*, 790–795. [CrossRef]

35. Koutchma, T. UV Light for Processing Foods. *Ozone Sci. Eng.* **2008**, *30*, 93–98. [CrossRef]
36. De Jesus Raposo, M.; de Morais, A.; de Morais, R. Emergent Sources of Prebiotics: Seaweeds and Microalgae. *Mar. Drugs* **2016**, *14*, 27. [CrossRef]
37. Martelli, F.; Favari, C.; Mena, P.; Guazzetti, S.; Ricci, A.; Rio, D.D.; Lazzi, C.; Neviani, E.; Bernini, V. Antimicrobial and Fermentation Potential of *Himanthalia elongata* in Food Applications. *Microorganisms* **2020**, *8*, 248. [CrossRef]
38. Jia, R.; Chen, H.; Chen, H.; Ding, W. Effects of fermentation with *Lacticaseibacillus rhamnosus* GG on product quality and fatty acids of goat milk yogurt. *J. Dairy Sci.* **2016**, *99*, 221–227. [CrossRef]
39. Rubio, R.; Aymerich, T.; Bover-Cid, S.; Guàrdia, M.D.; Arnau, J.; Garriga, M. Probiotic strains *Lacticaseibacillus plantarum* 299V and *Lacticaseibacillus rhamnosus* GG as starter cultures for fermented sausages. *LWT Food Sci. Technol.* **2013**, *54*, 51–56. [CrossRef]
40. Antelo, F.S.; Costa, J.A.V.; Kalil, S.J. Thermal degradation kinetics of the phycocyanin from *Spirulina platensis*. *Biochem. Eng. J.* **2008**, *41*, 43–47. [CrossRef]
41. Dutta, D.; Dutta, A.; Raychaudhuri, U.; Chakraborty, R. Rheological characteristics and thermal degradation kinetics of beta-carotene in pumpkin puree. *J. Food Eng.* **2006**, *76*, 538–546. [CrossRef]
42. Weemaes, C.A.; Ooms, V.; Van Loey, A.M.; Hendrickx, M.E. Kinetics of Chlorophyll Degradation and Color Loss in Heated Broccoli Juice. *J. Agric. Food Chem.* **1999**, *47*, 2404–2409. [CrossRef]
43. Wu, S.-C.; Wang, F.-J.; Pan, C.-L. The comparison of antioxidative properties of seaweed oligosaccharides fermented by two lactic acid bacteria. *J. Mar. Sci. Technol.* **2010**, *18*, 9.
44. Pereira, A.L.F.; Maciel, T.C.; Rodrigues, S. Probiotic beverage from cashew apple juice fermented with *Lacticaseibacillus casei*. *Food Res. Int.* **2011**, *44*, 1276–1283. [CrossRef]
45. Ricci, A.; Bernini, V.; Maoloni, A.; Cirlini, M.; Galaverna, G.; Neviani, E.; Lazzi, C. Vegetable By-Product Lacto-Fermentation as a New Source of Antimicrobial Compounds. *Microorganisms* **2019**, *7*, 607. [CrossRef] [PubMed]
46. Giulio, B.D.; Orlando, P.; Barba, G.; Coppola, R.; Rosa, M.D.; Sada, A.; Prisco, P.P.D.; Nazzaro, F. Use of alginate and cryo-protective sugars to improve the viability of lactic acid bacteria after freezing and freeze-drying. *World J. Microbiol. Biotechnol.* **2005**, *21*, 739–746. [CrossRef]
47. Reddy, K.B.P.K.; Awasthi, S.P.; Madhu, A.N.; Prapulla, S.G. Role of Cryoprotectants on the Viability and Functional Properties of Probiotic Lactic Acid Bacteria during Freeze Drying. *Food Biotechnol.* **2009**, *23*, 243–265. [CrossRef]
48. Varga, L.; Szigeti, J.; Kovács, R.; Földes, T.; Buti, S. Influence of a *Spirulina platensis* Biomass on the Microflora of Fermented ABT Milks During Storage (R1). *J. Dairy Sci.* **2002**, *85*, 1031–1038. [CrossRef]
49. Beheshtipour, H.; Mortazavian, A.M.; Haratian, P.; Darani, K.K. Effects of *Chlorella vulgaris* and *Arthrospira platensis* addition on viability of probiotic bacteria in yogurt and its biochemical properties. *Eur. Food Res. Technol.* **2012**, *235*, 719–728. [CrossRef]
50. De Marco Castro, E.; Shannon, E.; Abu-Ghannam, N. Effect of Fermentation on Enhancing the Nutraceutical Properties of *Arthrospira platensis* (Spirulina). *Fermentation* **2019**, *5*, 28. [CrossRef]
51. Dave, R.I.; Shah, N.P. Viability of yoghurt and probiotic bacteria in yoghurts made from commercial starter cultures. *Int. Dairy J.* **1997**, *7*, 31–41. [CrossRef]
52. Bianchi, F.; Careri, M.; Mangia, A.; Musci, M. Retention indices in the analysis of food aroma volatile compounds in temperature-programmed gas chromatography: Database creation and evaluation of precision and robustness. *J. Sep. Sci.* **2007**, *30*, 563–572. [CrossRef]
53. sherena.johnson@nist.gov Standard Reference Data. Available online: https://www.nist.gov/srd (accessed on 8 October 2020).
54. Ricci, A.; Cirlini, M.; Levante, A.; Dall'Asta, C.; Galaverna, G.; Lazzi, C. Volatile profile of elderberry juice: Effect of lactic acid fermentation using *L. plantarum*, *L. rhamnosus* and *L. casei* strains. *Food Res. Int.* **2018**, *105*, 412–422. [CrossRef]
55. Cirlini, M.; Dall'Asta, C.; Silvanini, A.; Beghè, D.; Fabbri, A.; Galaverna, G.; Ganino, T. Volatile fingerprinting of chestnut flours from traditional Emilia Romagna (Italy) cultivars. *Food Chem.* **2012**, *134*, 662–668. [CrossRef] [PubMed]
56. Goodner, K.L. Practical retention index models of OV-101, DB-1, DB-5, and DB-Wax for flavor and fragrance compounds. *LWT Food Sci. Technol.* **2008**, *41*, 951–958. [CrossRef]
57. Tanaka, T.; Yamaguchi, T.; Katsumata, R.; Kiuchi, K. Comparison of volatile components in commercial Itohiki-Natto by solid phase microextraction and gas chromatography. *Nippon Shokuhin Kagaku Kogaku Kaishi* **2003**, *50*, 278–285. [CrossRef]
58. Mena, P.; Cirlini, M.; Tassotti, M.; Herrlinger, K.; Dall'Asta, C.; Del Rio, D. Phytochemical Profiling of Flavonoids, Phenolic Acids, Terpenoids, and Volatile Fraction of a Rosemary (*Rosmarinus officinalis L.*) Extract. *Molecules* **2016**, *21*, 1576. [CrossRef]
59. Yamamoto, M.; Baldermann, S.; Yoshikawa, K.; Fujita, A.; Mase, N.; Watanabe, N. Determination of Volatile Compounds in Four Commercial Samples of Japanese Green Algae Using Solid Phase Microextraction Gas Chromatography Mass Spectrometry. *Sci. World J.* **2014**, *2014*, 1–8. [CrossRef]
60. Ricci, A.; Cirlini, M.; Guido, A.; Liberatore, C.; Ganino, T.; Lazzi, C.; Chiancone, B. From Byproduct to Resource: Fermented Apple Pomace as Beer Flavoring. *Foods* **2019**, *8*, 309. [CrossRef]
61. Dall'Asta, C.; Cirlini, M.; Morini, E.; Galaverna, G. Brand-dependent volatile fingerprinting of Italian wines from Valpolicella. *J. Chromatogr. A* **2011**, *1218*, 7557–7565. [CrossRef]
62. Ong, P.K.C.; Acree, T.E. Similarities in the Aroma Chemistry of Gewu1 rztraminer Variety Wines and Lychee (*Litchi chinesis Sonn.*) Fruit. *J. Agric. Food. Chem.* **1999**, *47*, 665–670. [CrossRef]
63. Yanagimoto, K.; Ochi, H.; Lee, K.-G.; Shibamoto, T. Antioxidative Activities of Fractions Obtained from Brewed Coffee. *J. Agric. Food Chem.* **2004**, *52*, 592–596. [CrossRef] [PubMed]

64. Babushok, V.I.; Linstrom, P.J.; Zenkevich, I.G. Retention Indices for Frequently Reported Compounds of Plant Essential Oils. *J. Phys. Chem. Ref. Data* **2011**, *40*, 043101. [CrossRef]
65. Brunton, N.P.; Cronin, D.A.; Monahan, F.J. Volatile components associated with freshly cooked and oxidized off-flavours in turkey breast meat. *Flavour Fragr. J.* **2002**, *17*, 327–334. [CrossRef]
66. Cirlini, M.; Mena, P.; Tassotti, M.; Herrlinger, K.; Nieman, K.; Dall'Asta, C.; Del Rio, D. Phenolic and Volatile Composition of a Dry Spearmint (*Mentha spicata* L.) Extract. *Molecules* **2016**, *21*, 1007. [CrossRef]
67. Schneider, H.; Gelpi, E.; Bennett, E.O.; Oró, J. Fatty acids of geochemical significance in microscopic algae. *Phytochemistry* **1970**, *9*, 613–617. [CrossRef]
68. Stoyanova, L.G.; Ustyugova, E.A.; Netrusov, A.I. Antibacterial metabolites of lactic acid bacteria: Their diversity and properties. *Appl. Biochem. Microbiol.* **2012**, *48*, 229–243. [CrossRef]
69. Höckelmann, C.; Jüttner, F. Off-flavours in water: Hydroxyketones and β-ionone derivatives as new odour compounds of freshwater cyanobacteria. *Flavour Fragr. J.* **2005**, *20*, 387–394. [CrossRef]
70. Do Nascimento, T.C.; Nass, P.P.; Fernandes, A.S.; Vieira, K.R.; Wagner, R.; Jacob-Lopes, E.; Zepka, L.Q. Exploratory data of the microalgae compounds for food purposes. *Data Brief* **2020**, *29*, 105182. [CrossRef]
71. Van Durme, J.; Goiris, K.; De Winne, A.; De Cooman, L.; Muylaert, K. Evaluation of the Volatile Composition and Sensory Properties of Five Species of Microalgae. *J. Agric. Food Chem.* **2013**, *61*, 10881–10890. [CrossRef]
72. Isleten Hosoglu, M. Aroma characterization of five microalgae species using solid-phase microextraction and gas chromatography–mass spectrometry/olfactometry. *Food Chem.* **2018**, *240*, 1210–1218. [CrossRef] [PubMed]
73. Achyuthan, K.; Harper, J.; Manginell, R.; Moorman, M. Volatile Metabolites Emission by In Vivo Microalgae—An Overlooked Opportunity? *Metabolites* **2017**, *7*, 39. [CrossRef] [PubMed]
74. Amàrita, F.; Fernàndez-Esplà, D.; Requena, T.; Pelaez, C. Conversion of methionine to methional by Lactococcus lactis. *FEMS Microbiol. Lett.* **2001**, *204*, 189–195. [CrossRef] [PubMed]
75. Amárita, F.; Martínez-Cuesta, C.M.; Taborda, G.; Soto-Yárritu, P.L.; Requena, T.; Peláez, C. Formation of methional by *Lactococcus lactis* IFPL730 under cheese model conditions. *Eur. Food Res. Technol.* **2002**, *214*, 58–62. [CrossRef]
76. Helinck, S.; Le Bars, D.; Moreau, D.; Yvon, M. Ability of Thermophilic Lactic Acid Bacteria To Produce Aroma Compounds from Amino Acids. *Appl. Environ. Microbiol.* **2004**, *70*, 3855–3861. [CrossRef] [PubMed]
77. De Melo Pereira, G.V.; da Silva Vale, A.; de Carvalho Neto, D.P.; Muynarsk, E.S.; Soccol, V.T.; Soccol, C.R. Lactic acid bacteria: What coffee industry should know? *Curr. Opin. Food Sci.* **2020**, *31*, 1–8. [CrossRef]
78. El-Gendy, S.M.; Abdel-Galil, H.; Shahin, Y.; Hegazi, F.Z. Acetoin and Diacetyl Production by Homo- and Heterofermentative Lactic Acid Bacteria. *J. Food Prot.* **1983**, *46*, 420–425. [CrossRef]
79. Milovanović, I.; Mišan, A.; Simeunović, J.; Kovač, D.; Jambrec, D.; Mandić, A. Determination of Volatile Organic Compounds in Selected Strains of Cyanobacteria. *J. Chem.* **2015**, *2015*, 1–6. [CrossRef]
80. Li, H.; Liu, L.; Zhang, S.; Cui, W.; Lv, J. Identification of Antifungal Compounds Produced by *Lacticaseibacillus casei* AST18. *Curr. Microbiol.* **2012**, *65*, 156–161. [CrossRef]

MDPI
St. Alban-Anlage 66
4052 Basel
Switzerland
Tel. +41 61 683 77 34
Fax +41 61 302 89 18
www.mdpi.com

Foods Editorial Office
E-mail: foods@mdpi.com
www.mdpi.com/journal/foods

www.ingramcontent.com/pod-product-compliance
Lightning Source LLC
LaVergne TN
LVHW071256170726
843515LV00006BA/1415

9783036539201